Innovative Renewable Energy

Series editor

Ali Sayigh
World Renewable Energy Congress, Brighton, UK

The primary objective of this book series is to highlight the best-implemented worldwide policies, projects and research dealing with renewable energy and the environment. The books are developed in partnership with the World Renewable Energy Network (WREN). WREN is one of the most effective organizations in supporting and enhancing the utilisation and implementation of renewable energy sources that are both environmentally safe and economically sustainable. Contributors to books in this series come from a worldwide network of agencies, laboratories, institutions, companies and individuals, all working together towards an international diffusion of renewable energy technologies and applications. With contributions from most countries in the world, books in this series promote the communication and technical education of scientists, engineers, technicians and managers in this field and address the energy needs of both developing and developed countries.

Each book in the series contains contributions from WREN members and cover the most-up-to-date research developments, government policies, business models, best practices, and innovations from countries all over the globe. Additionally, the series publishes a collection of best papers presented during the annual and bi-annual World Renewable Energy Congress and Forum each year.

More information about this series at http://www.springer.com/series/15925

Ali Sayigh
Editor

Sustainable Vernacular Architecture

How the Past Can Enrich the Future

Springer

Editor
Ali Sayigh
World Renewable Energy Congress
Brighton, UK

ISSN 2522-8927 ISSN 2522-8935 (electronic)
Innovative Renewable Energy
ISBN 978-3-030-06184-5 ISBN 978-3-030-06185-2 (eBook)
https://doi.org/10.1007/978-3-030-06185-2

Library of Congress Control Number: 2019933101

This Springer imprint is published by the registered company Springer Nature Switzerland AG
The registered company address is: Gewerbestrasse 11, 6330 Cham, Switzerland

Contents

About the Editor

Ali Sayigh is Chairman and Founder of the World Renewable Energy Congress and Council; Director General of World Renewable Energy Network (WREN); Chairman and Founder of the Arab Solar Energy Society; and Past Chairman of the UK Solar Energy Society. Dr. Sayigh was recently elected to chair the Iraqi Energy Institute, and he actively consults on renewable and sustainable energy issues for a number of international organizations, including UNESCO, ISESCO, UNDP, ESCWA, and UNIDO. Dr. Sayigh was Director of Solar Seminars at ICTP Trieste, Italy, from 1977 to 1995; Professor of Solar Energy at King Saud, Kuwait, and Reading Universities from 1969 to 1994; and Professor of Engineering at the University of Hertfordshire from 1994 to 2004. He was the founding expert in Renewable Energy at AOPEC. He is a Fellow of the Institute of Energy, a Fellow of the Institution of Electrical Engineers, and a Chartered Engineer. He has published more than 400 papers and has contributed to and edited more than 30 books. He has been Editor and Editor-in-Chief of several international journals including *Renewable Energy* and the *International Journal of Environmental Sciences and Technology* and Editor-in-Chief of the Major Reference Work *Comprehensive Renewable Energy*.

Chapter 1
Introduction

Ali Sayigh

> According to Foster: "**Vernacular architecture** can typically be understood to be a region's indigenous local building customs, materiality and the milieu in which it arises. **Vernacular architecture** is an **architectural** style that is designed based on local needs, availability of construction materials and reflecting local traditions. At least originally, **vernacular architecture did** not use formally schooled **architects**, but relied on the design skills and tradition of local builders".
>
> **Frank Gehry** began to come into his own as an architect. It was the time when … such sources as **vernacular architecture**, commercial imagery, and pop-art.

Ingels said that he wanted to break away from the identikit architecture created by the combined forces of Modernism and climate-control systems.

"The International Style of Modernism came with the advent of building services. In the end, the architecture became like a container space, essentially like a boring box with a basement full of machinery to make it inhabitable," he said. "As a result, buildings literally started to look identical all over the planet."

This global style failed to offer people choice and is forcing residents to adapt their lifestyles to the buildings, he said, adding that architects were finally realising that people around the world want to live in different ways, requiring different styles of architecture that could be localised and responsive to different environments.

And, he said that new technologies would enable architects to take back control of this design process.

"Today, we have sophisticated building technology, we can calculate and simulate the environments and performance of the building, the thermal exposure of envelop or the air flow through an urban space or structure".

"Rather than 'Architecture without Architects', it actually allows us to do 'Engineering without Engines'", he explained. "We can engineer a building and design a building with least reliance on active machinery to make it inhabitable."

> Virginia Fernandez explained what **Herzog** and **de Meuron** explained about the beauty of vernacular architecture:

A. Sayigh (✉)
World Renewable Energy Congress, Brighton, UK
e-mail: asayigh@wrenuk.co.uk

A. Sayigh (ed.), *Sustainable Vernacular Architecture*, Innovative Renewable Energy, https://doi.org/10.1007/978-3-030-06185-2_1

"Architectural discourse values the vernacular for its beauty, environmental performance, seamless connection to the landscape, use of local materials and the craft of its making. The renewed concern with the vernacular can be seen as a response to the overtly digital trends of the last decades, and the overall shift in the profession towards ecologically sustainable construction. There is an undeniable allure in a hand-cut mortise and tenon joint, and its reinterpretation by Herzog and de Meuron or Peter Zumthor. However, the definition of vernacular architecture as '…concerned with domestic and functional rather than public or monumental buildings' includes millions of informal dwellings spreading throughout Asia, Africa and South America. These buildings, especially in large cities, exhibit none of the qualities we typically associate with the vernacular; they are built with inexpensive prefabricated materials, radically modifying the landscape and often depleting and polluting natural resources. Many provide, by any measure, substandard living conditions. Yet, these buildings solve the urgent necessity of dwelling in the city, through an evolving tradition of building. This vernacular is based on the necessity to dwell, cultural aspirations and ingenuity to adapt to the surroundings".

Richard Rogers thinks ""**vernacular architecture**", real houses for real people. Rather … refusal in his various writings to **think** about the economy, or what the likely…, **Richard Rogers** says **architects** "have a duty to society"—but fears this has … and I would like to **think** it's an oath we **are** all required to make". He always includes vernacular architecture in his design.

Although **Zaha Hadid** began her remarkable **architectural** career in … As envisaged by Hadid, the Vitra Fire Station **would do** more than simply exist as an object in space. Rather, **she** used the building to **define** and **structure** the street on … the campus from its incongruously **traditional**, **vernacular** neighbours.

There are 17 chapters in this book from 14 different locations around the world. Dr. Neveen Hamza quoted several authors describing the subject, "She contests the idea that vernacular is to be mimicked and valorised. It seems the factors that lead to its definitions as 'sustainable' responsive to climate, economically affordable and occupants' needs are the exact factors that led to its demise".

Prof Khaled Al-Sallal described several examples from Yemen, Gulf Area, Iran, Egypt and North Africa by saying: "Vernacular architecture in many parts of the world has proven to satisfy the needs of human functions, which is guided by the sociocultural values of the inhabitants, along with high levels of thermal comfort based on least energy costs. The performance of vernacular architecture is mainly attributed to the smart strategies that indigenous adopted to cope with the harsh climatic conditions when technologies are not yet introduced".

Prof Manuel Guedes and his group have quoted several examples of vernacular architecture buildings in Portugal and they explained: "Vernacular architecture is an essential part of cultural heritage, and embodies the knowledge on how to design in harmony with nature, in a durable, healthy and sustainable way. The local design strategies emerge from centuries of empirical rationality, resulting in comfortable environments that passed the test of time. The knowledge derived from vernacular architecture is at the core of a truly sustainable design".

Dr. Mona Azarbayjani, in her chapter, seeking contemporary urban comfort through vernacular architectural principles in hot-arid climate referred to sustainability definition: "Looking at sustainability by its definition as longevity, analysing the architecture and urban patterns of traditional societies which overcame the harshness of environment over the course of time might bring passive solutions to both architectural and urbanistic scales of design. While the architectural scale was more of attention during past years, the purpose of this chapter is to look into urban features of hot-arid climate that has been used to bring urban comfort and to abate the harshness of the environment, and then to propose modern design alternatives using the benefit of contemporary advancement in technology and tools in architectural discourse".

Dr. Marta Szabo was describing one major building—a palace, in Hungary as a fine example of vernacular architecture: "The palace has a double-U-shaped design, and is surrounded by an enormous park. The building undergone several enlargements and modifications during the eighteenth century; its present shape being established in the time of the third generation of the Grassalkovich family. By then, the building had eight wings, and—besides the residential part—it contained a church, a theatre, a riding-quarter, a greenhouse for flowers and an orange orchard".

Dr. Salavatian and Dr. Malekjahan devoted their chapter about humid buildings in the north of Iran, "Houses in this area are of extroverted types (i.e. without central courtyard). No tangible border appears between buildings and their surroundings. Yards are the practical elements in vernacular houses of this region and could support various activities such as daily affairs, cooking in warm seasons, crops preparation, producing handicrafts, etc. They are a major linkage between private and public spaces meanwhile as a place for daily activities. A number of semi-open spaces for purposes such as crops storage, farm animals place, barns, etc. are created."

Architect Shaden Abusafieh from Jordan quoted that many authors praise the importance of vernacular buildings and described an old example of Al-Salt city. The author started rejecting foreign building forms and style which were alliance to the Arab buildings, then accepted some imported modification which has some technology needed in modern buildings in the region: "utilising the appropriate modern technology **to** express the traditional approach in a contemporary manner. It compromises between the modern technology and the neo-traditional trends. In this trend, followers understand, recruit and fuse the heritable values in traditional architecture as well as the aspects of the modern age to create the appropriate regional identity. The contemporary interpretation trend has an important challenge in forming this identity. This challenge appears in the attempt to respond to local climate and in the process of blending and obtaining the balance between the advanced generalities in the universal and the local or traditional aspects from the ancestors".

Prof Benoit Beckers and Dr. Elena Garcia-Nevado, from France, explained how urban planning enriched by its representations, from perspective to thermography. "Is it therefore worthwhile to go back in time in cities, and is it possible to draw lessons for our today megacities, so different in size, in operation, but also in location? Today the urban growth bulk takes place between the tropics, in areas of the world that were virtually uninhabited only two centuries ago. Aristotle considered that a city should not extend beyond what can be perceived at one glance, but there is not a high enough tower in Pudong from where one can only make an idea of the Shanghai urban area, with its sixty million inhabitants".

Dr. Antonella Trombadore concentrated on her chapter at the City of Mosul, Iraq and how architects kept to their tradition and culture in designing buildings. Dr. Trombadore quoted one of colleague Dr. Rifat Chadirji in creating yearly award to the best Iraqi architect in order to encourage young people to study architecture.

After Daesh destruction of Mosul and its liberation in 2017, a completion was announced to build the city and housing for displaced people estimated to be more than half million. The houses have to be traditional following the social customs of religion. To quote Dr. Trombadore: "the Mosul city suffers from a chronic housing shortage. The deficit in housing units in Nineveh is estimated to have reached 172,000 units in mid-2016, with a 53,000 units' deficit in Mosul alone. The major contributing factors to this shortage can be defined as: (1) the scarcity of tracts of land for new housing projects, and (2) the failure to update the city's 1973 master plan and create formal urban expansion zones for housing development".

Dr. Amina Batagarawa and her colleague discussed the vernacular architecture of the Hausa Traditional buildings in detail: "Knowledge of iron working was well developed early in Hausa culture. The most impressive Hausa iron products are the famous city gates, made of long strips of hammered metal joined together on sturdy frames and set on pivots instead of hinges. Otherwise, iron was used for complementary items mostly for nails with decoratively worked iron heads usually applied to the rails of the outer doors of houses.

THA construction in roofs, walls, columns, slab, beams, foundations, doors, windows, plastering and process of renovations are covered in the following sections".

Prof Marco Sala and his group described that vernacular buildings and villages become icon for tourism due to their sustainability and beauty. They sighted several Mediterranean villages.

"Through the years, a lot of small inner small towns, especially in Italy and south Mediterranean countries, had suffered heavy phenomena of exodus in order to supply the lack of development opportunities and possibilities. This had caused a deep demographic impoverishment, with serious negative consequences on the public building heritage and economic-cultural activities of historical settlements. This depletion, together with the inadequate of skills and services management, contributes to reduce further the touristic "attractiveness" of these areas, so that instead have great potentialities thanks to their "uniqueness", belonging to the extremely beautiful surrounding landscape and the artistic and cultural value of their small villages".

Prof Luca described the enhancement of vernacular architecture in a very useful example. He defined the vernacular architecture as: "Vernacular architectures spread throughout the world represent the result of an evolutionary process in which buildings' form and construction have been continuously refined with the aim of adapting to local climate and providing optimal living conditions for the human habitat. These building were built recurring to local materials and resources available on site".

Dr. Aliya Sanusi and her team considered that colonial architecture is vernacular and taken one example of colonial school and described fully: "This study focuses on the colonial building style, built between the seventeenth and twentieth centuries. Colonial building style is found in various building types such as residential, office buildings and school buildings. However, this study focuses on colonial school buildings, with aim to evaluate its viability in adapting to Malaysia climate and achieving thermal comfort among its occupants".

Prof Mohsen Aboulnaga and Architect Mona Mostafa explained how vernacular architecture not only in Egypt but globally enriched the architectural discipline and became very popular: "Vernacular architecture is representing native, domestic (local) and indigenous buildings in a country. It reflects the environmental, cultural and historical context in which it existed. Vernacular architecture uses available materials to fulfil the local needs. It is almost by definition ***'A sustainable architecture'*** since it utilised locally available resources without exhausting them. Also, vernacular architecture is described in a way that is the architecture that used to categorize methods of construction and traditions to address local desires. In addition, vernacular architecture tends to evolve over time to reflect the environmental, cultural and historical context in which it exists".

Dr. Judit Lopez-Besora, Helena Coch and Cristina Pardal from Spain too considered one element of the building (the roof) and described how vernacular architecture influences this part of buildings: "Vernacular architecture is as old as humanity. The primitive forms of habitation were designed by humans to fulfil their basic needs such as protection and security. At the beginning, caves and natural shelters were the most suitable places for this purpose, since humans did not have to carry out major modifications to make them habitable. After an extended period, humans left their shelters in caves and started to build small constructions such as huts, which gathered in permanent settlements. To reach a minimum standard of living, they used the means they had on hand, raw materials provided by the nature. These materials might be easily available and not far from the settlements, so that the construction and repair of the houses might not be complex".

Finally, Dr. Chiesa explained how the Italian building cooling technique in the past was fully influenced by vernacular architecture: "The air-conditioner market is growing fast with a huge increase in the number of installed units, both in industrialised countries and especially in emerging ones. For instance, the amount of installed units in China is growing exponentially and is reached to 120 million units in warm regions alone by 2017, as was underlined by an IEA study. Although an increase in the EER (Energy Efficiency Ratio) of air-conditioning units is evident, especially in industrialised contexts, constant growth levels of energy consumption in both industrialised and emerging countries are evident.

For this reason, alternative solutions are needed to reduce energy consumption in the summer season without sacrificing good comfort quality. The study of traditional archetypes, related to the specific climate regionalism of a location, shows that valid solutions are related with the inclusion of passive cooling techniques in buildings. Nevertheless, since these systems refer to a large toolbox, including different strategies and natural heat sinks, it is important to conceive their usage from early design phases in order to maximise their potential, increase their level of integration into building shapes and choose the best solutions for each climate and location. This chapter aims at translating traditional vernacular solutions for guaranteeing the indoor comfort in Italian climates into contemporary guidelines for architects".

Chapter 2
Contested Legacies: Vernacular Architecture Between Sustainability and the Exotic

Neveen Hamza

The Vernacular: Contested Sustainability

It is acknowledged that vernacular architecture was generally mastered by non-schooled architects. Vernacular architecture refers to an architectural construct built based on collective community responses to values and religious beliefs, local building material and years of trial and error (Oliver 2006; Weber and Yannas 2014). Vernacular architecture reflects cultural expressions with its programme of spaces and urban configurations. Academic discourses stand at cross roads with practices on the ground of conservation and people's perceptions of the feasibility of living in vernacular houses. Researchers and architects are often left wondering if they have to mimic the building practices or take a distant view of learning and archiving? Reconciling to the fact that vernacular architecture's aesthetics, form and land availability can be financially prohibitive and instead of an 'architecture for the poor' as leading Egyptian architect Hassan Fathi called it, it turns into an architecture for the very rich and singular buildings. How do we place a value on these buildings? Are they an architectural archive of building grandiose palaces and quarters that used to house the elite? Many of these today converted to Eco lodges with a sustainable label and a sometimes prohibitive price tag for even a one-night accommodation experience. Presented as environments where the contemporary occupants would experience the excitement of being away from modern day technologies and sterile interiors, and an ever-increasing demand for comfort. This chapter contests the idea that the vernacular is to be mimicked and valorized. It seems the factors that lead to its definitions as 'sustainably' responsive to climate, local economics and occupants' needs are the exact factors that led to its demise, leaving us today with contested sustainability ideas of vernacular architecture.

N. Hamza (✉)
School of Architecture, Planning and Landscape, The Quadrangle, Newcastle University, Newcastle upon Tyne, UK
e-mail: neveen.hamza@ncl.ac.uk

A. Sayigh (ed.), *Sustainable Vernacular Architecture*, Innovative Renewable Energy, https://doi.org/10.1007/978-3-030-06185-2_2

Interestingly, the protection and continuous publication of particular examples presents an underlying pattern of creating fascination with the architectural forms that could only to be afforded by the higher echelons of their respective societies. Forgetting that in hot arid areas, for instance, the timber roofs would be imported with a premium and their elaborate decorative elements depended on laborious craftsmanship that would be prohibitive for the society as a whole. The infamous vernacular houses of Islamic Cairo district of El-Moez in Fatimid Cairo, Egypt, Yazid Houses in Iran and the palace of Alhambra in Granada Spain are widely studied as examples of vernacularism, and in this chapter, the quarters of SanFang QiXiang, Fuzhou, China all represent vernacular housing for higher army ranks, wealthy merchants and heads of state, heavily decorated, gilded with gold and handcrafted timber screens and doors. Although in vernacular urban environments houses of households of various economic capabilities were housed next to each other but with varying plot sizes (Hakim 1986), what we mostly find today is the well-crafted and decorated vernacular of the rich. However, one underlying commonality is that all of these areas witnessed major upheavals in wars and social instability that led to their original wealthy owners often immigrating to safer havens, leaving their homes to be rented to poorer immigrants from rural areas seeking the opportunities of working in the city, and then overcrowding takes place and deterioration as well. Foruzanmehr and Vellinga (2011) draw attention to the widespread associations of vernacular architecture with poverty and underdevelopment; while this may well-play a part in the demise of the traditions, it is noteworthy that the compatibility of vernacular architecture with the current way of life of the people is rapidly changing. Few studies have investigated the vernacular architecture in various parts of the Middle East but most focus on 'one aspect of the morass of variables' (Roaf 1988) mainly concerned with descriptive architectural history, showing how the vernacular traditions concerned were built, used and imbued with meaning in the past. The goal of most of these studies has been the classification, listing and description of traditional house types and their characteristic features. Little attempt has been made to address the contemporary decline of the traditions, the difficulties in using traditional technologies in contemporary design or the opportunities they could offer to the architecture of today.

The UNESCO's list of protected and endangered heritage attempts to preserve cultural heritage and plays a role in our understanding and indeed romantic fascination with these built constructs. However, this raises three central questions. The first is a question of affordance, how can occupants shoulder the financial burden and accept the continuous need for maintenance and time that these buildings require?

Would occupants nowadays accept the temporal variations of room's thermal conditions that were in the past dealt with by moving furniture and changing patterns of daily use, such as sleeping on the roof in summer? Thirdly, if protected by government legislation, will this mean the need to evacuate in many cases occupants of low economic capabilities for these buildings to provide locations of touristic intrigue only to be experienced and endured for a limited time of exposure?

What we today experience as vernacular architecture is shaped by adaptive process that emulates biological evolution, and trials and errors of experimentation for thousands of years of following a Darwinian process of elimination. The construction that is frequently presented in the literature to offer its occupants climatically adaptive living environments that hold intrinsic qualities to support passive heating and cooling, and to use naturally resourced materials that in many cases have enabled ancient cultures to live in ways that are less energy and resource intensive than modern societies (Zhai and Previtali 2010). Rapoport (2006) and Hawkes (1996) introduce a cautionary note against the misleading 'environmental determinism' that reduces vernacular architecture to naïve copying of singular elements of form and material choice as the sole determinant of the physical environments. Rapoport expands on his discouragement of 'copying of certain formal qualities' of the vernacular architecture that are often based on a romanticized version of the vernacular into a valid approach of learning from vernacular design as a 'model system'. Elnabawi et al. (2013) using urban performance simulation of the main street of the Islamic Quarters in Cairo demonstrate how an environmental system of alleyways if leading to closed-end alleyways leads to hot spots but this heat is flushed if connected to the air paths of the indoor courtyards. Rapoport (2006) presents various examples where prevailing religious and cultural beliefs have prevailed over rational features such as orientation and size and positioning of openings for natural ventilation. The vernacular built environments are often shaped to respond to extreme weather conditions moulded by socio-economic values but also by a deep understanding and expectations of comfort on the streets as much as it is comfort indoors. Lingjiang et al. (2016), in their study of a vernacular house in Tibet, found that it was converted to a luxury hotel where occupants depended on electric heaters but were willing to endure some temporary discomfort for a short duration stay to experience the aesthetic and historic value of the interior environments.

The practice of adapting individual elements from vernacular architecture to inform contemporary design finds many examples between the 1930 and 1980s, at the time of political and social changes, oil import shocks and an awareness of a need for buildings to adapt to their local environment. Among other drivers were the claim to benefit from the superior adaptation of form and material selection of vernacular architecture to deliver thermal comfort to its occupants. A revival of the vernacular architecture also referred to as 'neo-vernacular' and following the bioregionalism philosophy is arguably based on placing a romantic value on this type of architecture. The discourse of what is considered vernacular architecture and how to learn from it is sometimes reduced to a nostalgia of a simpler mode of living, with a fewer population, planning systems that could cater for low densities and rich human intensities of interaction. No mention of the revival of symbols and forms of the vernacular architecture can ignore the architectural examples by Hassan Fathi, Charles Correa in India, and F. H. Kales (1899–1979) in designing the Wuhan University's main Campus buildings in China (1929–1936), blending local philosophies to modern construction methods. An attempt to revive local interpretations of identity through architecture language and pattern was also manifested in the adaptation of vernacular elements by the American architect Michael Graves in the five-star Al-Gona touristic village in Egypt (2007). All leads to the questioning of the authenticity of these projects and their actual response to social and economic expectations of performance (Fig. 2.1).

This chapter will review the current attempts of interpreting the value of the vernacular with its acclaimed sustainability. Sustainability indeed reaches out for

Fig. 2.1 Top left: Hassan Fathi New Baris Village; beautiful but expensive to build vaults and domes, considered crucial for the thermal and daylight performance on the buildings (courtesy, Google images). Top right: Charles Correa interpretation of the Indian vernacular symbolism in the British Consulate, Delhi, India (courtesy, Google images). Bottom left: F.H. Kales, Wuhan University old Library with its distinctive and expensive decorative blue vernacular construction and materials of tiles and timber decorative elements. Bottom right; Michael Graves, Elgouna Tourist Village, Red Sea, Egypt, a speculative development of five-star hotels and separate vernacular style villas exist for the enjoyment of the rich. ElGouna resort in Egypt designed by the international architect Michael Graves, as a speculative development that, earned from the architecture of Egyptian architect and advocate of the Nubian architecture style as architecture for the poor turns into this speculative ecotourism attraction where five-star hotels and separate vernacular style villas exist for the enjoyment of the rich at a premium

human engagement based on cultural ideologies, and a deep understanding of available climatic and natural resources. Regardless of the differing perspectives of looking at these particular case studies and their distinct architectural expression, it seems that the revival or even learning from the vernacular will lead to 'architecture for the rich' afforded only as a distinctive endeavour to differentiate oneself from the expression of modern construction techniques. Recently added vernacular buildings to the UNESCO's list of protected areas also show an endeavour by governments to valorize districts as open museums for an ever-increasing economy of tourism that seeks to provide a taste of a differentiated identity and culture through touristic attractions. The sustainability from an economic perspective then proves itself as a high return on investment to support the rich experiencing the exotic; sacrificing their comfort for the short excitement of living in these accommodations or even worse living in these houses and using mechanical means to provide comfort in a marketing endeavour to pitch these building as preserving the relics of the past but providing the modern amenities which might actually lead to increased levels of energy consumption as these buildings were built for a culture where occupants were mobile between various rooms depending on seasons and needs which might not be the mode of occupation expected from modern accommodation.

Contesting the Urban Vernacular: A System of Integrated Parts

Hakim (1986) explains the principles that govern the urban form in vernacular community responding to environmental principles of improving air movement or preventing rapid penetration of sand storms. The street layouts were formed based on animal movements and dimensions (3.5 m for one loaded camel and double this for main streets), and they were also laid out to respond to cultural norms of visual privacy of the indoors. In the case of Fuzhou to discourage lingering outdoors but in the case of Middle Eastern communities planning encouraged socialization in the alleys, creating shaded areas and built-in seats in external walls kown as the 'Sabat'. Elnabawi et al. (2015) simulated the microclimate of an outdoor urban form in Cairo, Egypt. Analysis of the Moez street (Fig. 2.2), measurements and simulation of the street highlight that the closed-ended streets that were designed to be more private streets away from the bustle of the main route through the Islamic quarter gates experience high mean radiant temperatures that couldn't possibly be comfortable during the peak day time, although narrow and shaded by the walls of the buildings (Elnabawi et al. 2015). These simulations show clearly that an integrated system of air movement between the micro-urban alley ways, the house entrances and courtyards could only achieve the required de-stratification of heat. However, this also raises the question on the land area required to maintain such a relationship and the building heights that translate to lower densification on land. The courtyard is a common and characteristic form of residential architecture in hot arid and humid climatic zone. It serves as a collector of cool air at night and a source of

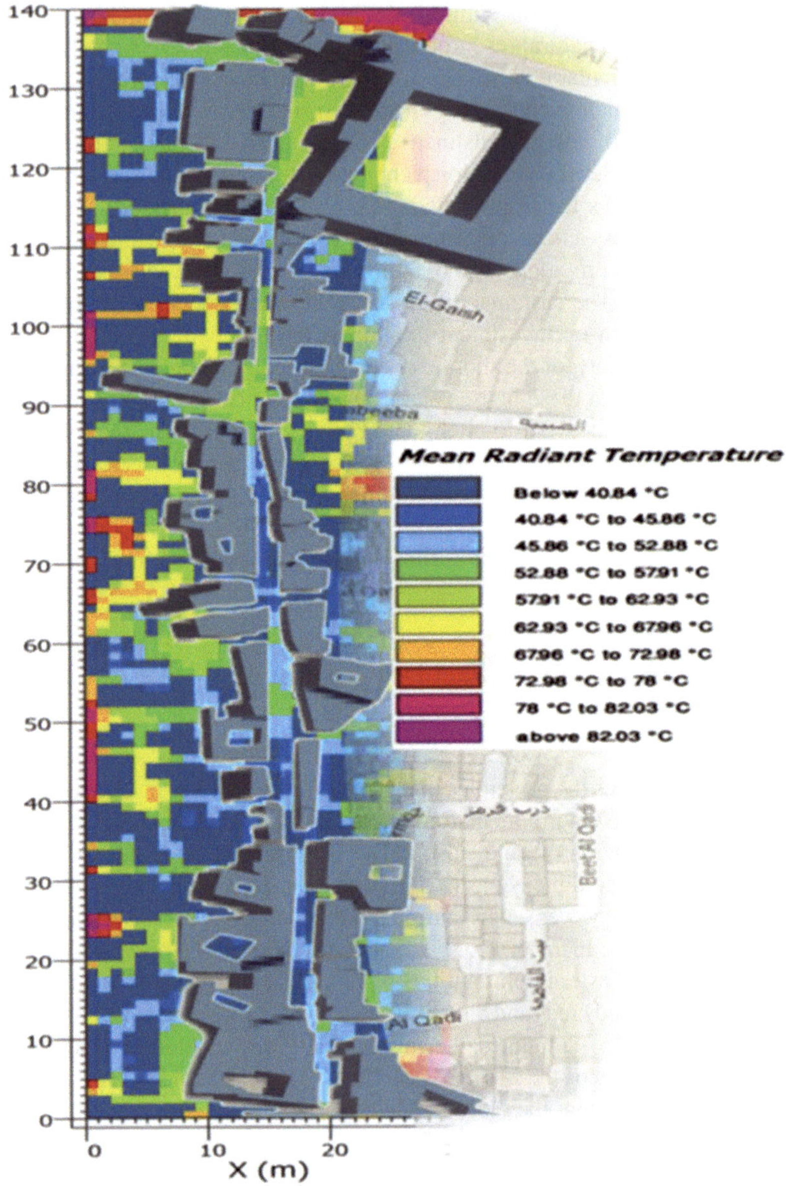

Fig. 2.2 The higher temperatures in alley ways when not connected to the ventilation system through courtyards, source: Elnabawi et al. (2013)

shade in the daytime, and also as a nucleus where families can meet to enjoy cultural activities of weaving, cooking and drying seasonal produce.

Foruzanmehr and Vellinga (2011) warn to not blindly follow a raft of technical performance research on vernacular elements such as the wind-catcher's performance in passive ventilation without understanding their social and cultural embodiment. Interestingly, their research showed that wind-catchers did not exist in all traditional houses. Large and elaborate wind-catchers were only for well-to-do families; they were expensive to build and maintain and hence ordinary households could not afford to have them. Although 40% may still be considered a significant number, this figure suggests that in spite of their popular representation, the wind-catchers may not have been an essential vernacular cooling system. Instead, they may have been auxiliary to the rest of the system. The research concluded that there were negative perceptions about having or maintaining these passive systems relating to the cost and their actual benefit to the indoor environment that would lead to their abandonment (Foruzanmehr and Vellinga 2011). Foruzanmehr and Vellinga (2011) concluded that their survey indicated that the decision to actively to maintain a vernacular feature like a wind-catcher or courtyard will not be taken on the basis of its energy efficiency or perceived association with the past, but will always be the result of a compromise, involving a variety of environmental and cultural factors and considerations (Foruzanmehr and Vellinga 2011).

These passive cooling systems are integral parts of a unified system and complement one another providing a diversity of temperatures throughout the house at different times of the day. The courtyards, wind-catchers, seasonal rooms, loggias, basements and massive structures all work together to provide comfortable conditions inside the house. What needs to be understood is that the air flow in vernacular urban developments is more complex than just the natural ventilation processes of the relationship between the courtyards, basements and sometimes wind-catchers. The outdoor street patterns and their shading played a role. These patterns not only respond to the need for movement of people and animals but also respond to social needs to engage or not, and protection from the onslaught of harsh climatic conditions.

Preserving the Whole: The Vernacular as Museum District (San Fang Qi Xiang of Fuzhou City, China)

San Fang Qi Xiang, or 'three districts and seven alleys', is located in Fuzhou city and covers an area of 0.39 km^2. The three lanes are Yijin, Wenru and Guanglu, and seven alleys are Yangqiao, Langguan, Anmin, Huangxiang, Taxiang, Gongxiang and Jibi. Its pattern of lanes and alleys shows the essence of South Chinese old cities, and makes San Fang Qi Xiang the representative of southern historic districts, including numerous National Designated Monuments like the historic residences of Yan Fu, Lin Congyi, Bing Xin and Lin Juemin. Because of its more than 400 rich, famous and powerful residents, this port city was known as 'Beverly Hills' of imperial China. Figure 2.3, shows the traditional densely packed houses with a main retail street and (Figs. 2.4 and 2.5) narrower alleys leading to stark walls with a main opening to various indoor courtyards and gardens.

Fig. 2.3 Left: the main commercial street with its elaborate timber decorated facades, right: the compact nature of the vernacular city with respecting the neighbour's building height, narrow alley ways that lead to sumptuous indoor gardens and courtyards

Fig. 2.4 Top left shows the elaborate and rich use of materials expressing the function of the main commercial road. Top right: the alleyways preserved by the UNESCO and the government protection protocol still lined by houses that are privately owned but with an ever-increasing number of craft shops and museums behind the white-washed walls. Right: the introduction on modern services of sewer networks and air conditioning units on the walls to provide a higher level of indoor comfort. Photos: author's personal collection

Fig. 2.5 Left: the lavish indoor garden for the family use, with its environmental function of providing outdoor shaded areas. Right: the handcrafted wooden screens to provide shelter from the hot summer heat awhile allowing for natural ventilation. Photos: author's personal collection

Being in the centre of the old city, San Fang Qi Xiang had been the aggregation of aristocrat and scholar-bureaucrat since Tang Dynasty. Until last century, eminent figures of the Chinese aristocrats and literary lived in San Fang Qi Xiang. Hundred dwelling houses and garden architectures which were built in Ming and Qin Dynasty still remain there, and San Fang Qi Xiang is also called 'the Museum of the buildings of the Ming and Qing Dynasty'. The government started its conservation programme in 1992 but this was stopped due to controversies of the methods of conservation. By then, the district has experienced major changes in its occupants as the civil war drove this social aristocratic class out of the region and sometimes out of the country, and landlords tended to rent the houses to carers or for cheap rent. By time, tenants couldn't afford the upkeep of these stately homes and started sub-letting to the poorer immigrants to the city. By the time the government started the conservation programme, 150,000 people lived there, and the density of population was 370,000 people/km^2; most of the residents didn't own the house they inhabited. It was a residential area with some commercial activities to serve its inhabitants, with a clear strategy to gentrify the area. In 2005, the conservation was restarted with government economic backing to relocate the building tenements who couldn't afford to buy the properties they occupied. The conservation of the vernacular adopted an understanding that houses have to be de-layered from the various changes on their fabric by occupants, and a more sensitive to the original design, materials and decoration was adopted to reinstate the individuality of these vernacular buildings (Hsiaoting 2012). Today, the vernacular district is advertised as one of China's 'Ten historical and cultural streets' since 2009. There are 20,000 visitors every day and 130,000 in midseason.

Currently, the government set a management committee as a mastermind for redevelopment, and a company responsible for all operations. All the investors who want to start their businesses have to be audited to achieve the standard of San Fang Qi Xiang. It keeps the special status of culture and promotes the value of culture, economy and society. On visiting the place in November 2017, it was a place that turned into an open-air museum for the preservation and showcasing the arts and

Fig. 2.6 The elaborate and expensive carvings in the courtyards. Top right: the maintenance of the fish pond, the symbol of prosperity but also an environmental feature to help cool the indoor shading courtyards. Bottom right: the virtual gaming machines, the new uses imposed on the traditional house to maintain 'spending time within' for visitors. Photos: author's personal collection

crafts of Fuzhou. It extended to show the food delicacies that the region offered and various tea shops. The whole site was a museum for intangible heritage housed in the vernacular conservation project. In some houses, digital and immersive games were housed in the historic houses to appeal to a younger generation (Fig. 2.6).

The Vernacular as a Backdrop for the Experience of the Exotic

Perceiving the vernacular as exotic has an influence on what we intend to learn from its construction today. The International Council of Sites and Monuments (ICOMOS 1999) describes as cultural tourism "any activity that allows visitors to experience discovering other people's lifestyles, allowing them to get to know their customs, traditions, natural environment and ideas and to have access to places of architectural, historical and archaeological interest". This pertains to travellers accommodation where there is a sense of self-fulfilment, experience and "good living". For those consumers the acquisition of experience of the "exclusive" and customized services are paramount. In an age of ever-increasing globalization, the protection, conservation, interpretation and promotion of the cultural heritage of each region is an important challenge.

Frangou et al. (2015) conclude that these are tourists seeking virtual experiences within staged backdrops of vernacular settings of an imitation of the 'old' but requiring modern facilities. These sought exotic experiences lead to the utilization of local architectural features in active traditional settlements, and a modern interpretation of the transaction between the interior and the shell that surrounds it. These seekers of the exotic search for opportunities are to engage with the manners and customs of a place, and finally the integration into the local community and everyday life (Frangou et al. 2015).

The International Cultural Tourism Charter (ICOMOS 1999) highlights that the majority of tourists today wish to visit areas of high environmental quality and strong local culture elements. ICOMOS, therefore, calls for 'the development of sustainable tourism responds to the needs of modern tourists and tourist areas while, at the same time, it protects and enriches the opportunities for tourism in the future. The sustainable tourism development leads to the management of all natural resources in a way that satisfies the economic, aesthetic and social parameters and needs, and preserves cultural diversity, basic ecological processes, and biodiversity and life support systems' (ICOMOS 1999).

Interior space in this case is easily disconnected from the natural environment and the architectural shell and follows its own style. So, it becomes an experimental design field for innovations and international styles. These interior developments acquire specific dimensions, incorporating lifestyle elements, body and beautification facilities, technological gadgets, lightings, etc. The interior's 'globalized image' removes the feeling of being integrated in the local community, and the differentiation between interior and exterior spaces causes confusion and disorientation. The principles of alternative tourism, adaptive reuse of existing buildings and complexes and their utilization as accommodation units, as well as the active and experiential participation in activities and local events provides the possibility of living in buildings and environments of local character. All these aspects leave their mark on vernacular Greek architecture and the potential use of buildings and architectural ensembles or parts of traditional communities, in seeking to provide an authentic tourism experience.

In Greece, the southern region of Peloponnese is known as 'Mani'. The Mani peninsula, Lakonia, is known for its turbulent history, rugged natural beauty and distinctive architecture providing visitors an unforgettable experience (Fig. 2.7).The Mani-Tower House designed by Z-level in 2011 is a manifestation of this trend to revitalize the 'vernacular' with the look-alike modern extension and services, encouraged by the government incentivization schemes since 1975 towards 'effective intervention to connect tourism development to the architectural heritage, leading to the modification of private traditional buildings into tourist accommodation, in more than 100 guesthouses. The state provided a subsidy for the modification of existing traditional buildings into tourist accommodation, with a 40% of total cost, according to the provisions of the development law'.

Fig. 2.7 Left: the rugged rocky topography of Mani making it an ideal location to watch over the seas and protect from the dangers of pirate raids. Right: Section showing the original construction of the house and the top extension. Photos courtesy: Babis Louzidis, source: https://www.archdaily.com/773124/mani-house-z-level/55e99faee58ecead28000199-mani-house-z-level-photo

Fig. 2.8 (Left) the new extension mimicking the eternal brick appearance, (right) the original massive stone walls. Photos courtesy: Babis Louzidis, source: https://www.archdaily.com/773124/mani-house-z-level/55e99faee58ecead28000199-mani-house-z-level-photo

The architects describe the site as initially an abandoned megalithic two-storey building dating from the eighteenth century, cut off from the road with access only over a stone-paved path; the ruin was located in a garden with olive trees, carob trees and prickly pear cacti, nestled in the embrace of the hillside with a cave. To the east, the yard abutted a sixteenth century tower, which is a historical monument, and alongside there was a Byzantine chapel with twelfth-century wall paintings (Z-level 2011). The intention of the design was for a spare renovation of the tower-house and its conversion into a holiday villa that exuded a primordial spirit of place. The residence was designed to function as a starting point and a place to familiarize oneself with the natural beauty of Mani as well as its cultural wealth, helping in this way to shed light on the traditional settlements in the area and try to bring them back to life, as they have slowly been abandoned with the passage of time (Z-level 2015).

The sectional drawings (Figs. 2.7 and 2.8) highlight the older stone-filled walls and the new extension of concrete-filled walls covered by the local stone. The original construction of the massive walls with their thermal mass means more

Fig. 2.9 The provision of modern indoor comfortable facilities to meet the expectations of tourists. Photos courtesy: Babis Louzidis, source: https://www.archdaily.com/773124/mani-house-z-level/55e99faee58ecead28000199-mani-house-z-level-photo

extraction and delivery of the local stone but above all finding the labour to construct it based on traditional methods is an added cost.

Figure 2.9 shows the architectural interventions to bring in high-level building interiors to provide an encounter with created exotic and staged interactions with the vernacular.

In Northern Egyptian dessert, 16 km from the charming town of Siwa lies Adrère Amellal. An ecolodge overlooking Lake Siwa and nestled at the foot of the mystical 'White Mountain'. Adrère Amellal is a newly built accommodation of 40 rooms, all built in the traditional Siwan style with indigenous salt-rock walls known as 'Kershef', and palm used for the roofs while doors, windows and fixtures are made of olive wood from annual tree trimmings, using traditional Siwan building techniques and styles which have a minimal impact on the environment.

The contemporary exotic of every room is unique, combining distinction and authenticity. The earthen buildings blend naturally into the landscape, and all furniture and crafts pay tribute to the talented local artisans. The facility claims that it attracts travellers who are looking for unique, high-quality, environmentally and socially conscious accommodation. The ecolodge's distinctive guests, which have included Prince Charles and the Duchess of Cornwall, are a testament to its success.

Contemporary plumbing is used throughout. Wastewater is first settled in self-contained sedimentation tanks, allowing the supernatant to flow through perforated pipes into a sealed wetland where indigenous papyrus plants are grown to complete the biodegradation and waste reduction process.

Adrère Amellal has been designed to operate as a low-profile structure with no lighting. Interference with the natural habitats of the area in both the construction and operation of the ecolodge was avoided. Adrère Amellal is not operated with electricity. Lamps and candles are used for lighting and on cold nights, coal-filled braziers are used for heating. Natural ventilation systems that rely on strategic positioning of doors and windows have been adopted, eliminating the need for air conditioning.

Fig. 2.10 Left the ruins of Shali village-Siwa, Egypt. The compact urban defensive configuration with their circular observatory towers built from the local materials creating the Kershef construction method. Photo courtsey: Prof.Hatem Altaweel. Right the vernacular contemporary interpretation in Hotel Adrer Amelal, Siwa, Egypt. The creation of the exotic encounter with local culture. Photo courtesy: Prof.Heba AbulFadl, July 2016

The architecture mimicking the original materials and getting some of its inspiration from the old ruins of the traditional village (Fig. 2.10) raises the question of who is the vernacular for?

Conclusions

This chapter highlights that the vernacular architecture, with its authentic references to cultural beliefs, social norms and the local climate, has worked to sustain communities of a different time and life style. Their intrinsic beauty, detail and most commonly referred to elements of vernacular bioregionalism might have been a symbol of wealth and power as well as their environmental performance that could not be afforded by all. Today, these conglomerations must be evaluated as a whole, an integrated system that works on a complex climatic, social and economic network between the urban morphologies and the indoor environments. It is evident that the created environments do need some support from mechanical systems to provide comfort by today's standards. The examples presented also point out to the romantic adoption of the stylistic elements related to passive architectural design are land intensive, difficult to find skills to build and need a considerable budget for their maintenance and upkeep. The examples presented in this chapter raise the question of whether by today's standards vernacular architecture is an architecture for the rich or the poor and if it will still require dependence on mechanical heating and cooling then how do we judge its sustainability?

References

Elnabawi M, Hamza N, Dudek S (2013). Use and evaluation of the envi-met model for two different urban forms in Cairo, Egypt: measurements and model simulations. In: 13th international conference of the International Building Performance Simulation Association. 2013, Chambéry, France

Elnabawi M, Hamza N, S Dudek S (2015) Numerical modelling evaluation for the microclimate of an outdoor urban form in Cairo, Egypt. HBRC J 11(2):246–251. open access www.sciencedirect.com

Foruzanmehr A, Vellinga M (2011) Vernacular architecture: questions of comfort and practicability. Build Res Inform 39(3):274–285

Frangou D, Georgiadou Z, Marnellos D (2015) The inhabitation experience as a means of promoting the local cultural heritage. In: 1st International Conference for Experiential Tourism IMIC 2015: Santorini island, Greece, accessed through researchgate, 12/6/2018

Hakim B (1986) Arabic Islamic cities. Routledge and Kegan Paul Plc, London

Hawkes D (1996) The environmental tradition: studies in the architecture of the environment. E&FN Spon, London

Hsiaoting Y (2012) Trends of preserving historic districts 48th ISOCARP congress 2012. Available on www.researchgate.com. Accessed 1 June 2018

ICOMOS (1999) International cultural tourism charter: managing tourism at places of heritage significance. http://www.international.icomos.org/charters/tourism_e.pdf. Accessed 25 Apr 2014

Lingjiang H, Hamza N, Bing L, Dava Z (2016) Climate-responsive design of traditional dwellings in the cold-arid regions of Tibet and a field investigation of indoor environments in winter. Energ Buildings 128:697–712

Oliver P (2006) Built to meet needs: cultural issues in vernacular architecture. Routledge, London

Rapoport A (2006) Vernacular design as a model system. In: Asquith L, Vellinga M (eds) Vernacular architecture in the twenty-first century, theory education and practice. Taylor & Francis, New York

Roaf S (1988) The wind catchers of Yazd, A. PhD thesis, Department of Architecture, Oxford Polytechnic, Oxford

Weber W, Yannas S (eds) (2014) Lessons from vernacular architecture. Routledge, London

Zhai Z, Previtali JM (2010) Ancient vernacular architecture: characteristics categorization and energy performance evaluation. Energ Buildings 42(3):357–365

Z-level (2011) House in 'mani' Greece, designed by Z-level architects in 2011. http://www.archilovers.com/projects/164149/maina.html#info

Z-level (2015) House in Mani, https://www.archdaily.com/773124/mani-house-z-level/55e99faee58ecead28000199-mani-house-z-level-photo. Accessed 12 Oct 2018

Chapter 3
Vernacular Architecture in the MENA Region: Review of Bioclimatic Strategies and Analysis of Case Studies

Khaled A. Al-Sallal and Meriem Rahmani

Introduction

The passive environmental control techniques in vernacular architecture demonstarted its effectivness in ensuring thermal comfort (Kim and Park 2010; Lee et al. 1996). A study conducted by Shan et al. concluded that applying passive environmental control techniques on modern houses in China helped reduce energy consumption and created comfortable indoor conditions (Shan et al. 2015). Toe and Kubota (2015) investigated the impact of potential vernacular passive cooling strategies (including roof/ceiling insulation, wall/window shading, night ventilation, courtyard/forced ventilation, microclimate and heat islands mitigation) on improving thermal comfort conditions in modern brick terraced houses located in hot humid climates of Malaysia. The strategies found in vernacular architecture include, but not limited to, connecting front entrance door to courtyards, linear planning of houses, orientation of the rooms based on the sun path and wind direction, the use of local materials, implementation of wind catchers, and courtyards (Chandel et al. 2016). In UAE for example, the use of thick walls, light colors, and compact urban fabric proved to provide good thermal performance in heritage buildings (Vine and Casey 1992). Several elements in the Emirati traditional house, such as the central courtyard with proper orientation of the spaces along with the reliance on *Bargeels* (wind towers) in some rooms, helped to create thermal comfort in the harsh hot humid climate of the coastal region of the UAE (Alkhalidi 2013). Other studies investigated how compact urban form influenced airflow and the potential for passive cooling (Al-Sallal and Al-Rais, 2011; Al- Sallal and Abouelhamd, 2018).

Based on comprehensive investigation of the vernacular architecture of old Sana'a, Al-Sallal (1996a, 2001, 2004, 2013) deduced that the morphology of the city fabric and the urban planning of the traditional housing clusters was the

K. A. Al-Sallal (✉) · M. Rahmani
UAE University, Al-Ain, UAE
e-mail: k.sallal@uaeu.ac.ae; 201570116@uaeu.ac.ae

A. Sayigh (ed.), *Sustainable Vernacular Architecture*, Innovative Renewable Energy, https://doi.org/10.1007/978-3-030-06185-2_3

main factors in achieving high levels of thermal comfort and satisfying the inhabitant's sociocultural values. Such a balanced morphology were demonstrated at four (4) levels of the city's architecture. The first was the morphology of the city in terms of its physical zones of urban forms and spaces developed over thousands of years. This includes the historical formations and functional zones of the city such as the Grand Mosque, the rainwater channels, the old markets, the roads, the urban gardens and squares, and the divisions of the housing quarters. The second level deals with the housing cluster morphology, which includes the clustering of the vernacular tower houses, the roads that serve them, the social square that serves the houses, and the urban garden on which these houses look upon to. The third level deals with the morphology of the individual house in terms of balancing between its form and space arrangements. Finally, the fourth level deals with the morphology of certain (passive) design features such as the openings.

The climate-responsive architecture developed by the indigenous people is a valuable resource of history-long experiences that we can learn from. Towards this end, this chapter provides a comprehensive review of the bioclimatic strategies and analysis of case studies in the vernacular architecture of the MENA region.

Traditional Communities

Improving the microclimate around buildings helps greatly to balance heat gains/losses and hence improving indoor comfort levels and reducing energy requirements. The climatically responsive site is one that helps to balance between needed environmental factors such as receiving less quantities of solar radiation and promoting exposure to desirable wind in the summer while this is reversed in the winter. The effective site must balance between the requirements of functions and the requirements of human comfort. Investigations reveal that many smart tactics were practiced by the inhabitants of old cities throughout their history to keep a balance between the influences of climatic/environmental factors and the sociocultural needs considering the daily and seasonally patterns of change.

The Historical City of Sana'a, Yemen

The historical city of Sana'a includes more than six thousand traditional tower houses that are still being used by their occupants. Previous studies by Al-Sallal (1996a, b, 2001, 2004, 2013, 2016a) and Al-Sallal et al. (1995) discussed in depth the morphology of the old city of Sana'a in Yemen (see Fig. 3.1), in terms of its streets' orientation and the achieved balance between building masses and urban or green spaces. This balanced morphology helped to offer sites with great opportunities for harnessing the solar radiation needed for winter heating while improving shading and passive cooling in the summer. The gardens of the old city of Sana'a

Fig. 3.1 Aerial view showing the traditional gardens in the old city of Sana'a (left image; Google Earth 2018) and a close-up view of an example of the traditional garden within the residential quarter (right image; reproduced from 26-SDNS 2011)

occupy 20% of the city's area, and 42% of each residential quarter; thus almost every house has an opportunity of having access to views into extensive gardens and the surrounding mountains through the windows that also provide access to the sun and wind. The clustering of the buildings helps protect the building from excessive solar radiation through proper shading and provides access to cool breezes in the summer while improving solar access and opportunities for solar heating in the winter.

Traditional Quarters of the Arabian Gulf Cities

There are three factors that helped to shape the desert architecture in the Arabian Gulf, namely the available materials, the hot climate, and the socio-cultural life of the people. This architecture is better known for its high density where buildings are attached, accessible by shaded narrow alleys (Sikka). The orientation of the alleys helped to limit the dusty winds and promote sea breezes in the coastal cities.

Previous studies on Al-Meraijah traditional quarter of Sharjah by Al-Sallal and Al-Zarooni (2000, 2001a, b) discussed how the compactness of the vernacular settlements maximized significantly the solar shading and minimized exposure of buildings' structure and occupants to the intense solar radiation. At the urban scale, the following methods were used for shading (Al-Sallal 2016a, b):

- Form-space proportion: Increase the density of the built-up area compared to the alleys and public squares.
- Size of the alleys: the width of the alleys was narrow and does not exceed 3 m.
- Alley's width and building height proportion: the alley's width was kept small compared to the building height (h/w = 2:1 – 1:1).
- Public social squares: the size and proportion of space helps the vegetation and surrounding buildings in the site to maximize solar shading.
- Covered alleys: Alleys in public zones (markets for example) were shaded by palm groves.

The alleys in the costal settlements of the Arabian Gulf, such as in Al Fahidi and Al Shandagha historical districts in Dubai, run from north or northwest at the sea direction and extend to south or southwest at the leeward side (Al-Sallal 2016a, b). This design helps the passage of the desirable north and northwest winds inside the alleys. Moreover, the direction of the alleys helped to create an optimized orientation for the houses that helps to catch the pleasant see breezes. This effective planning promotes passive cooling through:

- Convective cooling of the building structures through dispersing heat from the building masses.
- Convective and evaporative cooling by promoting ventilation to cool the building occupants via dissipating heat from their bodies.
- Stirring air currents to reduce the effects of high humidity levels.

Al-Sallal and Al-Rais (2011) carried out a study in Dubai to test human comfort in the traditional urban context by using computational fluid dynamics (CFD). Results indicate that narrow streets (with 4 m and less) perform better in terms of passive cooling. Narrow streets promote the wind speed circulation that passes through them which helps the wind to reach deeper parts of the narrow streets. In two other studies, Al-Sallal and Al-Rais (2012) investigated the outdoor human comfort in modern urban context of Dubai; and Al-Sallal and Abouelhamd (2018) compared between the planning patterns and the comfort levels resulted between the traditional and urban contexts of Dubai using CFD simulations of actual settings.

The Traditional Quarters in Bushehr and Yazd Cities, Iran

Bushehr city in Iran, located in a hot and humid climate, shows the importance of compact urban fabric to provide thermal comfort (see Fig. 3.2). Motealleh et al. (2016) investigated climatic responsive solutions in the vernacular architecture of the Bushehr city. The small area of surfaces exposed to the sun radiation, and the cool sea breeze, which penetrates the narrow deep alleys, are two main factors for climatic comfort in the city. The width of the alleys is no more than 1.5 m while the proportion of the alleys to the buildings height is almost 1:10. This design helps to shade almost the entire façades and alleys. The two factors work together to promote the wind circulation that helps maintain the required climatic comfort.

Another successful example on the relation between climatic comfort and vernacular architecture in Iran is Yazd city. The hot dry city responded to the harsh climatic conditions with a compact urban fabric penetrated by narrow streets and sidewalks directed from east to west direction only (see Fig. 3.3; Tavassoli 2016). The majority of the sidewalks are curved and narrow with high walls; some of the sidewalks are roofed (called Sabat in Persian) to provide extra shading. All these strategies are for the purpose of preventing the access of the hot air to the streets and thus the buildings (Keshtkaran 2011; Moradi 2005).

Fig. 3.2 The high urban compactness of the old Yazd, Iran (Google Earth 2018)

Fig. 3.3 The high urban compactness of the old Bushehr compared to the new settlements, Iran (Google Earth 2018)

The Traditional Quarters in North African Cities

The traditional quarters of Algiers and Tlemcen cities in Algeria are good examples that illustrate the compactness strategy (see Figs. 3.4, 3.5, and 3.6). The clustering of the urban form resulted in architecture that offered thermal advantages and satisfied the sociocultural needs of the inhabitants. In the clustered buildings configuration, buildings shared the boundary walls with no access to the outdoor enviroment. This necessitaed the inclusion of patios and courtyards to accommodate occupants' activities and provide access to daylight and ventilation. In this pattern, most rooms faced the courtyard with minimum or no openings to the streets.

In the Beni Izgen settlement in Ghardaia, Algeria, the housing units are encircled with a wall that provides security and safety against high velocity winds and sand

Fig. 3.4 The high urban compactness of the old neighbors in Tlemcen compared to the new settlements, Tlemcen, Algeria (Google Earth 2018)

Fig. 3.5 The high urban compactness of the old Casba compared to the new settlements, Algiers, Algeria (Google Earth 2018)

storms (see Figs. 3.7 and 3.8). Inside the settlements the housing units are grouped with narrow alleys to avoid the stormy winds and provide shading. The ratio between the street width and the building's height plays an essential role in shading the horizontal and vertical surfaces and create a comfortable microclimate during the day and at night (Bouchair et al. 2013) as shown in Fig. 3.9.

The traditional quarters of Cairo, Egypt, such as Khan Al Khalili (see Fig. 3.10) are also characterized with high urban densities and narrow alleys. Fahmy and sharples (2009) investigated the impact of urban form on thermal comfort in Cairo. This study used the H/W aspect ratio (AR; buildings height/distance

Fig. 3.6 View inside the old Casba of Algiers, Algeria

Fig. 3.7 The attached buildings penetrated by very narrow streets in Beni Izgen, Ghardaia, Algeria

Fig. 3.8 The high urban compactness of the old Beni Izgen compared to the new settlements, Ghardaia, Algeria (Google Earth 2018)

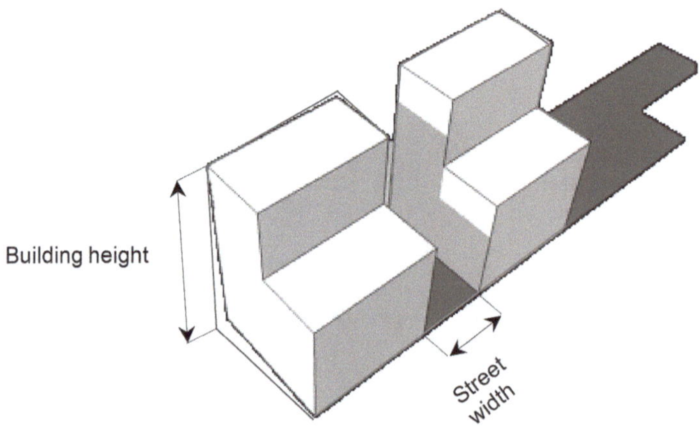

Fig. 3.9 Sketch showing the impact of high buildings with narrow streets on shading

between them), to determine how it influenced directly the incoming and outgoing radiations and the wind speed. The study investigated 3 case studies that reflect 3 different urban patterns: (1) high density, (2) medium density, (3) single dwelling arrangement. ENVI-met program was used to simulate the thermal comfort, 3 receptors were distributed in each location. The simulations were carried out at 9:00, 12:00, and 15:00 in June and July; which is the hottest period in the Cairo climate. The results revealed that the compact urban form had potential for

Fig. 3.10 The high urban compactness of the old Khan Al Khalili, Cairo, Egypt (Google Earth 2018)

better thermal comfort during the daytime because of the effect of solar shading. On the other hand, the wind flow is constricted, which could lead to health issues considering the high density of population in Cairo. The results of the second case (medium density) showed a potential for its arrangement to balance between solar access and wind speed. When the sun at high solar altitudes (i.e., around noon time), the limited depth of the street canyons allowed sun pentrations and hence it was recommended to enhance shading by implementation of vegetation shading. The wind flow in this arrangement was also constricted to pass through the cluster's courtyards due to the cluster form proportion. The results of the third case revealed that most building surfaces of the signle dwellings arrangement had exposure to high levels of solar radiation although this arrangement managed to bring more effective wind flows for passive cooling. Johansson (2006) measured the influence of urban geometry on outdoor thermal comfort in Fez, Morocco. The measurement study that endured for 1.5 years tested many cases of deep (9.7 AR) and shallow street canyons (0.6 AR). The results indicated the significance of deep streets' canyons on reducing air temperature by 10K cooler than shallow streets' canyons. This is attributed to the greater shading impact produced by the deep streets' canyons.

Bioclimatic Strategies in Vernacular Architecture

Designing a building with effective passive cooling requires examining all sources of heat gain and developing approaches to minimize their effect using various and integrated systems. The various categories of heat gains transmitting into a building can be listed into the following components (see Fig. 3.11):

- Heat gains by solar radiation, which includes two components: solar heat gain through windows' glazing ($Q_{\text{solar windows}}$) and solar heat gain through opaque surfaces such as walls and roof ($Q_{\text{solar walls/roof}}$).
- Heat gain by transmission through the building envelope due to difference in temperature between outside and inside air ($Q_{\text{transmission}}$).
- Internal heat gains generated by building equipment ($Q_{\text{equipment}}$), by lighting fixtures (Q_{light}), and by internal heat gain generated by building occupants (Q_{people}).

The total heat gain of a building is the sum of all these heat gains:

$$Q = Q_{\text{solar windows}} + Q_{\text{solar walls/roof}} + Q_{\text{equipment}} + Q_{\text{light}} + Q_{\text{transmission}} + Q_{\text{people}}$$

To minimize the total heat gain (Q), one needs to minimize each component. Reducing $Q_{\text{solar\_windows}}$ can be achieved by solar shading and minimizing the glazing area. Shading can be provided by natural elements like trees and other plants, by indoor fittings like curtains and venetian blinds, and by outdoor shading devices like overhangs or fins. Reducing $Q_{\text{solar walls/roof}}$ can be achieved by using low absorbance materials. This can be done by several methods: by painting the external surfaces of the building (walls and roof) with light colors, by cladding the walls and covering the roof with light-colored materials, and by using a green roof. Reducing $Q_{\text{transmission}}$ can be achieved by minimizing heat transmission through the building envelope surfaces. This requires reducing the U value, which can be achieved by increasing the thicknesses of different envelope components (x_a, x_b, x_c,…) and/or reducing their

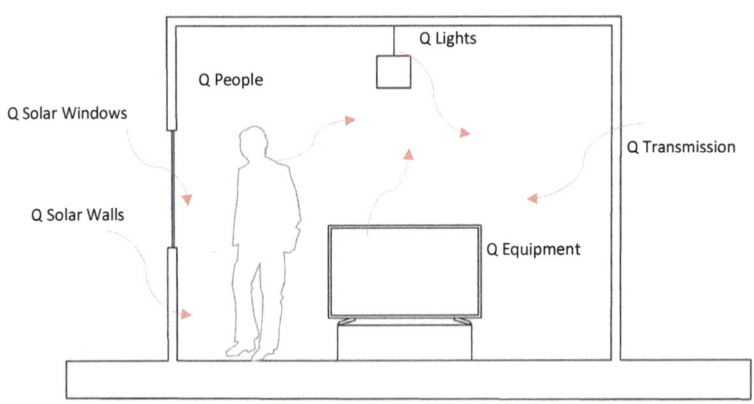

Fig. 3.11 External and internal heat gains of a building

thermal conductivities (k_a, k_b, k_c,…). Practically, this is done by adding thermal insulation to the envelope components and choosing materials with higher *R*-values. It can also be achieved by reducing the surface area (*A*) of the envelope components by proper design and by reducing the difference between the outdoor and indoor temperatures (T_{out} and T_{in}). T_{out} can be reduced by creating a better microclimate that helps to mitigate the outdoor conditions through building form design (e.g., courtyard) and/or landscape design. Finally, reducing $Q_{equipment}$ and Q_{light} can be achieved by more conscious use of equipment (use least heat generating equipment) and the consideration of energy-efficient lighting fixtures.

Cooling a building by passive means requires identifying the sources of coolness (or heat sinks) available in the natural environment where the building is located. The first natural source is the ambient water vapor, which controls the rate of evaporation into the air and evaporative cooling of moist surfaces. The second is the ambient air, which may be at a temperature low enough to reject heat by convection. The third is the upper atmosphere or the sky, which may be used by building hot surfaces to reject heat by thermal radiation. The fourth is wind, which may be in speeds high enough to maintain a desired airflow in the indoor space to promote convective and evaporative cooling.

Vernacular architecture in the hot regions has adopted many solutions to minimize heat gains and benefit from the natural sources of coolness. The following are some examples of vernacular architectural strategies found mostly in the MENA region (Al-Sallal 2016a):

- Clustering the buildings: This helps to reduce the exposed surface area as some are attached together and thus results in less transmission of heat into the indoor spaces. Moreover, it helps to reduce dust infiltration to the indoor spaces, a common problem in hot arid regions.
- Minimizing openings: When the size and number of openings are reduced, this helps to reduce the solar radiation transmitted through the glazing (i.e., $Q_{solar\ windows}$ and $Q_{transmission}$) and the building envelope ($Q_{transmision\ windows}$); hence, it helps to reduce heat gain. Limited openings can help in the reduction of dust penetration to the indoor spaces as well.
- Building with thick adobe walls: Using building materials with high thermal mass capacity such as adobe helps to store night coolness for several hours then release the stored coolness to the interior space during the hot daytime hours when it is needed.
- Creating a cool microclimate: Designing the form configuration of the building as introvert with a central courtyard that is planted with trees and water pool or fountain can help to create a cooler microclimate conditions than the site surroundings.
- Adapting the climate with sensible traditions: Indigenous people learned throughout history how to adapt their lifestyle to match climatic conditions/cycles. For example, the people in the harsh hot dry regions sleep on the roof of their houses when the outdoor ambient temperatures cool down while they move to the cool basements during the extreme hot daytime hours to feel more comfortable.

- Passive cooling systems: If the cooling system runs naturally without fuel or consuming energy, it is considered a passive system. Examples of vernacular passive systems comprise wind catchers, wind towers, curved/domed roofs, ice-makers, and water cisterns. Several reviews can be found in the literature; for example both Bahadori (1978, 1979, 1986) and Al-Sallal (2016a) presented comprehensive reviews about passive cooling systems. Fathy (1986) discussed the passive cooling systems used in the Arab houses of Egypt.

Form Configuration

A self-shading form helps to reduce exposure to the solar radiation. In hot dry climates, it is recommended to adopt compact forms since they have smaller areas of exposed surfaces to the outdoor conditions (high solar radiation and high ambient temperature) and as a result can minimize heat gains. Yet, compact forms are preferable during the daytime only. At nighttime, large exposed surfaces can help to lose heat via re-radiation (to the upper atmosphere) and by convection (to cool breezes). As a balanced solution, Givoni (1994a) has recommended to use a form that includes porches with closing insulated shutters in order to increase the surface area. Moreover, the form should be designed in a way that can improve cooling of the indoor by promoting cool breezes to pass through the building's openings. Generally, in hot humid climates it is recommended to adopt the spread-out forms, while in hot dry climates the introvert form configuration with a central courtyard is more effective (Al-Sallal 2016a). Environmental influences should play an essential role in the spatial zoning of the building. The zoning should help utilize the desirable environmental factors (e.g., cool breezes) while preventing the undesirable ones (e.g., hot and dusty wind).

Thermal and Sociocultural Benefits of the Courtyards

The courtyard design is a sustainable design strategy that can balance between many design requirements: functional, sociocultural, and bioclimatic. The courtyard is an essential element in the vernacular architecture of many hot regions across the world. *Al-Hawsh* (meaning courtyard in Arabic) can be found in the houses of many traditional quarters of the Arab cities or villages (e.g., Khan Al-Khalili in Cairo, Egypt; Zebid village, Yemen; Old Damascus, Syria; Bastakia in Dubai, UAE; Sidi Bu Said, Tunis; Qasba in Algiers, Algeria). It has the potential to decrease excessive sun exposure and mitigate the harsh climate of the desert hot arid regions because the rooms overlooking the courtyard are protected from the heat and the windstorms and sand (Keshtkaran 2011; Meamarian 1999). The courtyard also plays an essential climatic role as a wind generator tool when the cooler air from the surrounding rooms replaces the hot ascended air. It also operates as a tank

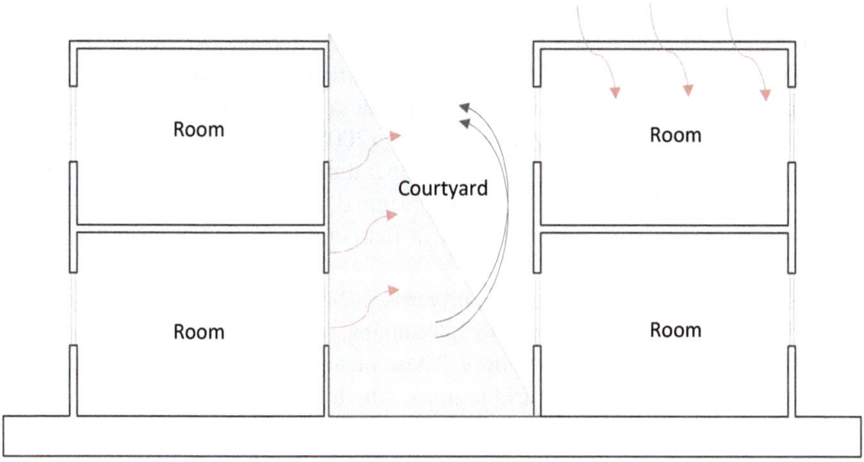

Fig. 3.12 Thermal performance of courtyards

to accumulate the air that was cooled by emission (re-radiation) from the building roof to the night's dark sky. This process helps to create a pleasant microclimate for the residents of the house.

According to Muhaisen (2006) and Al-Hemiddi and Al-Saud (2001) the thermal performance of courtyards depends on two strategies: natural ventilation and protection from solar radiation (shading). Solar shading helps to limit solar heat gains through the walls and in the meantime promotes a pressure differential between the shaded and sunny areas which eventually creates air movement (see Fig. 3.12; Agha 2015). Also, the difference in temperature between the surfaces of the building that absorbs heat and the cool air accumulated in the rooms (by the assistance of the wind scoops) creates convective currents, which provide comfort during the day. The accumulated heat in the building structure is emitted to the clear night sky at night causing radiant cooling. The coolness is stored in the thermal mass of the building, which helps to keep the house cool for several hours in the next day (Al-Sallal 2016a). Moreover, the introvert building form of the courtyard configuration usually uses limited windows to the streets and thus it helps reduce solar heat gain. It also helps reduce the noise levels coming from the outside, and hence creates a calmer atmosphere (Bouchair et al. 2013). The design target when considering courtyard configuration in a building is to maximize solar shading in the summer, while minimizing solar exposure in the winter (Muhaisen 2006; Muhaisen and Gadi 2006; Soflaei et al. 2016). Courtyards remain cold during specific daytime, due to the shading provided by the surrounding attached rooms. After this the temperature increases gradually when the surfaces are exposed to the solar radiation. During nighttime, the surfaces release the stored heat to the sky (Al Jawadi 2011). This generates natural ventilation caused by the heat differences between the hot air in the courtyard and the cold one in the surrounding spaces. Hence, the courtyard works as a channel (or tank) to replace the hot air by the cold air. As a result this operation helps to reduce the building temperature as a whole (Al Jawadi 2011; Moosavi et al. 2014). A study conducted by Al-Masri and Abu-Hijleh (2012) in the UAE tested the difference in energy consumption between a closed building and a multistory courtyard one. It revealed that the building with the courtyard consumed less energy (6.9%) compared to the one with no courtyard. Manioğlu and Yılmaz (2008) in Turkey carried out a similar study testing the difference of room temperature between two houses; the first had a courtyard while the second was designed as a modern house without it. Results showed a significant reduction in temperature (11 °C) attributed to the rooms in the courtyard house.

To increase the performance of courtyards, it is recommended to add means of evaporative cooling such as basins, jars, fountains, or even more primary means like sprinkling water on the courtyard floor. These means proved to promote the evaporative cooling especially in hot arid regions, which helps maintain thermal comfort in the courtyards (Al-Azzawi 1984; Bahadori 1978; Beazley and Harverson 1982). Givoni (1998) explained the process of evaporative cooling when the lower air of the courtyard contacts the water, as this helps moisturize the air. Warm and dry air absorbs water better because they are both equal in the temperature and vapor pressure. Water generally loses its characteristics as a liquid and changes to gas when exposed to the dry air. Hence, the air temperature decreases until it reaches the same level of water temperature. However, it is important to mention that the surfaces should be protected from solar radiation because uncovered surfaces will help equalize the temperature of the water to that of the ambient air. Hence, shading the courtyard is a significant factor that should be taken into consideration (Abdulkareem 2016).

Wind Towers as Passive Cooling Devices

The wind tower is another effective strategy for passive cooling in regions with hot climate especially when it is coupled with the courtyard as this combined arrangement can increase its effectiveness considerably. The wind tower also creates a charismatic architecture with its beautiful vertical form that creates nice contrast with the low horizontal lines of the buildings and a dynamic skyline in the district. In Iran, wind towers are also designed to respond to the climatic conditions. High-speed winds to severe sandstorms are usually common in Yazd city. Therefore, the vernacular wind towers in this region were built to consider both the favorable and the unfavorable winds (Keshtkaran 2011). According to Meamarian, there are two types of wind towers: purely functional or symbolic and functional wind towers. In the second type, the wind tower is used to indicate the financial status of the landowner in addition to its climatic function. He described the wind tower as comprising the following elements from down to top respectively: stokehole, stalk, chest, chain, and shelves (Meamarian 1999).

Thermal Benefits of Curved Roofs

Generally, curved roofs are used in large spaces that require extra height. Traditionally, this central space is usually used for living and socializing. The building roofs tend to loose heat via convection more than radiation. Even though flat and curved roofs absorb the same amount of solar radiation, curved roofs, which have larger surface area, lose heat more effectively compared to the flat ones. The majority of the absorbed heat by the roof is lost by convection. The hot air that is generated inside or outside the house transfers by convective currents to the upper level, which is trapped just under the curved roof. The trapped hot air helps to limit heat transmission from outside to inside by the generated thermal buffer (see Fig. 3.13). When the wind flows over the curved roof, its velocity increases while its pressure decreases. The apex of the curved/domed roof usually includes an opening that permits the hot air to flow out. The curved roof has two forms: a vault (i.e., cylindrical form) or a dome (i.e., spherical form). In cases when wind comes predominantly from one direction it is suggested to use the vault, while the dome is better in cases when the wind comes from several directions.

Building Materials

Building materials played a significant impact in vernacular architecture. In Oman for instance, mud/baked bricks were used to build thick walls (300-450 mm) to help prevent the impact of solar radiations. Roofs generally differ in terms of the house size; for large houses it was common to use flat earth roofs that were supported by

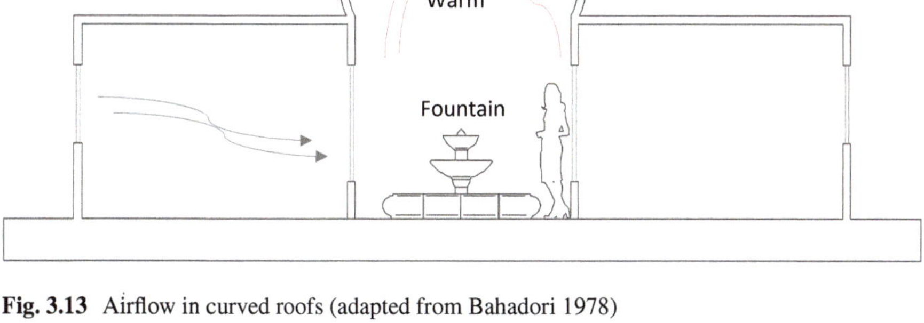

Fig. 3.13 Airflow in curved roofs (adapted from Bahadori 1978)

mangrove poles or palm trunks. Simple mud-brick houses, on the other hand, have pitched palm frond (barristi) roofs. Houses in Oman can reach up to a height of three (3) stories. (Majid et al. 2012). The strategy of thick walls and roofs was used in the UAE as well. Thick walls proved to delay the heat flow inside the thermal mass, and thus reduce its impact in the indoors (Alkhalidi 2013). At daytime, the heat flow passes through the wall to the surface of the indoor wall and this helps to keep the walls cool. At nighttime the operation reverses; the heat flow goes towards the surface of the outdoor walls to keep the indoor surface of the wall cool. The cool wall surfaces absorb the heat from the air and this helps to maintain a comfortable air temperature indoors (Coles and Jackson 1975). The Khaimah (meaning tents in Arabic) and the Arish (or Barasti, meaning palm fronds' huts in Arabic) are two common designs in the vernacular architecture of the Arab Gulf region. The palm tree's trunks were used to construct the frames of the roof, while the branches were used to weave solid mats for the house walls after removing the leaves. The loosely woven mats were used to permit the penetration of the cool breeze in summer, while, in winter, an extra layer was needed to protect the house from the cold (Kansara 2016).

In Iran, and specifically in Yazd, mud and molded materials (especially adobe) are the common construction material used to build houses, wind towers, and even sidewalks. Mud and adobe, besides their availability, have strong resistance against the solar radiations, which made them practical and feasible materials for the climatic conditions of the city. Moreover, the high thermal capacity of the mud and adobe helped to slow down the heat from reaching the indoor spaces. Adobe for instance can help delay the heat transfer through the wall by 8 h; hence the accumulated heat during the day will reach the indoors during nighttime when the temperature is already increased (Meamarian 1999; Moradi 2005).

In hot regions of Algeria, specifically in M'zab settlements, residents used local materials such as lime, stones, and mud to build thick walls and roofs. These thick walls work as an insulator from solar radiation during morning and as a reservoir that releases heat during night (Bouchair et al. 2013). The importance of using local materials in vernacular architecture is also related to the embodied and operational energy in maintenance, construction, and transportation. Modern architecture as it uses non-local materials consumes more energy compared to the vernacular architecture (Chandel et al. 2016). A study conducted by Shukla et al. (2009) tested the embodied energy from construction of a house that uses low energy materials such as cow dung, soil, and sand. Results indicate that vernacular houses proved to save a significant amount of embodied energy and thus CO_2 emissions compared to conventional houses.

Openings

Most of the developed heat in building's indoor comes from solar gain transferred through the building's glazing and opaque surfaces. Hence, controlling solar gain by effective distribution of solid and glazing walls/roofs should be a priority in all hot climates. Generally, in hot dry climates it is recommended to avoid large openings especially on the west and east orientations due to the low solar altitude. On the other hand, large openings are recommended in hot humid climates to help promote cool breezes into the building, but they have to be shaded by extended porches or shading devices. South orientation openings require external horizontal shading devices, while west and east orientation openings require vertical external shading devices. Egg-crate shading devices, which help prevent all kinds of solar radiation including the direct, the diffuse, and the reflected components, are a suitable solution in very hot climates. Cutting down the heat gain by manipulation of the solar-optical properties of the glazed surfaces is another important strategy that needs to be considered when daylighting is a design requirement.

In Iran it was common to minimize openings to the streets and place them on the courtyard to offer higher protection from sun radiation. When windows were used on the streets, they were placed at high level on the wall to protect the house from excessive sunlight and dust (Meamarian 1999). Similarly, Algerian houses in hot regions followed the same pattern; these houses were constructed with small or no openings towards the streets but towards the central courtyard. In some rare cases, one can find windows overlooking the public roads, yet with horizontal overhangs to provide shading and minimize glare. Moreover, windows were placed high on the walls to reduce the heat gain and increase the light penetration (Bouchair et al. 2013). The location of windows in vernacular architecture was also highly influenced by the sociocultural requirements. These requirements usually did not favor the use of external openings due to the significant considerations of privacy. As a result, this could reduce the effectiveness of natural ventilation (and thus the overall thermal performance) due to the role of external windows to discharge the hot air inside the rooms and allow the fresh cold air from the courtyard to enter the room. As a solution to this issue, many alternatives have been used. For instance, in Baghdad small high openings overlooking the streets are created in the ground floor, while, in the first floor, larger windows covered with timber lattice screen or "Mashrabiya" helped to promote cross ventilation and air movement (Al-Azzawi 1984). In other places, such as Diyarbakir, windows are placed in the upper and lower levels. The upper level windows are for allowing the hot air to exit, while the lower level windows are for introducing the cold air to the adjacent spaces (Bekleyen and Dalkiliç 2011). Vernacular windows helped also in promoting evaporative cooling by the help of other means (e.g., porous water jars) to moisturize the air and thus reduce its temperature (Abdulkareem 2016; Aboul Naga 1990). In addition to ventilation and passive cooling, vernacular windows were also used for the provision of natural lighting.

Color of the Envelope

The color of the envelope's surfaces has proved to have a remarkable impact on the building's heat gain and the indoor temperature resulted. Examples from several traditional settlements in hot climates, such as the houses of the Sidi Bu Said village

Fig. 3.14 Minimizing heat gain via white color of the building's envelope as demonstrated by the traditional architecture of Zebid, Yemen (ERYJ 2016)

on the Mediterranean in Tunis and the houses of the Zebid village on the Red Sea in Yemen (see Fig. 3.14), demonstrate the appropriateness of using white (or other light colors) to minimize heat gains in hot regions.

Generally, it is recommended to use white surfaces for the building envelope in hot climates. An experiment carried out in Haifa by Givoni (1994b) tested the difference of using roofs that were made of the same material (i.e., Ytong or lightweight concrete 600 kg/m^3) in three different thicknesses (7, 12, 20 cm), but differ in their colors (white and gray). The results demonstrated a significant difference between the two cases. The external surface temperature for the white and gray roof was equal to 27.5 and 69 °C, respectively. The temperatures in the indoor ceiling was about 25.5 °C for all tested thicknesses of the white roof; and 45 °C, 39 °C, and 33 °C for the gray roof with a thickness of 7, 12, and 20 cm, respectively. In another similar experiment, Givoni tested walls made of the same material (solid concrete and Ytong, 12 and 22 cm thick) but in two different colors, white and gray. The results of the gray wall case showed an increase in the average indoor temperature (4.5 °C) above the average outdoor temperature and increase in the maxima indoor temperature (8 °C for the 12 cm) above the maximum outdoor temperature. On the contrary, the results of the white wall revealed that the average indoor air temperatures were below the average outdoor temperature most of the daytime hours, and the indoor maxima were about 2 °C below the outdoor maximum.

Case Studies

Coastal Architecture in the Arab Gulf

The most common building form configuration in the Arab Gulf region is "the introvert plan," in which the rooms are arranged around a central courtyard. This plan responds to both social and climatic requirements. Privacy of family members (especially women) and flexibility of the use of spaces to suit extended families are two examples of sociocultural requirements (Al-Sallal 2016a). The reliance on bioclimatic ventilation openings such as the Dreeshah (main window) and the Masqat (wind puller) with different shapes (square or rectangle) was common in the vernacular UAE houses (Alkhalidi 2013; Vale and Vale 1991). Windows were used in two parallel walls to provide ample cross circulation airflow in addition to other vents that promote escape of the hot air. A special decorative pattern is usually used to secure those openings without blocking the ventilation (Alkhalidi 2013; Vale and Vale 1991). The Leewan, which is a big space used generally as a living room, is another means for ventilation provision (Alkhalidi 2013). It is covered from the road side with wood screens while the other side is left open to the court. The main axis of the Leewan takes South-North to avoid excessive heat gains and sun penetration (Steele and Fathy 1997).

The vernacular house relied on some strategies to avoid the excessive heat of the climate and provide thermal comfort. Several methods were adopted to provide effective solar shading. These are as follows:

- The walls and vegetation in the courtyard help to maximize shading during the day hours.
- The Masqat (air-puller wall) provides shading through its solid walls.
- The wooden shutters of the *dreeshah* (the traditional window) are flexible shading devices that can provide partial or full shading.
- The verandahs (arcades) around the courtyard provide considerable shading to the rooms' windows and walls.

For passive cooling, other methods were adopted. These are as follows:

- The courtyards played an essential role as a wind generating tool. They helped to promote convective cooling, where the hot air goes up and the cooler air replaces it from the surrounding rooms. The generated air currents result also in evaporative cooling by the passage of air through and around the vegetation in the courtyards and thus help in lowering the air temperature.
- The wind tower helps to cool the rooms through its four triangular vertical tunnels catching air from all directions. It is usually used in the major living room of the house with a vertical form taking a rectangular shape comprising arched openings pointed to the outside. The openings allow the fresh air to enter the shaft of the tower then to the living spaces below. In summer, the apertures are covered with a metal mesh to prevent pigeons and dirt from entering, while in winter they are covered completely with wooden shutters (Coles and Jackson 1975). In order to take advantage of the breeze coming from any side, the Barageel (airshaft) in the UAE has four sections shaped as separate triangles when seen in cross section of the tower shaft. These sections were built this way to face the four cardinal points. The thermal mass of the walls absorbs the heat from the incoming airflow (through exchange of the heat with the night coolth stored in the tower walls) to promote a comfortable cool air in the indoor space compared with the air outside. In the hot dry

regions, it is common to place water in the bottom of the Barageel to humidify the air (Alkhalidi 2013). Hence, wind towers were used as an evaporative cooling means.
- The air-puller wall permits the circulation of airflow through the rooms and removes hot air through convective cooling. Its solid wall helps to minimize solar gain to the indoor space while it maintains the dwellers' privacy.

Vernacular Tower House of Sana'a, Yemen

Al-Sallal (2001), through his investigation of the vernacular houses of Sana'a, explained how that the vertical form of the traditional tower house of Sana'a has bioclimatic advantages with regard to solar access and shading. In summer, when the solar altitude angles are high, less surfaces of the vertical form are exposed to the perpendicular solar radiations compared to the horizontal form, which results in less heat gains. In winter, on the other hand, when the solar altitude angles are low, more vertical surfaces are exposed to the solar radiation, compared to the horizontal ones, especially on the south façade (east and west façades are generally blocked by attached buildings). This helps to promote higher heat gains (see Fig. 3.15).

The lower floor of the tower house is frequently double story in height and used for keeping animals and for storage (Al-Sallal 2013). The higher floors are residential quarters and include rooms that are functionally polyvalent and nonspecific, rooms that can be used interchangeably for eating, sleeping, recreation, and domestic tasks. This flexible use of living space is reflected by the absence of cumbersome furniture. The tower house walls are built of 50 cm square, black, volcanic ashlar stone on the lower levels (i.e., up to approximately 6–10 m above street level) and 40 cm baked exposed brickwork above that. The roof consists of a frame of wooden beams set 50–60 cm apart, covered by branches and twigs, on top of which lie layers of finely sifted earth, wet and compact, up to a thickness of 30 cm.

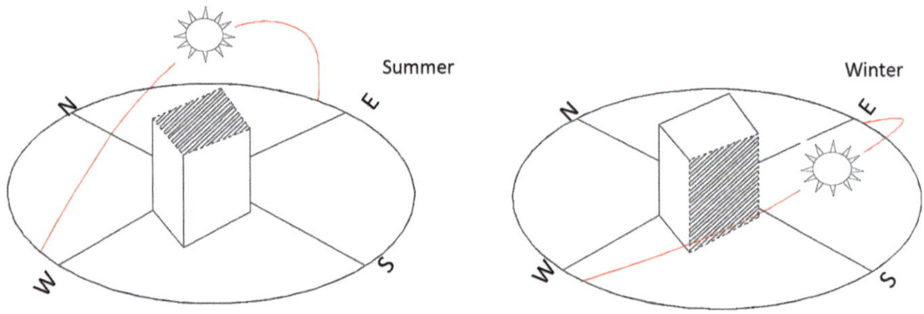

Fig. 3.15 Sun path during summer and winter and its impact on exposed surfaces of vertical buildings

According to Al-Sallal (2001, 2013, 2016a), four (4) conditions must exist to achieve a controlled thermal environment. First, the implementation of a good thermal insulation for the structure to reduce heat transfer by conduction and minimizing air infiltration must be considered. Second, the size and orientation of the window area used for solar collection should be adequate. Third, the thermal mass should be distributed properly to ensure absorption of energy generated from solar radiation without increasing air temperature in the indoor space. Finally, shading of the solar collection glazing with attention to drain out the unneeded accumulated heat should be considered during the times when energy collection is not needed. The Sana'a tower house satisfied these four conditions, which can be explained as follows:

- The first condition (energy conservation) is satisfied by the selection of low conductive construction materials for the vernacular walls (e.g., burnt mud bricks with U value = 0.72 $Wm^{-2}K^{-1}$) and roofs (e.g., earth filling with U value = 1.83–2.86 $Wm^{-2}K^{-1}$). These vernacular materials are more energy efficient and provide better thermal comfort than most of the modern construction materials used today in Yemen (Al-Sallal, 1993).
- The second condition is satisfied by the proper orientation of the building's glazing for solar collection, which is towards the south. The rooms that have higher frequency of occupancy, which usually exist in the higher floors, are provided with larger windows. This helps to improve opportunities for access to the garden views and provision of thermal comfort and ventilation. In this regard, the vertical zoning of the tower house (arrangement of activities vertically) helped to match between the thermal comfort requirements and the socio-cultural needs.
- The third condition is achieved by the use of highly massive materials on the walls and floors that helps in securing a comfortable indoor thermal environment with stable internal temperature during the summer and winter (Al-Sallal et al. 1995). The walls in the upper floors, where the occupants live, were constructed by burnt-bricks, which has a relatively low heat transfer coefficient (U=0.72 $Wm^{-2}K^{-1}$) with high thermal mass. The solid stone material that has a relatively high heat transfer coefficient (U=1.47 $Wm^{-2}K^{-1}$) with high potential of thermal mass is used in the lower floors whose spaces are mainly utilized for seasonal storage of food and raising animals (chicken or sheep) (Al-Sallal, 1993). This approach of distributing the proper wall material as per the needs of each floor helped to maximize indoor thermal comfort for the occupants, provide cool spaces for seasonal storage, strengthen the structure of the building (the stone walls carry the brick walls but not the opposite), and protect the building structure at the street level from potential erosion caused by flooding of rain water and daily traffic.
- The fourth condition is achieved by the provision of effective shading devices such as the overhangs (or *kunnah* in the local dialect of Sana'a) and natural ventilation through the small vents located near the ceilings (or *shaqoos* in the local dialect of Sana'a) to prevent space overheating (Al-Sallal and Cook 1992).

Vernacular Courtyard House of Baghdad, Iraq

The traditional house of Baghdad can demonstrate the effectiveness of the courtyard configuration and how the courtyard integrates with the wind scoops to minimize heat gains and provide passive cooling. The courtyard collects all the warm air being pushed out from the rooms by the wind scoops and get rids of the heat by radiation to the sky and by convection to the ambient air, especially during the night hours (Fardeheb 1987; Al-Sallal 2016a). Generally, designing courtyards requires a comprehensive study of the courtyard space configuration and how it integrates with the affecting factors and pattern environmental strategies (Al-Hafith et al. 2017). Al Jawadi (2011) tested the difference between the outside temperature and the temperature in a house courtyard in Iraq. A decrease of 9 °C was recorded in the courtyard compared to the outdoor temperature.

Wind scoops operate in a similar way to the wind towers. They can be found in the Baghdadi vernacular house of Iraq. The main differences between the wind scoop and the wind towers are in the size and the facing direction; wind scoops are generally smaller and shorter than the wind towers and they face only one direction. Figure 3.16 illustrates an example of a wind scoop in Baghdad. The device has the following dimensions: 90–120 cm for both width and height over roof level opening, depth equals to 60 cm, and finally, inlet opening height 90–120 cm (Al-Azzawi 1969; Fardeheb 1987). The top part of the roof is made of wood, brick, or metal and includes inlet openings covered with a 45° inclined top surface in a design that promotes wind capturing. At night, the coolness is stored in the high thermal mass walls by the channeled cool wind. During daytime, the cool walls capture and cool the wind, which helps to push the warm air from the rooms to the courtyard. The

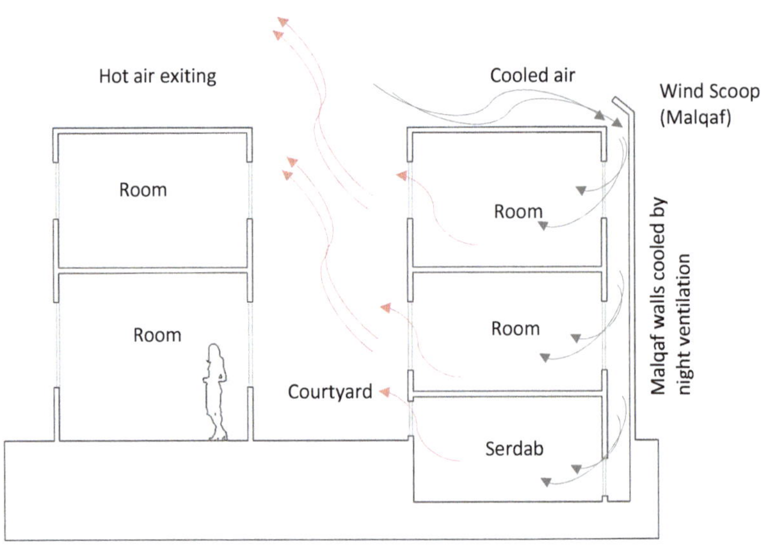

Fig. 3.16 The wind scoop of the traditional house of Baghdad (adapted from Al-Azzawi 1969)

courtyard consequently collects all the warm air and discharges the heat by convection to the ambient air or radiation to the sky, especially during the nighttime. Each room has one wind scoop and the number increases in the cases of large rooms. The malqaf can help to cool the room temperature by 5 °C compared to the outside (Fardeheb 1987; Lezine 1971).

The Baghdadi traditional house depended on wind scoops to ventilate and cool rooms and basements. The *serdab* (meaning basement in Arabic) is the coolest space in this traditional house (see Fig. 3.17). A previous study monitored five different rooms in a traditional Baghdadi house, one of which was in the serdab. A peak reduction of 21% in the temperature was recorded in the serdab at 3:00 pm. Wind catchers can also promote dehumidification. A measurement study was conducted to test the performance of the serdab (in terms of air temperatures and relative humidity), with and without wind catchers. In the case without wind catcher, air temperature and relative humidity were equal to 25–28 °C and 40–55% (during the day) and 60–65% (at night), respectively. In the case with wind catcher, air temperature was unexpectedly higher (27 °C at midnight to 32 °C in the afternoon) while humidity was lower (25% at 4:00 pm to 35% at 2:00 am) (Danby 1984; Fardeheb 1987).

Vernacular Architecture of Hot Arid Regions in Iran

Trees and plants along with high shading around the building's site and in courtyards maintained indoor and outdoor thermal comfort in the vernacular architecture of the hot arid regions of Iran. Adding trees, shrubs, and flowers can create a self-sufficient microclimate as it increases the relative humidity and provides shading (Soflaei et al. 2016). According to Safarzadeh and Bahadori (2005) and Soflaei et al. (2016) the passive cooling of the courtyard has four (4) features. These are as follows:

1. The shading provided by the courtyard walls on the ground, south-facing wall and windows.

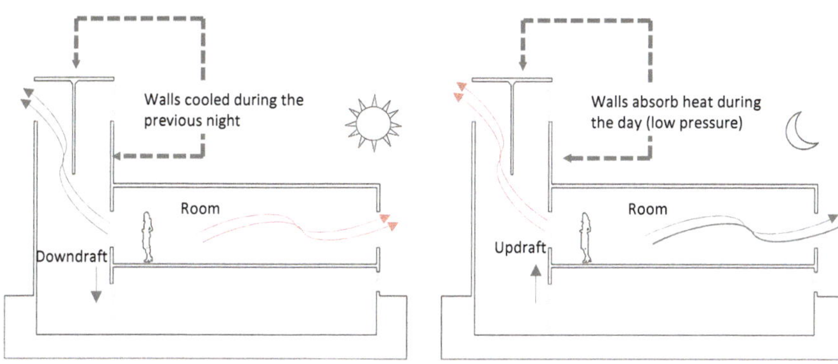

Fig. 3.17 How wind conditions affect wind tower (Baud-Geer) modes of operation

2. The shading provided by trees on the ground, south-facing windows.
3. The effects of the lawn, shrubs, flowers, and pool.
4. The wind-shading effects provided by the courtyard walls and trees on the infiltration rate of air through the building.

Safarzadeh and Bahadori (2005) investigated numerically the passive cooling effects of a courtyard in a small building in Tehran, which was shaded by the courtyard walls and two large trees (of various shapes) planted immediately next to the south wall of the building and had a pool, a lawn, and flowers in the courtyard. Choosing appropriate types of trees and distributing them properly on the site can help to minimize solar radiation and provide shading. Trees and other plants can also help to reduce the ambient temperature through the process of evapotranspiration and create areas for outdoor activities. Using limited water bodies such as pools, fountains, and water channels shaded by plants can improve evaporative cooling and create beautiful outdoor environments. Outdoor landscaping encourages occupants to spend some hours outside the building and hence helps to reduce the cooling load inside. The study concluded that these features can reduce the cooling energy requirements of the building, but they cannot maintain thermal comfort by themselves during the extreme hot summer hours in Tehran.

The wind catcher (also called Baud-Geer in Persian) is a cooling and ventilation device, developed in the traditional architecture of Iran and some Arab countries. It operates by creating change in air density in and around the tower, which helps to create a draft that pulls the air either up or down through the tower. Its operation is influenced by the wind conditions and time (see Fig. 3.13; Bahadori 1978), which can be explained as follows:

- Absence of wind at nighttime: When there is no wind at nighttime, the wind tower functions as chimney. During the daytime, the tower walls are exposed to the sun; they absorb the solar heat and transfer it to the inside. The heat warms the air layers next to the walls. This creates difference in air temperature and a natural convection is created. The hot air rises and exits the tower while it pulls the air from the windows in the lower levels.
- Absence of wind during the day: When there is no wind during the day, the wind tower functions as a reverse chimney. The upper part of the tower is cooled down at night and remains cool during the early hours of the daytime; as a result the air layers next to the walls are also cooled down. This creates natural convection but in a reverse direction; the difference in air temperature pushes the cool air down and distributes it into the rooms. The cool air from the tower replaces the hot air in the rooms. When the temperature of the tower reaches that of the ambient air, the tower begins to operate again like a chimney.
- Existence of wind at night: When there is wind at night, the tower works as a wind catcher, by catching the wind at the top opening. Higher towers are more effective than lower ones since they can reach to the air layers that have higher wind speeds. Lower wind towers are less effective because of the wind resistance caused by the ground and building masses that result in slowing down wind speeds.

- Existence of wind during the day: When there is wind during the day, the rate of air circulation is increased as a result of combined effects of wind catching (similar to the wind nighttime operation) and solar chimney (similar to the no-wind daytime operation).

In the vernacular architecture of Iran, wind towers combine evaporative and sensible cooling. The following methods were invented for this purpose:

- By implementing a small pool with a fountain at thc bottom of the wind tower.
- Increase the distance between the tower and the building (approx. 50 m), in addition to the implementation of a ground tunnel that connects between the building basement and the bottom of the wind tower (see Fig. 3.18). Planting the ground above the tunnel was used to keep the soil moist and hence can help to cool the air passing through the tunnel evaporatively and sensibly.
- Wind tower in conjunction with underground stream (see Fig. 3.19): The velocity of the air increases when it gets into the tower and then slows down, and its pressure decreases when it moves from the vertical tower to the door opening (smaller in cross section). This point is near to the outlet opening of the stream shaft inside the building and as a result the pressure decreases. The pressure continues to decrease until it creates sufficient pressure difference between this point and the top outer opening of the stream shaft. This results in the sucking of ambient air from the outer opening of the stream shaft. The air is channeled through the stream shaft to pass over the water stream and is cooled evaporatively before entering the building.

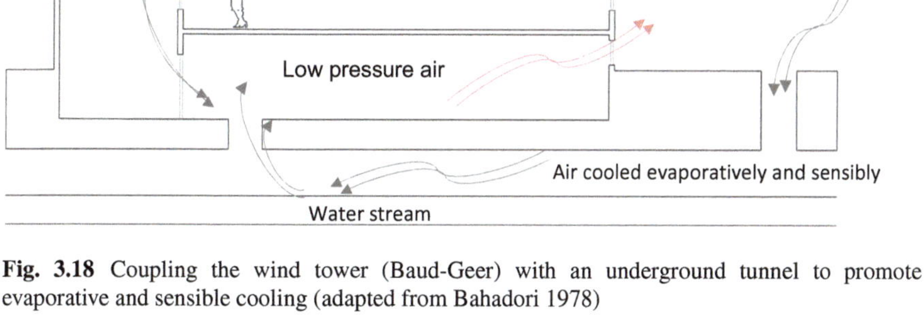

Fig. 3.18 Coupling the wind tower (Baud-Geer) with an underground tunnel to promote evaporative and sensible cooling (adapted from Bahadori 1978)

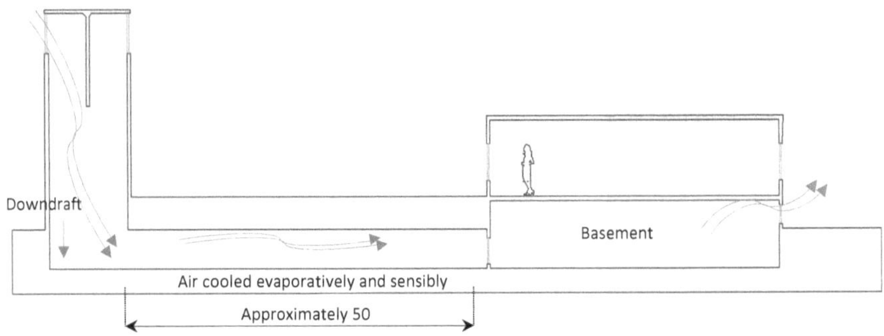

Fig. 3.19 Coupling the wind tower (Baud-Geer) with a water stream to promote evaporative and sensible cooling (adapted from Bahadori 1978)

Vernacular House of Cairo, Egypt

The Malqaf (wind catcher in Arabic) is a device that was used in the traditional architecture of Cairo, Egypt, to provide passive cooling and ventilation. The same concept of the device was found in different countries in the Middle East under different names. The wind catcher was used to cool down the indoor spaces by catching the cool wind from upper levels where the air is less dusty and cooler. In some cases windows cannot provide sufficient cooling airflow for many reasons; for instance, windows are designed to serve other requirements such as daylighting and view, which require specific window size (larger or smaller than required for ventilation), and this might conflict with the cooling requirements. Hence, wind catchers became very popular and useful in bringing cool and less dusty air to indoor spaces in these kinds of urban environments.

The Qa'a of Mohib Addin Ashaafi Al-Muwaqqi in Cairo is a good example of the traditional Egyptian houses cooled by the Malqaf. This wind catcher was built in the fourteenth century A.D. It operates by the air movement caused by the difference of pressure (Fig. 3.20). This design employed the stack effect to produce convective cooling during the daytime when the wind velocity is slow (Fathy 1986). The increase of the pressure at the entrance of the malqaf causes passage of cool breeze from the north through the "lwan," an intermediate zone that connects to the "Qa'a" or the central space. From there, the air flows to the "Dur-qa'a" or the upper part of the central space; then it exits through the shaded high-level windows (close to the ceiling). The design of the roof that covers the central space promotes airflow using the Bernoulli effect. The top part of the roofs helps to generate stack effect that is caused by the difference in temperature between the cooler inlet (caused by solar heat gain through the roof) opening and the exit point, which improves the airflow performance. The well-known architect, Hassan Fathy, attempted to develop the traditional design to improve its performance (see Fig. 3.21). He added wetted baffles and attached the wind tower to a wind-escape in order to promote the airflow and increase the evaporative cooling (Fathy 1986).

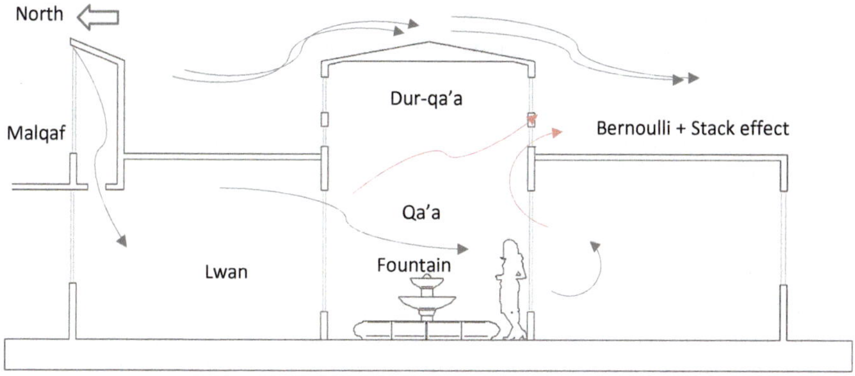

Fig. 3.20 Operation of the Egyptian malqaf in the Qa'a of Mohib Addin Ashaafi Al-Muwaqqi (adapted from Fathy 1986)

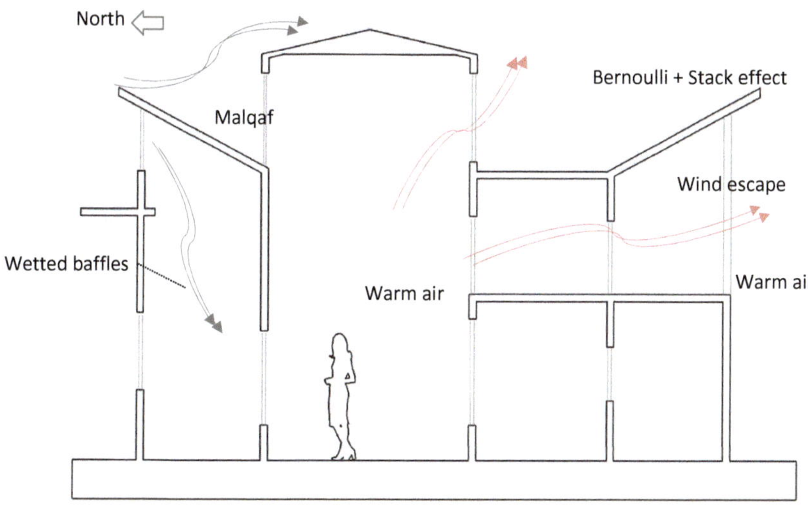

Fig. 3.21 Improving the performance of the Egyptian malqaf by adding wetted baffles and a wind-escape to maximize evaporative cooling and promote airflow (adapted from Fathy 1986)

Conclusion

Vernacular architecture will always remain as a model to re-explore how it managed to keep a balance between human functions, sociocultural values, and human comfort. It was based on history-long practical experiences of the indigenous people, and hence proved its robustness against harsh climatic conditions before modern technologies were not yet introduced. The strategies in vernacular architecture were applied from the larger scale of the urban massing and city spatial arrangement to other smaller scales of the buildings' clustering and building and passive systems design. These all worked together to provide a comfortable microclimate and satisfy the functional and sociocultural needs. Effective passive cooling relies on minimizing all sources of heat gain along with strategies to improve heat loss by structures and people. The vernacular strategies adopted in the MENA region can be summarized into the following: Clustering the buildings, minimizing openings, building with thick adobe walls, creating a cool microclimate, adapting the climate with sensible traditions, and finally using passive cooling systems.

At the urban level, the morphology of the city was shaped to respond to climatic requirements. In Sana'a, Yemen, for instance, it helped to enhance harnessing of solar radiation for passive heating in the relatively cold winter of the highlands region, along with solar shading and passive cooling strategies in the summer. The compact urban fabric was a common strategy in many hot climate locations found in the Arab Gulf, Iran, Algeria, Morocco, and Egypt. Compactness and buildings' attachment helped to reduce the exposure of the building surfaces to the harsh climate and hence improve thermal comfort and provide efficient spaces.

At the architectural level, the self-shading form helped minimize heat gain into the building. The vernacular courtyard house served as the heart of the functional and sociocultural activities while in the meantime served a major role as a passive cooling system. Wind towers were effective solutions for passive cooling in hot climates especially if they were combined with courtyards. Local building materials such as mud and backed bricks were used to build thick walls to modulate the extreme hot temperatures. The windows in the hot dry regions were small, shaded, and properly oriented to avoid solar gains. In the hot humid regions, windows were shaped and sized to promote cool breezes. Finally, light colors were used in the external skin of buildings to mitigate the impact of solar radiation.

The bioclimatic solutions once practiced in the vernacular architecture of many settlements in the MENA region provide very valuable knowledge base from which we can learn theory and application of sustainable design solutions.

References

26-SDNS (2011) 26 September Daily News Site, The Directorate of the old Sana'a is going out of its way on Friday in support of legitimacy, Friday April 1st, 2011. http://www.26sep.net/news_details.php?sid=72682. Accessed 16 Apr 2018

Abdulkareem HA (2016) Thermal comfort through the microclimates of the courtyard. A critical review of the middle-eastern courtyard house as a climatic response. Procedia Soc Behav Sci 216:662–674. https://doi.org/10.1016/j.sbspro.2015.12.054

Aboul Naga M (1990) Natural ventilation and cooling by evaporation in hot-arid climates. PhD thesis, doctoral dissertation. Retrieved from http://etheses.whiterose.ac.uk/4043/1/uk_bl-ethos_278300.pdf

Agha R (2015) Traditional environmental performance: the impact of active systems upon the courtyard house type, Iraq. J Sustain Dev 8(8):28. https://doi.org/10.5539/jsd.v8n8p28

Al Jawadi M (2011) Model of house design responsive to hot-dry climate. J Hous Sci 35(3):171–183

Al-Azzawi SH (1969) Oriental houses in Iraq. In: Oliver P (ed) Shelter and society. Barrie and Rockliff, London, pp 91–102

Al-Azzawi SA (1984) A descriptive, analytical and comparative study of traditional courtyard houses and modern non-courtyard houses in Baghdad in the context of urban design in the hot dry climates of the sub-tropics. PhD thesis. Retrieved from http://ethos.bl.uk/OrderDetails.do?uin=uk.bl.ethos.311467

Al-Hafith O, Satish BK, Bradbury S, Wilde PD (2017) The impact of courtyard parameters on its shading level an experimental study in Baghdad, Iraq. Energy Procedia 134:99–109. https://doi.org/10.1016/j.egypro.2017.09.539

Al-Hemiddi NA, Al-Saud KA (2001) The effect of a ventilated interior courtyard on the thermal performance of a house in a hot–arid region. Renew Energy 24(3–4):581–595. https://doi.org/10.1016/s0960-1481(01)00045-3

Alkhalidi A (2013) Sustainable application of interior spaces in traditional houses of the United Arab Emirates. Procedia Soc Behav Sci 102:288–299. https://doi.org/10.1016/j.sbspro.2013.10.743

Al-Masri N, Abu-Hijleh B (2012) Courtyard housing in midrise buildings: an environmental assessment in hot-arid climate. Renew Sust Energ Rev 16(4):1892–1898. https://doi.org/10.1016/j.rser.2012.01.008

Al-Sallal KA (1993) The role of indigenous fenestration to reduce energy requirements in the Sana'a house. In: Proceedings of fourth Arab international solar energy conference, vol 2. Royal Scientific Society, Amman, pp 875–888

Al-Sallal KA (1996a) Traditional methods in new forms: insight to achieve energy conservation in the modern Sana'a house. In: Proceedings of world renewable energy congress IV in Denver

Al-Sallal KA (1996b) Solar access/shading and building form: geometrical study of the traditional housing cluster in Sana'a. In: Proceedings of world renewable energy congress IV in Denver

Al-Sallal KA (2001) The balanced synthesis of form and space in the vernacular house of Sanaa: bioclimatic and functional analysis. Archit Sci Rev 44(4):419–428. https://doi.org/10.1080/00038628.2001.9696922

Al-Sallal KA (2004) Sana'a: transformation of the old city and the impacts of the modern era. In: Elsheshtawy Y (ed) Blurring identities: the middle eastern city in an age of globalization. Routledge, London

Al-Sallal KA (2013) Vernacular tower architecture of Sana'a: theory and method for deriving sustainable design guidelines. In: Sayigh A (ed) Sustainability, energy and architecture: case studies in realizing green buildings. Academic Press, London, p 257

Al-Sallal KA (2016a) Passive and low energy cooling. In: Low energy low carbon architecture. CRC Press, Boca Raton, pp 57–102

Al-Sallal KA (2016b) Learning sustainability from Arab gulf vernacular architecture. In: Sayigh A (ed) Mediterranean green buildings & renewable energy. Springer, Cham, pp 885–896

Al-Sallal KA, Abouelhamd A (2018) Natural ventilation in hot seaside urban environments. In: Sayigh A (ed) Seaside building design: principles and practice. Springer, Berlin

Al-Sallal KA, Al-Rais L (2011) Outdoor airflow analysis and potential for passive cooling in the traditional urban context of Dubai. Renew Energy 36(9):2494–2501. https://doi.org/10.1016/j.renene.2011.01.035

Al-Sallal KA, Al-Rais L (2012) Outdoor airflow analysis and potential for passive cooling in the modern urban context of Dubai. Renew Energy 38:40–49

Al-Sallal KA, Al-Zaroony S (2000) Climatic and social impact on physical forms of the vernacular architecture in Sharjah. In: Proceedings of American solar energy society SOLAR'2000 conference in Madison, Wisconsin, 16–21 June

Al-Sallal KA, Al-Zaroony S (2001a) Climatic and social impact on physical forms of the vernacular architecture in Sharjah. In Abstracts of 2nd annual UAEU research conference, Al-Ain, 17–18 April

Al-Sallal KA, Al-Zaroony S (2001b) Vernacular approach for low energy architecture: the case of Al-Murija district in Sharjah. In: Proceedings of Sharjah solar energy Conference 2001 in Sharjah, 19–22 February

Al-Sallal KA, Cook J (1992) Sana'a historical windows: integration between comfort and aesthetics. In: Burley S, Arden ME (eds) Proceedings of 17th national passive conference of American solar energy society, vol 17. American Solar Energy Society, Boulder, pp 197–202
Al-Sallal KA, Ayssa AZ, Al-Sabahi HA (1995) Thermal performance and energy analysis for Sana'a vernacular house. In: Ben Ghadi S (ed) Proceedings of applications on renewable energy in Yemen workshop, Aden University, Yemen
Bahadori MN (1978) Passive cooling systems in Iranian architecture. Sci Am 238(2):144–154. https://doi.org/10.1038/scientificamerican0278-144
Bahadori MN (1979) Natural cooling in hot arid regions. In: Sayigh AAM (ed) Solar energy applications in buildings. Academic, Cambridge, pp 195–225
Bahadori MN (1986) Natural air-conditioning systems. In: Böer KW (ed) Advances in solar energy. Plenum Press, New York, pp 283–356
Beazley E, Harverson M (1982) Living with the desert working buildings of the Iranian plateau. Aris and Phillips, Warminster
Bekleyen A., Dalkiliç N (2011) The influence of climate and privacy on indigenous courtyard houses in Diyarbakır, Turkey. Scientific research and essays. https://doi.org/10.5897/SRE10.958
Bouchair A, Tebbouche H, Hammouni A, Lehtihet M, Blibli M (2013) Compact cities as a response to the challenging local environmental constraints in hot arid lands of Algeria. Energy Procedia 42:493–502. https://doi.org/10.1016/j.egypro.2013.11.050
Chandel S, Sharma V, Marwah BM (2016) Review of energy efficient features in vernacular architecture for improving indoor thermal comfort conditions. Renew Sust Energ Rev 65:459–477. https://doi.org/10.1016/j.rser.2016.07.038
Coles A, Jackson P (1975) A wind tower house in Dubai. AARP, London
Danby M (1984) The internal environmental aspects of the traditional Islamic house and their relevance to modern housing. In: Bowen A (ed) Proceedings of PLEA (1984). Pergamon Press, New York, pp 664–671
ERYJ (2016) Embassy of the republic of Yemen in Japan. http://www.yemen.jp/images/zebid-01.jpg. Accessed 13 Apr 2018
Fahmy M, Sharples S (2009) On the development of an urban passive thermal comfort system in Cairo, Egypt. Build Environ 44(9):1907–1916. https://doi.org/10.1016/j.buildenv.2009.01.010
Fardeheb F (1987) Examination and classification of passive solar cooling strategies in middle eastern vernacular architecture. Passive Solar J 4(4):377–417
Fathy H (1986) Natural energy and vernacular architecture: principles and examples with reference to hot arid climates. University of Chicago Press, Chicago
Givoni B (1994a) Passive and low energy cooling of buildings. Van Nostrand Reinhold, New York
Givoni B (1994b) Building and urban design guidelines for different climates. Van Nostrand Reinhold, New York
Givoni B (1998) Climate considerations in building and urban design. Wiley, Baltimore
Google Earth Pro (2018) https://earth.google.com/web
Johansson E (2006) Influence of urban geometry on outdoor thermal comfort in a hot dry climate: a study in Fez, Morocco. Build Environ 41(10):1326–1338. https://doi.org/10.1016/j.buildenv.2005.05.022
Kansara T (2016) Transitional zone design in the desert environment of Abu Dhabi: vernacular vs. modern. Sustain Cities Soc 23:59–67. https://doi.org/10.1016/j.scs.2016.02.003
Keshtkaran P (2011) Harmonization between climate and architecture in vernacular heritage: a case study in Yazd, Iran. Procedia Eng 21:428–438. https://doi.org/10.1016/j.proeng.2011.11.2035
Kim T, Park J (2010) Natural ventilation with traditional Korean opening in contemporary house. Build Environ 45(1):51–57. https://doi.org/10.1016/j.buildenv.2009.05.016
Lee K, Han D, Lim H (1996) 96/04277 passive design principles and techniques for folk houses in Cheju Island and Ullung Island of Korea. Energy Fuels 37(4):296. https://doi.org/10.1016/0140-6701(96)82552-5
Lezine A (1971) La protection contre la chaleur dans l'architecture Musulmane d'Egypte. Bull Etud Orient 124, Damascus Syria: 7–17

Majid NH, Shuichi H, Takagi N (2012) Vernacular wisdom: the basis of formulating compatible living environment in Oman. Procedia Soc Behav Sci 68:637–648. https://doi.org/10.1016/j.sbspro.2012.12.255

Manioğlu G, Yılmaz Z (2008) Energy efficient design strategies in the hot dry area of Turkey. Build Environ 43(7):1301–1309. https://doi.org/10.1016/j.buildenv.2007.03.014

Meamarian GH (1999) Residential architecture of Iran. The University of Science & Industry, Tehran, pp 239–347

Moosavi L, Mahyuddin N, Ghafar NA, Ismail MA (2014) Thermal performance of atria: an overview of natural ventilation effective designs. Rcnew Sust Energ Rev 34:654–670. https://doi.org/10.1016/j.rser.2014.02.035

Moradi S (2005) Environmental control system. Ashiyan, Tehran, pp 34–41, 55–57, 128–134

Motealleh P, Zolfaghari M, Parsaee M (2016) Investigating climate responsive solutions in vernacular architecture of Bushehr city. HBRC J. https://doi.org/10.1016/j.hbrcj.2016.08.001

Muhaisen AS (2006) Shading simulation of the courtyard form in different climatic regions. Build Environ 41(12):1731–1741. https://doi.org/10.1016/j.buildenv.2005.07.016

Muhaisen AS, Gadi MB (2006) Effect of courtyard proportions on solar heat gain and energy requirement in the temperate climate of Rome. Build Environ 41(3):245–253. https://doi.org/10.1016/j.buildenv.2005.01.031

Safarzadeh H, Bahadori M (2005) Passive cooling effects of courtyards. Build Environ 40(1):89–104. https://doi.org/10.1016/j.buildenv.2004.04.014

Shan M, Wang P, Li J, Yue G, Yang X (2015) Energy and environment in Chinese rural buildings: situations, challenges, and intervention strategies. Build Environ 91:271–282. https://doi.org/10.1016/j.buildenv.2015.03.016

Shukla A, Tiwari G, Sodha M (2009) Embodied energy analysis of adobe house. Renew Energy 34(3):755–761. https://doi.org/10.1016/j.renene.2008.04.002

Soflaei F, Shokouhian M, Shemirani SM (2016) Investigation of Iranian traditional courtyard as passive cooling strategy (a field study on BS climate). Int J Sustain Built Env 5(1):99–113. https://doi.org/10.1016/j.ijsbe.2015.12.001

Steele J, Fathy H (1997) An architecture for people: the complete works of Hassan Fathy. Thames and Hudson, London

Tavassoli M (2016) Urban form and architecture in the hot environment zone of Iran. Urban structure in hot arid environments the urban book series, pp 101–106. https://doi.org/10.1007/978-3-319-39098-7_5

Toe DH, Kubota T (2015) Comparative assessment of vernacular passive cooling techniques for improving indoor thermal comfort of modern terraced houses in hot–humid climate of Malaysia. Sol Energy 114:229–258. https://doi.org/10.1016/j.solener.2015.01.035

Vale B, Vale R (1991) Green architecture: design for a sustainable future. Thames and Hudson, London

Vine P, Casey P (1992) The heritage of Qat. Immel, London

Chapter 4
Vernacular Architecture in Portugal: Regional Variations

Rui Nogueira Simões, Inês Cabral, Fernando Cerqueira Barros, Gilberto Carlos, Mariana Correia, Bruno Marques, and Manuel Correia Guedes

Introduction

The globalization process and technological evolution brought complexity as well as ambiguity to local identities and also a rupture on man's secular relationship with nature. There is still a growing demand for high-tech energy-consuming mechanisms to support artificial indoor environments, resultant from perverted—and unsustainable—comfort expectations.

Vernacular architecture is an essential part of cultural heritage and embodies the knowledge on how to design in harmony with nature, in a durable, healthy, and sustainable way. The local design strategies emerge from centuries of empirical rationality, resulting in comfortable environments that passed the test of time. The knowledge derived from vernacular architecture is at the core of a truly sustainable design.

This chapter aims at providing an overview of the regional differences in vernacular architecture in Portugal. It builds upon a major reference work: the National Survey on Regional architecture, conducted during the 1950s—a document that should always be an inspiration to the present and future generations.

The National Survey on Regional Architecture

The goal of making a national survey of regional architecture during the 1950s was seen as a tool for modernizing Portuguese architecture. One of the main promoters of this idea was the Brazilian architect Lúcio Costa, collaborator of Le Corbusier and

R. N. Simões · I. Cabral · F. C. Barros · G. Carlos · M. Correia
Faculty of Architecture, Universidade Lusófona, Lisbon, Portugal

B. Marques · M. C. Guedes (✉)
Higher technical Institute, University of Lisbon, Lisbon, Portugal
e-mail: manuel.guedes@tecnico.ulisboa.pt

A. Sayigh (ed.), *Sustainable Vernacular Architecture*, Innovative Renewable Energy, https://doi.org/10.1007/978-3-030-06185-2_4

coauthor of the Brasilia project with his student Oscar Niemeyer. In southern Portugal, the initiative would be led by Keil do Amaral, who was the president of the Portuguese Architects Association (AAP), at the time of the national survey. Fernando Távora would lead the initiative in the northern regions of Portugal (AAP 1961).

Two reasons have prompted the making of such a large survey that would cover all regional architecture across the entire country: the first was proving that vernacular (traditional) architecture presented a variety of differences and a richness that contrasted with the uniform model propagated by the Nationalist regime at the time (the "Estado Novo"), during the 1940s; the second motif had to do with the timing, since the survey coincided with the end of the 1950s, so it provided data on traditional models that were integrated in the recent modern architecture style in Portugal.

The survey was initiated in 1955 with the sponsorship of the Ministry of Building and was led by the Portuguese Architects' Association. The national territory was divided into six zones: (1) Minho, Douro Litoral, and Beira Litoral; (2) Trás-os-Montes and Alto Douro; (3) Beira Alta and Beira Baixa; (4) Estremadura, Ribatejo, and Beira Litoral; (5) inland Alentejo; and (6) Algarve, southern Alentejo, and litoral Alentejo, allocated to six teams. Each team had three architects. Fernando Távora was the coordinator for the northern regions, Francisco Keil do Amaral coordinated the southern regions, and the remaining regions were studied by architects such as José Huertas Lobo, Nuno Teotónio Pereira, Francisco Silva Dias, Frederico George, and Celestino de Castro.

It is current among critics of the survey that the survey was an excuse for the teams to find a clear relationship between the chosen models and the modern architecture. In fact, Leal (2009) argues the survey facilitated a modern look on popular architecture, which is acknowledged even by some architects involved in the survey. Teotónio Pereira wrote that "we were very happy, very satisfied, when we found expressions of popular architecture that had similarities with what we thought modern architecture was. That happened when we discovered houses with a mono pitched roof or walls with blind gables, thus similar to expressions that we tried to use in our architecture. We were really happy when we saw a building that seemed to be modern, which could have been made by one of us" (Pereira et al. 1961 in Leal 2009).

At the end of the 1950s, during the Survey and while preparing the publication in 1961, a new challenge was under discussion. By then there was a clear reaction to some controversial application of the *international style* premises in Portugal, without the necessary adaptation to the national context. This idea is clear in a statement by Nuno Portas in 1958: "Let us remember that if the International Style served us… to rebel against neoclassicism and pseudo-traditionalism, the unconditional and long-term acceptance of this type of architecture is becoming dangerous, because it lacks bonds to our backgrounds (…)". It is this insistent attitude that dictated in part, the making of the National Survey on Regional Architecture as a contribution for the Portuguese Pavilion at the Universal Exhibition in Brussels by Pedro Cid (Ramos 2012). So we can assume that when the teams were in the field, rather than the concern to find a relationship between traditional models and modern architecture, they wanted to discover how to make modern Portuguese architecture.

Regional Variations: Context Considerations

According to Fernandes, Orlando Ribeiro, a geographer by training, was the first researcher looking for a full assessment of patterns across the Portuguese territory, based on its geographical and historical expertise (Fig. 4.1) (Ribeiro 1968).

Orlando Ribeiro describes "a northern civilization made of granite, articulated with rough morphological values, adapted to local climate and to an agro-pastoral livelihood, where settlements could be described as midway between the strong density of Trás-os Montes and the dispersion settlements in Minho." He also describes "the South as a civilization relying on clay, located on flatlands and plateau areas of the meridional region where natural light is plentiful, and which is bonded to soft materials transformed by fire and painted with lime, as a response to a life more open to the outdoors, more grounded on earth, and settled in dense villages which are interlinked."

This description shows two different geological and climatic regions in Portugal: the Atlantic ones (based on granite) and the Mediterranean ones (based on clay). Furthermore it shows the diverse historical and cultural contexts (the Portuguese Celtic, Sueve Christian culture, and the Romanesque and Moorish Portugal) and how its complexity impacted the settlements.

It is later known that there was a third geoclimatic region articulated with the Spanish inland culture, a bigger diversity of building materials (such as schist and wood), and that archeology led to older references.

According to Fernandes (1963), the survey on the regional Portuguese architecture provided knowledge about other subregions of the vernacular architecture. The work has emphasized the links between the cultural and functional context, between northern and inland regions with the neighboring regions of Galicia and Meseta, and reconnects the south of the river Tejo traits with the Andalusian and Spanish Extremadura. As such we can state the inquiry had its own international dimension.

Minho

The region of the Atlantic Northwest is characterized by its temperate climate, cool weather, and high precipitation. It is a region predominantly granitic (despite some vestiges of schist), which is a stone associated with different types of construction. This region is characterized by irrigated agriculture, forming a green landscape with dense and diverse vegetation. It presents diverse crop fields, such as cereals, pasture, vegetables, vineyard, olive trees, as well as maize, a crop introduced during the sixteenth and seventeenth century, with its cultural and social implications, such as an extreme partitioning of the land, which resulted in a characteristic minifundio system.

The Minho landscape mosaic is subdivided in two major areas: the valley—"ribeira" (in the vicinity of the main rivers: Minho, Lima, Cávado, Ave), and the

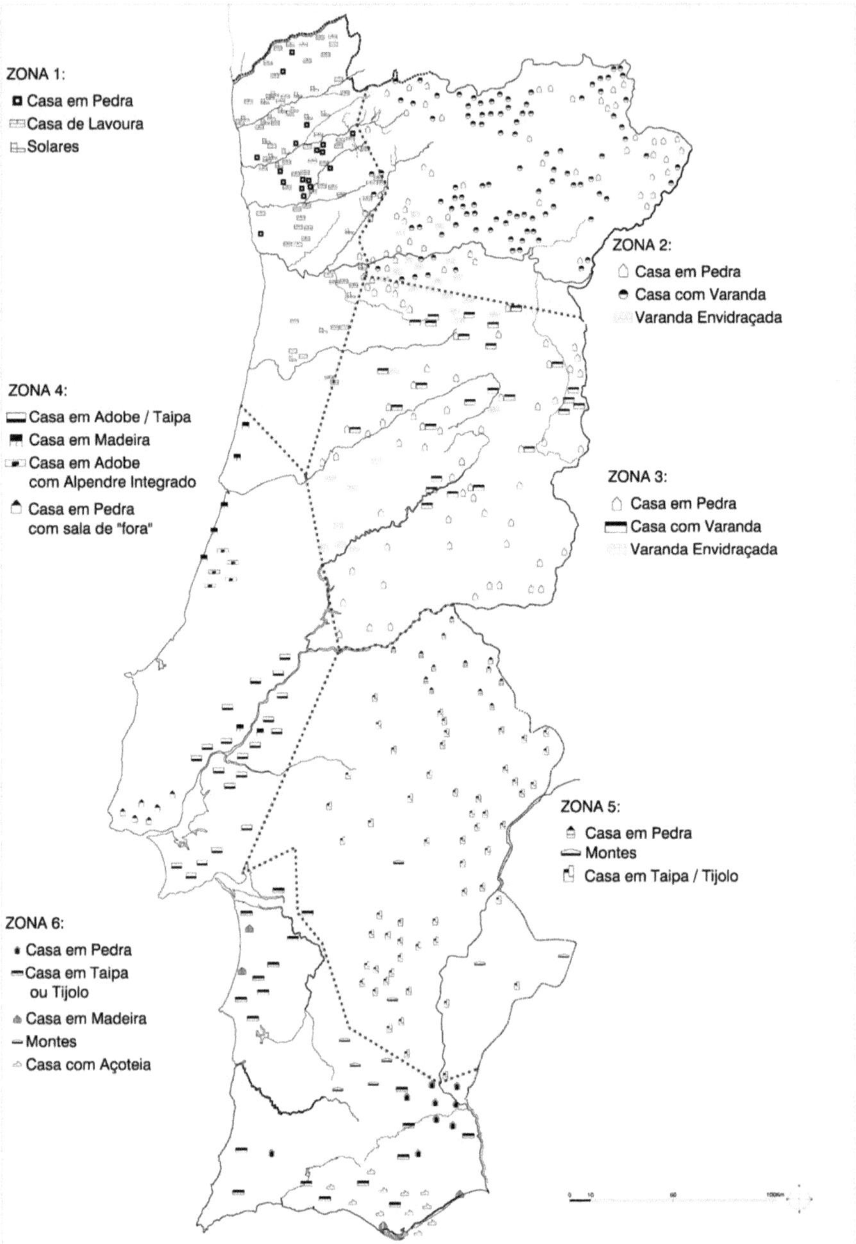

Fig. 4.1 Map with the surveyed regions (1–6) and respective typologies (Maia et al. 2013)

mountain—"serra" (located in the peripheral regions of the river and associated with the main mountain ranges of the region: Arga, Peneda, Amarela, Gerês, Cabreira).

In the valley, settlements are dispersed, with land being cultivated between forest patches (called soutos), including on hills, predominantly "matos" (mainly gorse), that once provided the bed for the animals and constitutes the main fertilizer. The houses are scattered, or in small clusters, located between these cultivated lands, and so the notion of agglomerations or dense "villages" does not really exist. In the small valleys, among the scattered dwellings, there is a predominance of small private properties, which stand out as larger farms belonging to wealthy farmers, some of them members of the local nobility, living in manor houses. In addition to these, the church stands out with its tower, churchyard, and cross, where the cemetery and the parish house are usually implanted, thus marking the core of each parish. However, the dispersion of houses is such that in many cases, "the last houses of one place are confused with the first ones of the next one" (Oliveira and Galhano 2003), and it is therefore difficult to perceive the territorial limits of these small units of colonization of the Minhoto territory.

The cultural and economic aspects of the mountain range areas present some aspects that are similar to the valley areas (such as the importance of maize). Both societies share the importance of pastoralism in the mountainous and high mountain areas, which in some cases depend on communal pastoral systems (vigias or vezeiras) practiced in some cases in a transhumant regime (Peneda, Amarela, and Gerês mountain ranges). This mountain settlement is similar to that of the adjacent Trás-os-Montes region, thus concentrated: houses clustered in a compact and dense way, contiguous to each other and sometimes sharing walls, forming small, irregular and narrow streets. The concept of "village" prevails in the mountain areas, which is surrounded by crop fields, forest areas, brushwoods, and common lands (and in some cases, include complementary settlements for seasonal use—corrals or summer villages, called brandas).

In this region prevails the house of 2 stories, where the ground floor holds stables, corrals, cellars, and warehouses, and in the upper floor all the spaces are used by people: kitchen, living room, bed rooms, and the balcony. The floors have independent access (second floor is accessed by an external staircase, in many cases with a covered porch or balcony). The most common building material is stone (granite or schist, depending on local availability, and economic reasons), usually also having some parts built of wood. The walls are commonly made of exposed stone (especially in the mountainous zone), and in some cases painted with lime or plaster (when it is, usually the skirtings, wedges, and sill trimmings appear in gray or other dark color).

The balconies, which in many cases constitute additions to the original constructions, are made of wood, or in large blocks of granite (perpianho), and are often protected by a roof eave. These structures are used as a corridor or as storage room, often working as indoor/outdoor space (or transitional), connected to a staircase. Their height and opening size are often interlinked with local climate conditions (being higher and open in the valley zones, and lower and closed in the mountain range). Usually, there are divisions at the end of these balconies (rooms, or partition for loom). This balcony-corridor, common in rainfed farm regimes, could be the origin of other construction types of the Entre-Douro-e-Minho, such as the great verandas (varandões) and sun drying flooring (sequeiros), characteristic of the region of Baixo Minho.

In Entre-Douro-e-Minho region the kitchen is the core of the house, given the importance of the fireplace, and the multifunctionality associated with this space. The living room (which in many cases works as a bedroom) was the most elaborated division, and it has an important ceremonial function (hosting the pascal compass, weddings, funerals, etc.). The bedrooms are often functional, and small in size, and in many cases lack windows (alcoves). When the house is small, all divisions are located in the same space. However, in the case of more affluent houses (and therefore of larger size and care in their distribution), certain dependencies (especially those associated with animals and agricultural work) are located in independent divisions and buildings, which in many of the cases are arranged around a courtyard.

In the mountainous areas the houses are typologically similar to those of the valley zone, although of smaller dimensions and use rustic features. Given the characteristics of the clustered mountain settlement, in many cases these houses do not have any external space of their own, communicating the stairway directly to the public space. When they present this space, it is generally a small courtyard (or quinteiro) that works as a transitional space between the public and the private ones, allocated for agricultural and pastoral activities of the family. In the steepest areas the house is implemented in a dialogue with the slope, and in many cases there is no external staircase, since it can be accessed directly on the upper floor by the highest part of the terrain. Unlike valley areas, where roof tile prevails (on roofs of 3 or 4 surfaces), in the mountainous zones it is usual for the roof to be a pitched roof (having persisted in the most remote villages until recent times, the use of thatched roofs).

The importance of the introduction of maize crop in the region has manifested itself in the structuring of the territory, through the partitioning of land, irrigation structures, and from the architectural point of view, the proliferation of storage elements such as dry flooring, haylofts, and granaries. These are indeed a highly developed and perfected element, which had great prominence in Arquitectura Popular em Portugal (AAP 1961). In the valley areas, the granaries are built mainly in wood (or mixed granite-wood structure), being located in the vicinity of the dwellings, or agricultural annexes of their own, dry flooring or threshing floor, etc. As for the mountainous areas, the granaries become more robust, and sometimes are built exclusively in granite (Soajo), implemented around communitarian spaces, as communal threshing floors, in large groups of three, four, or five dozens (Soajo, Lindoso, etc.).

At the northern end of the municipality of *Arcos de Valdevez*, nested in the initial section of the valley of the river Vez, and the abrupt slopes of the northwest slope of the mountain of Peneda, is located the parish of Sistelo, marked by strong contrasts between the humanized landscape of the valley of Vez and a more open landscape with wide horizons looking at the plateau of Peneda (Fig. 4.2). The variances in steepness gave rise to distinct microclimates within the parish, which are directly associated with the different elevations and the availability of water and insolation,

Fig. 4.2 Sistelo: valley landscape with terraces to the maize, and mountain landscape in a summer village (credit: Barros 2013)

Fig. 4.3 Cachena cattle in a summer village, and buildings in Branda da Gêmea at 1000 m (credit: Barros 2015)

which clearly distinguish the deep areas of the valley of Vez and adjacent streams (corgas), and areas of mountain and high mountain, that approach the plateau of Peneda.

The breeding and grazing of Cachena cattle has been one of the most important pillars of the family economy of all the mountain communities of Peneda, associated with the smallness of the fertile valley spaces, and the strong availability of high mountain pastures, which could be exploited (especially in the summer season), was punctuating the saw of corrals (Fig. 4.3) and "brandas" (summer villages), laying the foundations for the structuring of a transhumant and seasonal use of space, between the low valley areas—used throughout the year (and where the villages are located)—and high mountain areas—where livestock goes up, between March and April, and where they remain until September or October.

Another key point to understand the settlement was the introduction since sixteenth century of the maize. The need for irrigation moulded the slopes of the valley into monumental terraces, thus creating horizontal levels that hold the water and allow irrigation. These are associated with the granaries (canastros), intended for

Fig. 4.4 Terraces in Valley of Vez between Padrão and Porta Cova, and Granaries of Padrão (credit: Barros et al. 2016)

Fig. 4.5 Granite houses in Padrão's village, and houses and pastoral shelters constructed in corbelled dome structure at Gêmea's summer village (credit: Fernando Cerqueira Barros)

the storage and drying of corn, built in granite and wood, elevated in order to avoid soil moisture, and protected from climbing rodents. Within dense villages, which characterize the concentrated settlement of the mountain, the granaries are clustered, oriented for maximizing exposure to sun and winds, to better conserve the cereal (Fig. 4.4).

As granite is a primordial material, villages contain robust buildings, with few openings, defending themselves against the harshness of the climate, in some cases enclosed around private courtyards, which are associated with haylofts, made out of wood. These granaries were added and clustered and implemented in communitarian lands. Alongside the rivers one can find the mills as well as interesting bridges, which allowed the passage of the waterways, in important territorial connections (Fig. 4.5).

The mountain region of Peneda is located within the National Park where landscape features granite mountains alternating with oak forest and rivers in the valley.

Fig. 4.6 A house and a street in the winter village of Pontes (credit: Cabral and Chalfoun 1998)

Until recently *Castro Laboreiro* settlements depended on a transhumant livelihood. In this region shepherds spent their summertime in higher grounds (above 1000 m), while in Winter they were forced to move the cattle to lower grounds (under 800 m) to avoid the long snowy season. Therefore the Castro people built houses in both sites. This type of settlement has created three different types of villages: the summer village or Branda, the winter village or Inverneira, and permanently inhabited village or "lugar fixo" such as the town of Castro Laboreiro (Cabral 2009).

Most villages had less than 30 houses, a common oven and a common water tank. The houses were often located in the least productive land, sometimes on the rockbed mingling with it. Surrounding the village there was a communal land used for grazing and where some oak trees and birch trees were harvested for building the houses. Small vegetable gardens were close to the houses. Due to the uneven precipitation regime, water was scarce in the summertime and demanded the construction of levees that in some cases were also used during wintertime to prevent the fields from freezing (Campos de Lima) (Fig. 4.6).

The traditional Castro house is characterized by a unique typology with small variants. According to Viana (1999) the castro house has two stories, where the ground floor is used for sheltering animals (cows, sheep, and goats) and the first floor is residential. The outer walls are commonly assembled with crossed over granite stones, stabilizing the shell. Wood walls confine the divisions of the living area. The stones are usually of irregular shape for houses built before 1960. The pitched roof structure is made of local oak wood; once thatched it is now covered with ceramic tile. In both cases there was no chimney over the fireplace. The floor slab around the fireplace is made out of stone and it is called "lar." In the bedrooms the floor is replaced by oak wood (Fig. 4.7). The exterior walls can be as thick as 80 cm on the ground floor and 60 cm on the first floor. This high thermal inertia of these walls provides comfort in summertime but not in wintertime, since openings are very small, reducing solar gains. The small openings for windows avoid the harsh winter weather but it is also due to the stone's low resistance to horizontal loads (Cabral et al. 2013)

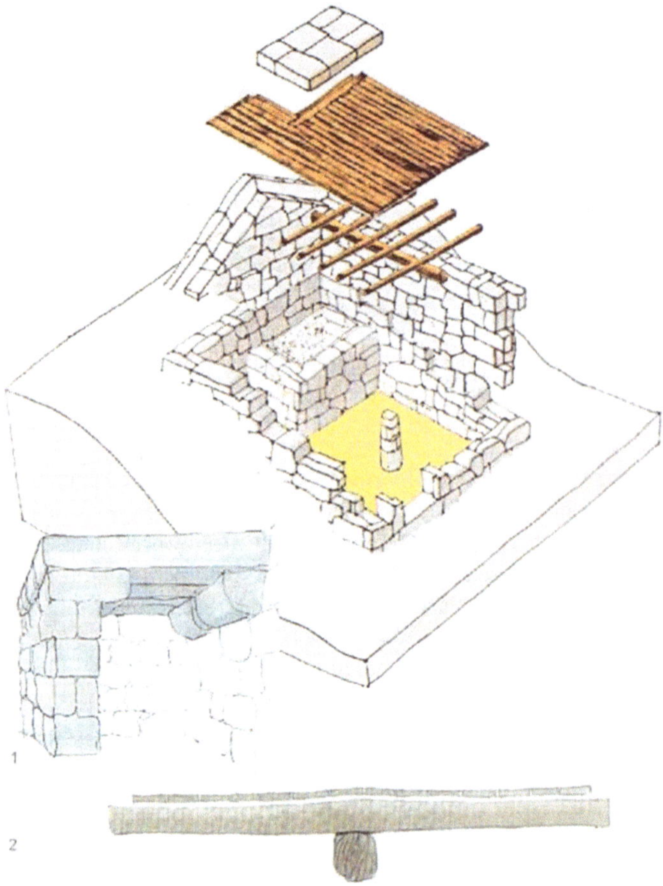

Fig. 4.7 Typical building system in Castro (Credit: Viana 1999)

Trás-os-Montes and Alto Douro

The region of Trás-os-Montes and Alto Douro is separated from the Northwest Atlantic coastal region by a sequence of mountains (Gerês, Cabreira, Barroso, Alvão, and Marão). These mountains determine the climate in the Transmontana eastern inland region, which is mostly continental as opposed to the western Atlantic coastal climate. The landscape includes several plateaus, and extensive cereal crop fields in the highlands (rye and wheat, sometimes fallow), as opposed to the deep valleys carved by the river Douro and its main tributaries (where vines and olive plantations are traditional, as well as almond trees).

Predominantly a schistous area (although large granitic areas are also present), this region is commonly divided into two subregions due to its subclimates: the "Terra Quente" (associated with the river Douro valley and its main tributaries) has

Mediterranean geographical features, and the "Terra Fria" (associated with the extensive plateaus located along a west-east axis) has a more continental climate. Climate in both cases is harsh, with its cold winters and abundant rainfall, snow and frost, alternated by scorching, dry summers and high temperatures: the so-called "Iberia Seca" (Oliveira et al. 1969). The vegetated groves are scarce, and fragmented in patches of chestnut or oak forests, whereas in the "Terra Quente" the olive and the cork oak forest is more common.

The economy depends on two separate activities: the pastoral and communal livelihood of the plateaus of "Terra Fria" (with its common pastures and meadows, where livestock herding is shared by the community in a vezeira regime; and ovens, mills, and communitarian mills are also shared, under a system of self-governance—such as the "Concelho" of Rio de Onor) and a family-owned activity, particularly in the large vineyards of Port Wine, typical of "Terra Quente."

The population density of the region is lower than in the coastal area, and settlement is mostly concentrated. The communities are located near water springs, in sheltered and sometimes rocky areas, allocating most of the surrounding productive land for both extensive cereal crops and livestock grazing. The concentrated settlement system is related to the historical origins of the settlement in this region, and to a communal management of the territory (commons and communitarian livelihoods), which until recently governed these societies (Sampaio 2009).

The villages are compact and dense consisting of rustic aligned houses forming streets. In the center, the Parish Church (or Chapel) stands out among the buildings. Besides this massive built area, there was usually a large central area where a communal threshing floor was located. Surrounding the village, in the nearest plots, and especially in those best irrigated, one can find the vegetable gardens, and further away the areas of extensive cultivation, one can find grazing fields. It was verified that the houses are mostly (especially in the Terra Fria) built in raw stone, without lime plaster, with prevailing thatched roofs or roofs in slate.

The popular house, which generally does not differ much from other northern regions (stone house, two floors with different functions, and outside staircase and balcony), presents in Trás-os-Montes a singularity, especially in the larger houses: the patio (or curralada). This large compartment, located in the central areas, or adjacent to the dwelling, has important functions in the family economy, and is associated with farming activities.

It is around this patio that all divisions are distributed and accessed. The access to the patio is generally made by a large carriage door (porta carral), which opens to the street, and that door is usually covered by a porch, sometimes becoming a division for haystack (designated in some areas as "theater of hay"—"teatro do feno").

The smaller houses, which are surrounded by streets, sometimes have thick walls, shared with adjacent houses, have no patio, and alternatively a small yard in the back, where they tanned manures. Sometimes these smaller houses may not have any outdoor space, and the manure would be tanned in the corrals of the ground floor, or alternatively stacked in the street, immediately in front of the house.

The most used construction material in the whole region is stone (schist or granite, according to local resources), complemented by wood (balconies, pavement, roofing structures). The roof varies between the thatched roof, the slate roof, or ceramic tile (the latter clearly a more recent solution, not available in most remote areas until the middle of the twentieth century).

A primordial element in the Transmontana housing is the balcony, which, having no place properly defined, presents itself in different typologies and functions, according to the type of house in which it was installed. In the streetfront houses, it is located in many cases in the main façade, towards the street, functioning as an extension of the stairs landing. In the courtyard houses, it usually appears along the various sides, varying between a balcony with one front (when the patio is on one side of the house), or a balcony on three or four sides (when the patio is in the center). It should also be noted that, in some cases, the balcony can be installed in the rear façade of the house, especially if it has a backyard, or in cases where the rear façade reveals better sun exposure (since the balcony was also a place to dry the cereals).

From a functional point of view, the balcony can present similarities, for example, with the balcony of the Minho region, since it is also an auxiliary space for family activities, agricultural activities, also for sun drying cereals, storage, as well as meeting or eating space.

Alongside the rural cases, the balcony is also present in Trás-os-Montes urban dwellings. In the house of Tras-os-Montes, the kitchen is the essential division of the house, as it is located close to the fireplace, used for hosting visitors, and where the owners perform the most important tasks of the family life. Contrary to other regions, in many cases, the kitchen had no oven, since, due to the persistence of communal customs in this region, the communal oven was shared by the villagers.

In many cases these communal ovens present singular and important architectural forms. These are solid constructions of reinforced materials, built with large blocks of granite, linked to its inside space, exterior covered spaces or porch (such as the one of Santo André, which appears in prominence in "Arquitectura Popular em Portugal" (AAP 1961)), and that besides meant for baking, were also used as a place for dating and meeting, and sometimes even for an overnight stay for travelers and beggars.

Tourém and *Pitões das Júnias* are two villages located in the municipality of Montalegre, in the most northwestern region of Trás-os-Montes, and close to the border with Galicia (Spain). These villages are located in the Mourela plateau, bordered by the mountain of Gerês (on the west side), Barroso (south side), and Larouco (east side), and the valleys of the rivers Cávado and Rabagão (south side) and Salas (north side). This region, generally called Barroso, was one among the case studies in the Regional Architecture survey within zone 2 (see Fig. 4.8).

Looking westward one can find the mountain range of Gerês (vastly depopulated) and the northern area of Galicia, where most villages from Barroso show similarities with the west regions, i.e., the mountainous areas of inland Alto-Minho with some Atlantic characteristics that have penetrated this region through the valleys of the rivers Cávado and Rabagão. The area where Pitões and Tourém are located corresponds to a granite zone (its constructive tradition may be included in what Orlando Ribeiro called "The Granitic Civilization of Northern Portugal" (Ribeiro 1961)).

Fig. 4.8 Pitões das Júnias—aspects of the village (Source: AAP 1961)

Fig. 4.9 Tourém's houses (source: Barros 2013 and AAP 1961)

From the economic point of view, the population relied on a typical mountain livelihood, where cattle grazing prevails, as the "Community's Ox" ("Boi do Povo") became a symbol of fertility and reason for pride for each community. The main crops are rye, which is associated with livestock, and more recently also potato (maize is an exception in the Barroso area). The best land is for cattle grazing and grain growing. The poorest lands and common lands were meant for small livestock (goats and sheep).

The two villages are built along irregular paths and alleys, intersected by crosswalks (canelhas), and their dwelling houses are clustered like small neighborhoods in a rustic urban space. This matrix opens in small common spaces, where one can find a fountain or a community oven, the chapel, or the church and its churchyard.

The traditional house can be described as three different main types: (a) the elementary house only has one division per floor; (b) the intermediate type has a central block in granite where a staircase is carved, has a balcony and/or outdoors division in with wooden pavements; (c) the third type is the most affluent houses of larger dimensions, corresponding to the house-patio typology (Fig. 4.9).

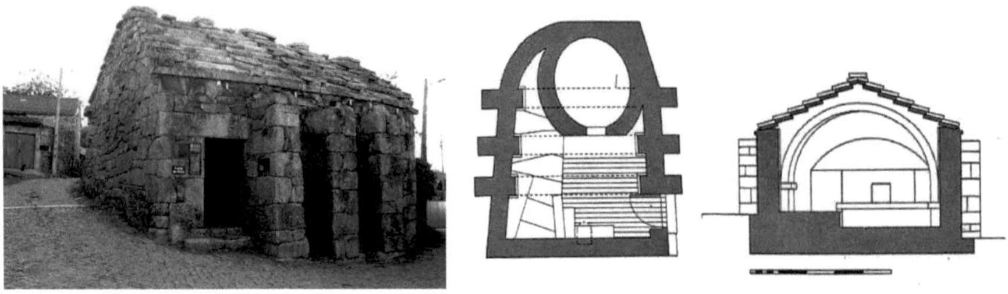

Fig. 4.10 Tourém—Community oven (source: Barros 2013)

At the time of the survey, there was a clear predominance of thatched roofs in most buildings (similar to the Alto-Minho mountain areas). The use of this roofing and its binding (windbreaker stones, wooden sleepers, etc.) is explained in Arquitectura Popular em Portugal (AAP 1961), which was disappeared in part due to a fire hazard.

Community ovens are considered evolved architectural forms (AAP 1961). Unlike most dwellings, the furnace is covered by granite slabs, for structural reasons and to prevent fires. In Tourém these are supported by three granite arches, reinforced by buttresses (Fig. 4.8). In some cases we can see the resemblance with a temple, both as a space for making bread and as a community building, a space open to all villagers. The ovens were used in turns, as various families had different roles (e.g., supplying wood to light the fire) (Fig. 4.10).

At the moment these villages are less isolated and remote than in former times, but very much affected by migration. In 1971 these settlements became part of the National Park of Peneda-Gerês, and later also part of the World Biosphere Reserve of UNESCO. The material and immaterial heritage of this region has been preserved and managed by the Barroso Ecomuseum.

Beira Interior

The green hills of Gardunha and the extensive plains of Idanha and the Natural Park of the International Tagus are the main features of this region. In the county of Fundão, at the foothills of the Gardunha mountain and belonging to the so-called Cova da Beira, is located the largest producer of cherry. But other products such as cheese and olive are also an important part of the local economy.

In terms of historical heritage, one can find well-preserved villages, part of the Historic Villages Network, such as the villages of Castelo Novo (in Fundão), and Idanha-a-Velha and Monsanto (Idanha-a-Nova). More recently, the network of Schist villages covers villages like Baroque and Janeiro de Cima and Foz do Cobrão.

Regional architecture in Beira Interior is characterized by compact houses made of exposed stone walls and small windows. Buildings are two stories high with storage

Fig. 4.11 The traditional glazed porch in Cova da Beira (source: AAP 1961)

rooms in the ground floor and porches with exterior staircases sometimes glazed in order to protect from the strong winds (see Fig. 4.11). Chimneys are omitted in these houses so heat is not lost during cold winters. Often there is a patio in front of the house for some agricultural works (see Fig. 4.8). Locally available materials like granite, schist, and clay are largely used for walls and roofs. Thatched roofs were also common. Pine and oak wood are the most common materials used in structure and partitioning.

The *Quinta da França* is located on the foothills of Serra da Estrela (Continental Portugal's highest mountain) and close to Covilhã. It has about 500 ha and has diverse activities like agriculture, animal husbandry, and forestry. The territory is limited by the river Zêzere on west boundary where you can find three small dams (Cabral and Chalfoun 1998).

There are about 24 rural houses covering a total area of 4800 square meters. The houses are very diverse in size and are scattered over the farm due to their historical occupation and different types of farming, forming 12 clusters.

The Casa dos Eucaliptos (Fig. 4.12) is located on the top of a small hill, facing SE and protected from winds by a dense tree line of eucalyptus. There are also deciduous trees on the south side shading the patio. The house is close to a levee along which there are several trees. Farther south there is an irrigation pond. The house geometry is a 30 m long rectangle rotated 43° towards the east side.

There is also a small construction on the east side that is slightly below ground where there used to be three pigsties (Fig. 4.10). The house is bermed from the north side. It is settled in a rocky basement. This allowed the nonexistence of foundations and also permitted the use of a sloped piece of land less favorable for agricultural use. This earth berm is also a way of diminishing the shadow cast by the house on its north side. The berm maintains the wall at constant temperature, its inertia reducing seasonal seasonal weather changes and daily temperature swings.

Fig. 4.12 The House of Eucaliptos in Quinta da França with three pigsties and a patio (credit: Ines Cabral)

Estremadura and Ribatejo

The so-called region 4 was defined as transitional and contrasting, as the team realized the region combined feature from north and south vernacular architecture. However we can say that in the area there are mostly white washed homes, stone or adobe and rammed earth in South, but some exceptions stand out. And these exceptions are many and diverse and also well mapped in the survey. From the chapel to the seashore to wells and cisterns, wineries and windmills and watermills, they all show how important the production and storage of items were for the supply of the city of Lisbon.

Also in the exemplification of housing types, the survey team choose a few cases that are strict to same confined areas. These are the "avieiros," wooden constructions on the riverbanks of the river Tagus or the houses with front porches "casas alpendradas" near Leiria, both endangered.

Fig. 4.13 Group of "saloio" houses, Ar, Sintra" (AAP 1961)

Another example of housing in the outskirts of Lisbon is the "Saloia" *House* (Figs. 4.13 and 4.14), which stands out in the landscape for being one of the few two-story cottages in this region. This type of housing unlike the aforementioned types still has a few examples in good condition and has been the investigated post-survey.

It is today consensual that the regional type of housing does not stop just in the northern riverbank of the river Tagus, as shown in the survey, but there is also part of the south riverbank. João Vieira Caldas in his study called "rural houses in Lisbon suburban areas" refers to the evolution of the definition of zoning, and states that it is the very architecture that bears witness to the presence of the "Saloio" people also on the South riverbank of the Tagus, pointing examples that show this. Fernandes and de Lurdes Janeiro already in their work on the "architecture of the Saloia region" defined this region to be located between two places of pilgrimage: "Nossa senhora da Nazaré" and "Nossa Senhora do Cabo" in Espichel, two places quite symbolically as there are two chapels with cubic shape (Caldas and Correia 1999).

The origin of the people who inhabit this region is not easy to track because they "should never have been a distinct population group, physically individualized or differentiated from the Portuguese". However most of the authors assign its origin to Arab ancestors and Christians who have left Lisbon during the "Reconquista" (term defining the conquest of Portugal to the Moors by Afonso Henriques), to occupy the surrounding lands and living off agriculture. Fernandes and Janeiro (2008) date the origin of the Saloio as far back as the invasion of the Iberian Peninsula by the Arab peoples, accompanied by people from the Berbers tribe, who stay from the start in these border regions, while the Arabs occupied the town. The

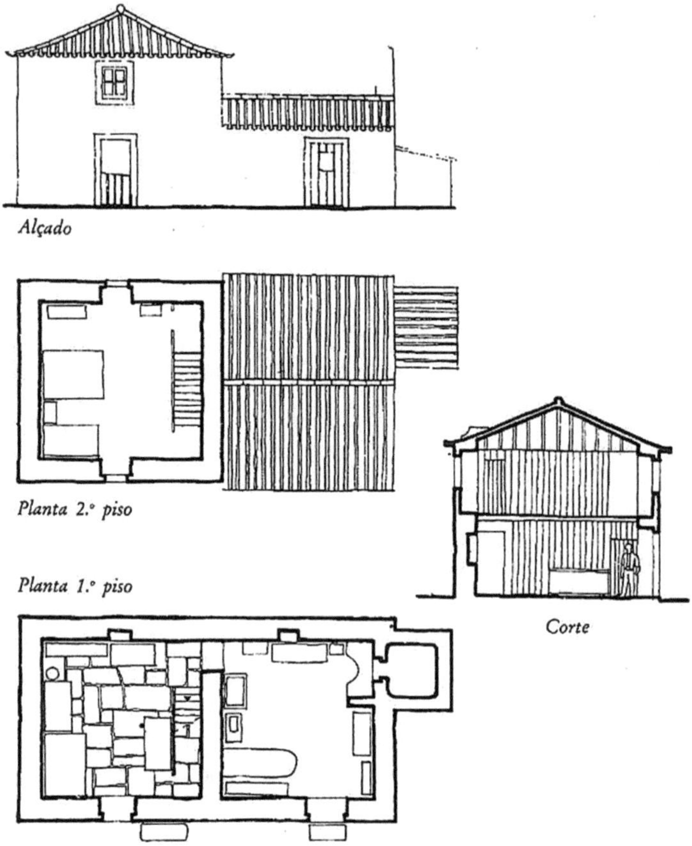

Fig. 4.14 "Saloio" house—plans, elevation, and section (AAP 1961)

miscegenation of race that occurred then is hard to distinguish these populations along with its manners.

Unlike the people, the house "saloia" is perfectly detectable, because its characteristics are easily recognizable among numerous dwellings of this region, as far as its dimensions which are repeated in various cases even though they are far from each other. The "saloio" House is part of the generic model of the South, Mediterranean architecture, plastered, whitewashed, and tiled. However the existence of two floors in most cases suggests influences from Northern Portugal which according to Vieira Caldas (…) should become a transitional group. Were single houses of rural character initially isolated from surrounding farmland (Fig. 4.15) which later given way to other buildings, so building an urban structure as a "hard process given its rustic and simple expression of single House"?. These two authors agree on the characterization of the types of the "saloio" home:

Fig. 4.15 "Saloio" house, Assafora, Sintra (AAP 1961)

Type 1—house with two or more divisions and four-pitched roof.

Type 2—identical to type 1, but with a cubic volume of two stories at the end. This volume has a four-pitched roof and a lower roof with two or three pitches. The second floor of the cubic body was the bedroom, and in the first floor is the living room "quarto de fora" that communicates with the existing kitchen in the body only with a story.

Type 3—two-story house resulting from duplication of cubic volume with plant similar to the previous but with two bedrooms upstairs.

The use of geometry in the volume of these houses and especially the presence of the cube refer to an Islamic influence. J.M. Fernandes and M.L. Janeiro suggest a proportional ratio of 1 to 2 is used in the dimensions of these volumes. This proportion can be found in some "saloio" rural houses and curiously in the "Kaaba", in Mecca, Islam's sacred building. Not being a perfect cube the relation between the width, depth, and height is the same as those of the "saloio" home. Other factors are the small size of the living spaces and the openings J. Vieira Caldas also commented that houses with windows seats, with shutters or crates, have similarities with typologies from the Islamic culture.

However it is the roof that has more interpretations. The characteristic design of the four-pitched roof, and the curvature along the cornice "saqueado" ending with the almost horizontal tiles, is very particular and it appears to be a singular feature of Portuguese architecture. However this type of roof is not restricted even to vernacular architecture or to the region of Lisbon and can be found in some areas of the Algarve, as in Tavira for example. According to J. Vieira Caldas, it appears mainly in mansions outside the region of Lisbon and the Algarve, not being frequent in the vernacular architecture of other regions.

It must be pointed out that these two studies that we've been referencing not confined only to the popular models of the local region. In the case of J. Vieira

Caldas is made a complete survey and characterization of erudite rural housing in this region and the relationship with vernacular models. In the study of J.M. Fernandes and M.L. Janeiro is treated in depth the spread of the "Saloio" house in other regions, refer: Azores, mainly on the island of St. Mary, Madeira Island, and the Canary Islands.

Finally, the survey refers to the two most well-known examples of a dwelling group, based on the "saloio" house: The row house for rural workers in the "Picanceira" near "Mafra" known by the name "bairro dos Ilhéus" dated at the end of the nineteenth century, and the dwellings from pilgrims in the Sanctuary of "Nossa Senhora do Cabo" in Espichel from eighteenth century.

The "Bairro dos ilheus," in the island of S. Miguel, Azores, is the repetition of a very simple module that, taking advantage of the slope, has two-story dwellings with entries in two levels. The module has a cylindrical volume on the ground floor accessing the oven (Fig. 4.16). Downstairs would be the kitchen with a patio and upstairs the living room (casa de forra) with street entrance and a room where the stairs are.

The sanctuary of "Nossa Senhora do cabo" in *Cabo Espichel* consists of a set of buildings: the church dated at the beginning of the eighteenth century and the two long buildings of dwellings supported in a simple arcade, which defines a gallery, space of transition between the square and the interior spaces. In the center develops a vast space of a rectangular square open to the east (Fig. 4.17). The housings for pilgrims in "Saloio" architecture, of great interest and monumentality, were built in the first half of the eighteenth century, by a group of pilgrims "sirios" from the "saloio" region, as indicates a plaque at the site.

Each module has a square plan and consists of a ground floor with part of the gallery and a shop, and the second floor comprises the hostel room where pilgrims could stay overnight. The main facades have a very interesting fenestration scheme - missing a window in the space of the stairs and causing an interruption in the row of windows.

Fig. 4.16 "Bairro dos Ilheus", Picanceira, Mafra (AAP 1961)

Fig. 4.17 "Nossa Senhora do Cabo," Espichel, Setúbal (Ordem dos Arquitetos)

This relates to the presence of the geometric square. The asymmetry caused by the different lengths of the two sets of rooms creates a very interesting tension in the square. This group of buildings was classified as heritage in 1950.

The fishermen called *Avieiros*, as they were originally from Praia de Vieira (in Leiria), used to migrate to the riverbanks looking for less strenuous fishing conditions in the coastal areas during the winter months. As river fishing was similar to their traditional livelihood, this community moved to the banks of Tagus and Sado rivers (Moreira 1987). During the migration period, the Avieiros generally lived in very poor conditions, either in their boats or in precarious constructions made out of the local available material, such as reeds and branches (Palla and Gaspar 2015).

This population settled in the banks of the Tagus river, from Alhandra to Santarém, organized in stable communities that lived on fishing shad (fish that, like the salmon, used to migrate to rivers to spawn).

Initially the Avieiros lived in their boats on the riverbanks. Later looking for better conditions, the Avieiros built permanent palaphytic villages (Fig. 4.18), mostly in the higher sites of the riverbanks, protected from the regular floods of the Tagus river (Fig. 4.19). These spaces served as a storage place for fishing gear as well as improved shelter for the family than the boats (sometimes only the children lived there).

Today these palaphitic constructions and settlements are threatened by abandonment as younger generations tend to migrate looking for jobs in nearby cities, as the fishing activity is declining due to depleted fish stocks. Moreover the wood buildings do not provide living conditions according to modern standards.

Fig. 4.18 House in Avieira's village of Palhota

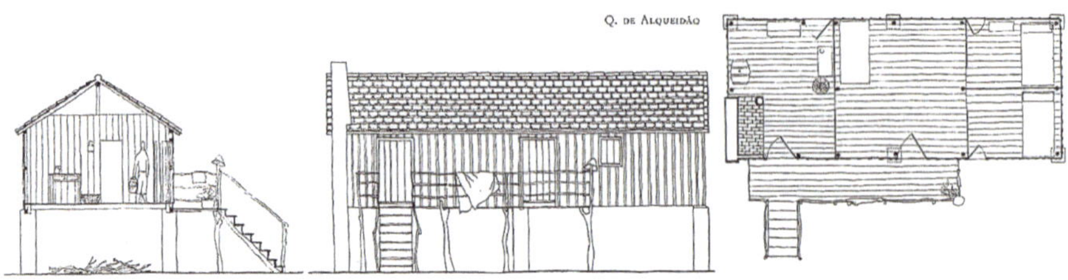

Fig. 4.19 Typical palaphitic house in the Tejo river (source: AAP 1961)

Inland Alentejo

Although the Alentejo region is usually addressed as having a very homogeneous building culture, a more thorough look can expose a rich and diverse architectural heritage.

The inland Alentejo is usually classified as a southern Mediterranean region, with hot and dry climate half a year. Despite the cold winter and mild autumn, the low precipitation is a strong feature, contrasting with the littoral region where the influence of the Atlantic Ocean attenuates the temperature swing. The relief presents a distinctive smooth and undulating configuration, in a dynamic combination of wide valleys and low plateaus. The traditional forest called Montado includes three species: cork trees, holm oaks, and olive trees. Silviculture provides complementary income for both farmers and shepherds.

The settlements of Alentejo are the most disperse and less dense of the country, and its population is decreasing and aging. The economy of the region is based on extensive cultures of cereal, a legacy historically attributed to the roman presence in the Iberian Peninsula. The twentieth century national policies accentuated the monoculture regime, forcing the optimization of the dry-land farming and establishing

Alentejo as the "granary of Portugal." Therefore, the landscape is divided in large properties, speckled with dispersed small settlements and isolated rural units, called Montes, also known as "Inland meridional" settlements (Ribeiro 1963). A high percentage of land has thin, poor, and rocky soil, thus not productive. The Alentejo inland subsoil is mainly composed of no crystalline schist, with specific areas where marbles, granites, and limestone can be found.

Despite the availability of many materials, until the middle of the twentieth century, the dominant building technique was rammed earth (Correia 2007). The vast use of this technique by the communities with less economic resources helps us understand its strong presence in rural houses conditioning most of the vernacular typologies of Alentejo's inland.

The architecture is based on structural façades of load bearing earth walls. The material determines a rectangular shape of the floor plan and limits the height of the walls as well as the number and size of its openings. Due to the structural limitations of the material and the historical record of seismic occurrences (of low intensity) it is rather usual the integration of several structural reinforcement elements like buttresses, relieving arches, stone benches, and tie rods in the traditional solutions (Correia and Carlos 2015). The Alentejo traditional architecture can also resort to adobe, fired brick, or stone masonry (usually schist or granite), but all of these techniques are commonly used as complementary to the rammed earth construction system or applied as principal material in rather particular places and circumstances.

Lime, wood, reed, and woodwaxen constitute the most common transversal elements of the vernacular typologies applied in cladding and finishing solutions. These materials provided a protective coating and coverage against weathering, for indoor spaces and for the main building components, constituting a rather light solution that does not compromise the building structural performance. The external lime wash painting, made of several layers, constitutes another distinctive aspect of the region's traditional architecture. This solution is not only fundamental for protection of the rammed earth walls, from natural erosion, but also contributed to minimize the thermal inertia of the exposed facades in summertime. The most simple, and therefore common, roofing system relies on the transversal support of wood beams, of circle section, placed directly under the pitched roof. Additionally one can find the vaulted ceiling, considered one of the most interesting solutions of the Alentejo traditional construction. It is made of bricks (abobadilha) and according to the national survey, this solution expresses the high expertise of the Alentejo's masons as these low vaults were made without using a cymbal, relying entirely on the quality of the used lime mortar and the technical skill of the worker.

As easily inferred, most of the identified vernacular typologies of the Alentejo inland are related to farming activities. The rural complex, the "Monte," constitutes the most common example, where one can find both the residential building, and agricultural and/or livestock-related constructions. The more elaborated Montes present a central enclosed space. This space called patio usually results from the placing of linear sections of walls, connecting the different buildings of the complex. Besides the one-story house, there is a barn, a cow shelter, and a pig shelter as the most common complementing buildings. The collective bread oven was an

Fig. 4.20 Monte da Boleja (credit: Gilberto Carlos)

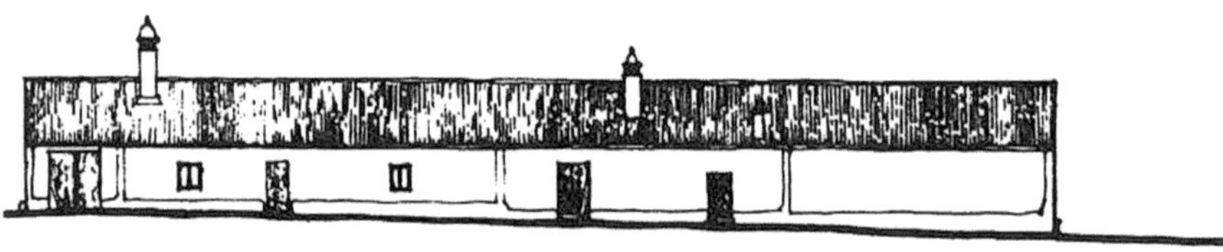

Fig. 4.21 Monte da Boleja front elevation (credit: AAP 1961)

important addition to the domestic life. It is very frequent to find this element attached to the end of the main building or even detached but close by, covered by a dome. Besides the different types of openings, all of the aforementioned complementary buildings used the building technology of the main construction.

As it can be observed in the following example (Figs. 4.20 and 4.21), located in the Baleizão, the rural traditional house is usually structured with a single volume of low and horizontal expression. The kitchen constitutes the central space, where the entrance is placed, and also functions as a living room. The small bedrooms, the "alcovas," are directly connected to this space. In the lower inner land area, rammed earth is combined with schist elements, namely in the execution of the buttresses, reinforced corners, and in the framing elements of the openings.

The *house of Monte da Boleja*, located in Baleizão, is composed of a single ground floor. It has 60 cm deep rammed walls, on the top of a shallow foundation of schist. It integrates several and different big buttresses, made of schist masonry, except in the front façade, where most of the openings are located. The front façade is exposed to north-east in order to avoid the most aggressive sun exposure. The plastered facades are carefully lime washed, contrasting with the traditional blue color of the wainscot of the wall, the soco, and the openings shutters, the postigos. The application of strong colors like green and yellow on this region, circumscribed to the mentioned transition elements, is usually related with the original intention to repel the insects. The inside partitions are executed in fire brick masonry. The pitched roof, with a slight inclination, is directly supported by simple round wood

Fig. 4.22 Reguengos de Monsaraz (credit: Gilberto Carlos)

beams and is covered by traditional "canal" tiles. In this case the pavement is covered with thin square mosaic brick, the local "baldosa," but in many buildings the floor could be simply made of compacted earth and the areas of more intense use could resort to river pebbles. The tall and thick chimney, intersecting the roof, acquires a significant expression, reflecting the importance of the kitchen, but not as much as in the houses located inside the compact settlements.

The second example represents the Alentejo's townhouse in *Monsaraz* (Fig. 4.22), typically urban defined by more compact structures. In this particular dwelling the construction is developed in two floors, although one-story houses are also common. The building is more compact, and the inside partition presents a less regular matrix. Although plastered and lime washed rammed earth continues to be the prevailing construction system, in these examples, granite, schist, and fired brick masonry are also common. In all the examples studied, the lime mortar is a determinant element (Fig. 4.23).

The coverage system resembles very much the previous example, but here the channeling of the rain water plays an important role. Here some of the pitched roofs combined the gutters and eaves conducting rainwater to an underground cistern, shared by the locals. The support of the higher level floor is provided by simple wood beams, but in most of these cases, they are combined with stone arches (or vaulted ceilings), centrally placed in the lower divisions, thus helping bearing the added weight. The lower floor usually contains two divisions of collective use. The entrance leads directly to the kitchen. The upper stairs divide the kitchen from the rear living room from which one can usually access a small patio with some small

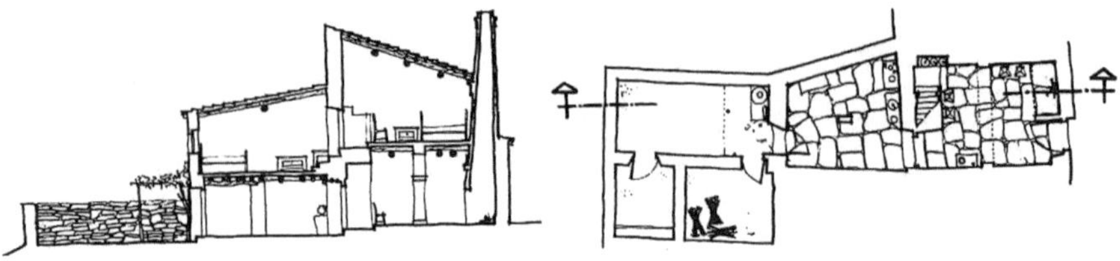

Fig. 4.23 Reguengos de Monsaraz: floor plan and section (AAP 1961)

storage or animal shelter. The upper floor contains bedrooms on opposite sides, articulated through the stairs, with openings to the street and to the rear patio, respectively. It is frequent to use vineyard trellis as a coverage solution for the transition areas with the exterior. Due to its abundance in the region, most of the ground pavements are covered with schist slabs. The chimney acquires a relevant position, tall and thick, emerging as an outgoing element, extending from the street facade over the roof lines.

Coastal Alentejo, Lower Alentejo, and Algarve

The so-called zone 6 which is larger than the others (cf. Fig. 4.1) is comprised of the entire region of the Algarve, Alentejo coastside (from the river Sado to the Algarve), and a portion of Lower Alentejo that borders the Algarve. Furthermore the national survey team divided the province of the Algarve into three distinct zones because they understood that these had characteristics of settlement, and economy differentiated in addition to differences in the building materials.

These 3 zones are the lower Algarve, the Barrocal or Algarve (where limestone is abundant), and the Higher Algarve with its mountain area. This natural division also results from topography and geology in three zones which are still easily identifiable today by the different types of settlement, geographical, and architectural features, and more recently by different economic development levels.

The south of Portugal presents unitary characteristics in traditional construction, linked to clay and lime, possibly as a result of its Roman past when it was integrated in Lusitânia and Moçárabe land as part of the Al-garb Andaluz kingdom, with small variations in each region.

The Alentejo coast and the southern zone of the lower Alentejo present an architecture close to the one existing inland, which is characterized mostly by buildings in simple construction with earth walls not very high, may have buttresses, and pitched roofs with reduced slope covered with tiles. This configuration and the construction materials used resulted in an architecture inserted in the landscape that often stands out only for the high and generally large chimney in the Alentejo.

The simple spatial division was mostly characterized by a large central space with fire where all the functions of domestic life—eating, living, and sleeping—were performed, according to Ramos (2010); this characteristic was quite common

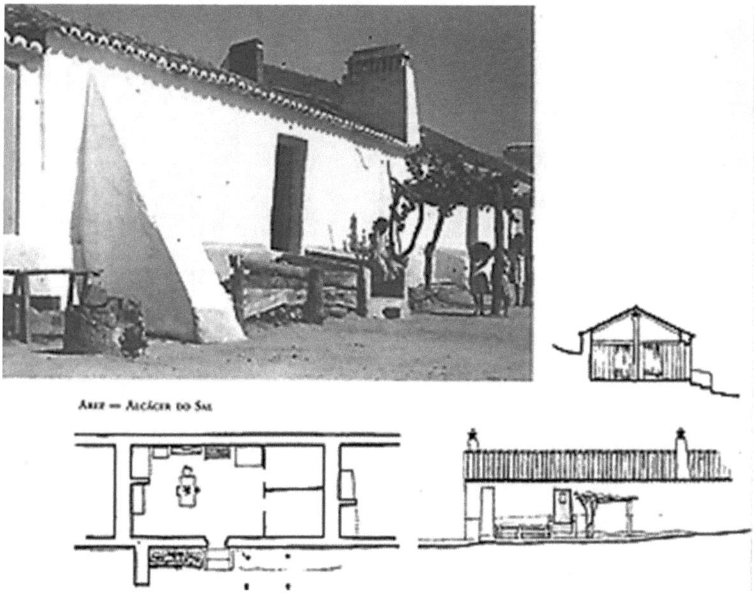

Fig. 4.24 House in Arez, Alcaçer do Sal (AAP 1961)

in the oldest rural dwellings: "The rural house of popular origin can present a simple organization, usually based on a single compartment, and may have light partitions that protect the sleeping place" (Ramos 2010).

The town houses, in the rice fields of the Sado Valley for example, are referred to in the survey as single room dwellings with a fireplace, with wood partition walls in sleeping zones that appear to be of later construction (Fig. 4.24).

These low-wall dwellings with buttresses are also notable for their large chimneys perpendicular to the façade and the fact that there is no other type of opening beyond the entrance door. This open door and high chimney ensured the necessary cross-ventilation for the single central space. In the survey that Mariana Correia made on the "taipa" dwellings in Alentejo, these were located in the same zone of *Alcácer do Sal*, and with the same characteristics (inquiry 35 and 37): houses with a central rectangular space, with fireplaces, and few windows. Some of these windows according to direct testimony were later opened. This simplicity in the typology of housing in the river Sado valley is referred to in the survey as "a synthesis of man's minimum needs" (AAP 1961) (Fig. 4.25).

Also near Alcácer do Sal but in the sandy areas near the *Comporta* beach, we can find the houses of fishermen with completely different constructive characteristics. These are constructions with a wooden structure of reduced section, as much in the walls as in the finishings that are filled by two layers of stem and topped in the walls by horizontal boards generally covered with clay (Fig. 4.26). The floors are made of rammed earth. The rectangular plants sometimes had two or more bedrooms with a living room in the center that might have a fireplace. In some cases, a dwelling

Fig. 4.25 Houses in Carrasqueira, Alcaçer do Sal (source: AAP 1961)

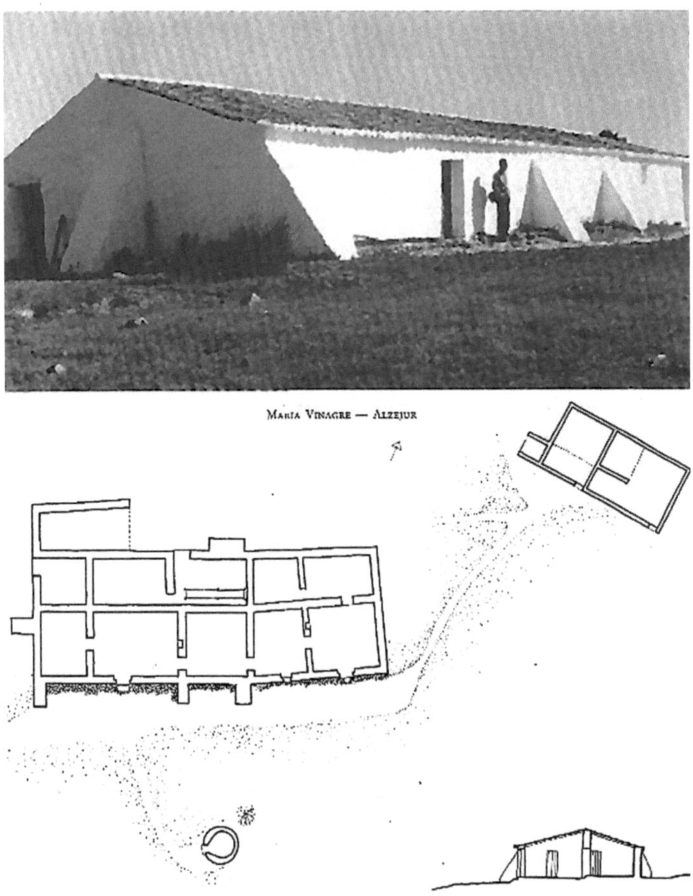

Fig. 4.26 House in Arez, Alcaçer do Sal (AAP 1961)

consisting of two huts may be found, one with a fire hut, in order to isolate the flammable fireplace from the rest of the house.

Algarve—Team 6 has conducted in a very detailed study of the area of the Algarve: addressed the urban structure, housing types, and construction materials used, and, unlike other chapters, also the relationship between climate and architecture. The natural division of the Algarve into three regions, as already mentioned, also corresponds to different types of architectures and construction materials. While in the lower Algarve the use of clay is the most frequent, in the mountainous areas it is the stone masonry, varying according to the region. In the barrocal or limestone Algarve, the stone masonry of the region is the most used building material.

The dwellings of the lower Algarve have characteristics close to the Alentejo houses: low walls with two-story roofs, but usually have a more quadrangular plan and are less extensive than most Alentejo houses. Another distinguishing characteristic is the small chimneys if we compare with the Alentejo and with a network with decorative elements for the extraction of the fumes. The plants of the interior, more subdivided than in the Alentejo dwellings, have the kitchen separated from the reception area outside. Compartments dedicated to animals or agricultural stowage appear next to the dwelling or in the same set.

In the dwellings of the places closer to the Alentejo such as in the *house of Maria Vinagre* in the Aljezur area, these differences are diluted. This house was one of the models raised in the survey, as it is an interesting rare example in the Algarve architecture. Its plant has a master wall in the longitudinal direction, which divides the zone of the residential space in the south of the space for animals and storage in the north (Fig. 4.26). This example is also notable for containing "Moirões" (buttresses) rare in the region of the Algarve as in (Fernandes and Janeiro 2008), who also refers this house in his work on popular Algarve architecture: "it would be very important to safeguard what remains of this authentic relic of Portuguese vernacular architecture—a rare and beautiful vestige of the use of the "moirão" in the construction of rural Algarve.

In the housing of "*quarto estradas*" for example, also referenced in the survey, enclosures for animals are located next to the house hidden by the terrain slope (Fig. 4.27). This house is a very interesting example of this type of architecture because it has three volumes of different finishes but with a common layout which suggests different times of construction. Outside the building in front of it, a living space spreads a terrace flanked by benches and small whitewashed walls topped by a grapevine that allows the shade in summer days. These walls, which extend the ground floor creating terraces next to the houses as a form of appropriation of the landscape, sometimes include a cistern or a tank, a popular feature in the Algarve, though conversational benches along the exterior walls are typical in different regions of Alentejo.

In central Algarve, the most remarkable example and also the most known is one that articulates a volume with a slanted coverage of one single roof with another covered by a roof terrace (açoteia). This one, which generally is oriented to the street, is topped by a parapet, almost always with decoration, serving as a balcony on the roof, very characteristic in this region. These houses may be isolated or

Fig. 4.27 House in Quatro estradas, Lagos (AAP 1961)

clustered. The survey finds this type of housing in settlements between Porches and Tavira, but mainly in the "barrocal" area. In his thesis "Algarve Buildings" Agarez (2013) disassembles the possibility of this type of parament (Fig. 4.28) being genuinely vernacular, dating back to the nineteenth century. Existing originally only in terraces for users' protection, it began to be used in all types of roofs from 1872, when a law enforced drainage in pitched roofs to avoid run off in public spaces. The spatial organization of these houses is very similar to the lower Algarve, with a square plan in rural houses and while urban homes feature a deep rectangular shape with the smaller side towards the street (Fig. 4.29).

We should also refer to the urban housing group of *Olhão and Fuzeta*. These are town houses of fishermen with long rectangular plants with "patios" in the back. In the case of Fuzeta these "patios" are the ceiling of a lower floor, possibly due to the topography of the terrain. In Olhão the typical building is usually a story high. In both cases the dwellings have flat roofs or açoteias, accessible by a ladder outside.

The origin of these terraced houses is questioned by zone 6 authors in the survey, citing António Sérgio, as he claims it to be the result of a recent Northern Morocco influence (nineteenth century), due to permanent contacts between the populations of these two regions and across the Atlantic. Another hypothesis would be this type of roof deriving from the weather, as local climate differs from others with clear distinguishable architectural patterns. Also Ricardo Agarez raises the issue of the origin of açoteias in Olhão and its relationship with Morocco, citing Leite de Vasconcelos. This author presents another reason besides the direct relationship between the population of these towns and the north of Morocco but adds up that examples of these styles in other locations in the Central Algarve, like Estoi, Moncarapacho, or Tavira, would justify this theory itself.

Finally, the houses of the *"Caldeirão" mountain and the "Guadiana" valley*, both with very simple volume and plant with a single-pitch tiled roof. This roof that accompanies the slope of the terrain is sometimes creating gaps inside the housing, in cases where they are implanted in accentuated slopes. With very simplified plant and sometimes just with the door as the only opening, these stone

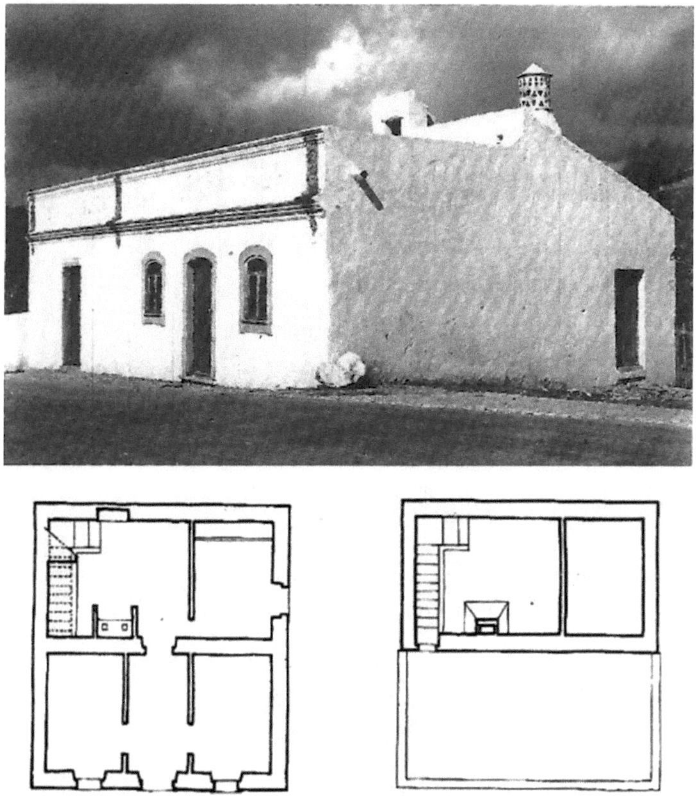

Fig. 4.28 House in Fonte do Bispo, Tavira (AAP 1961)

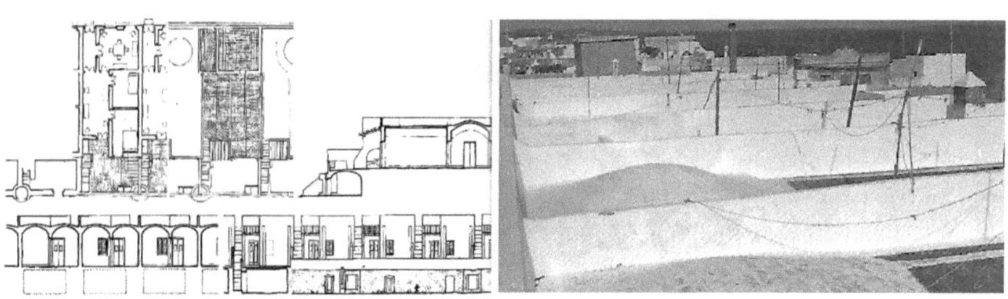

Fig. 4.29 Row houses in Fuzeta, Olhão—plans, sections, elevation, and "Açoteias" (AAP 1961)

houses feature an outdoor fireplace to cook in the summer and for heating the interior of the houses during winter. Smoke naturally through flows through the tiles in these constructions with no chimney (Fig. 4.30). In the national survey of regional architecture several examples portray this type of rural house but always isolated dwellings. However in a thorough study about mountain houses of these two regions, conducted by Costa (2014), most of the cases studied are clusters of

Fig. 4.30 Houses in Alcoutim—plan and section (AAP 1961)

dwellings or dwellings grouped with facilities for animals and agricultural support. Some of these provided a community life (Milheiro 2012).

In the area of the river "Guadiana" the "montes" were a cell junction, some are housing, some with farm use, and some with dual function. The system of these villages was a living system that was going to engage new cells when it was necessary, creating an organic system quite adapted to the terrain. In "Beliche de Cima" (Fig. 4.31) for example there are several single clusters of houses, but every construction of these can have one or more cells for homes, and in some cases a family could dwell in a cell of a building, but having the agricultural compartments in another building.

Architecture and Climate To conclude this reflection about the Algarve, it must be stated that the zone 6 team—coast and far southern Alentejo and Algarve—have a section dedicated to the climate and its interaction with the Architecture, relating design with thermal performance in Summer and Winter.

In this chapter the architects of the Team 6 made a geographic characterization of the region, with large areas of south-oriented slopes favoring both the settlement of villages as single housing. Maps of the region's temperature variations and precipitation are presented.

Various architectural aspects are pointed in the studied cases to improve the climate-comfort issues: the privileged South orientation; the protection from the Atlantic winds; the fixed shading (balconies, porches, terraces, etc.). As says João Santa-Rita "In Algarve, preferring the southern Quadrant for their house and creating heat transition spaces: courtyards, Porches, trellised vines, etc."

To the end, the team no. 6 shows several detailed examples of water collection systems, through the sloped roofing and guttering systems that guide the water to cisterns or tanks – a way of storing rainwater in regions where it is rare in Summer. "In the short terms one can say, like the Algarve team, that the entire elements put all together set—the most perfect conditions of living and using" (Santa Rita 2012).

Vernacular Revival

The vernacular legacy has been largely forgotten since the dawn of the "Mechanical Era." To many, it was associated with poverty, lack of living conditions, and rural environments; instead, modern building techniques represented greater value and efficiency. Local references have become subdued by a mass culture, turning architecture into a "fast consumption" product (Sampaio 2009).

The need for a sustainable architecture arrived in Portugal in the last two decades. Teaching and research on bioclimatic design began in the turn to the twenty-first century—bringing with them a new interest on vernacular design.

Today, new generations of architects start (or restart) to perceive vernacular as a fundamental source of knowledge, bringing about a shift in architectural paradigms. Vernacular is no longer viewed as poor and inefficient—instead, it is a rich source of renewed inspiration, and vernacular buildings are very efficient. Efficiency is no longer a mere attempt to mimic mechanized uniform environments such as those

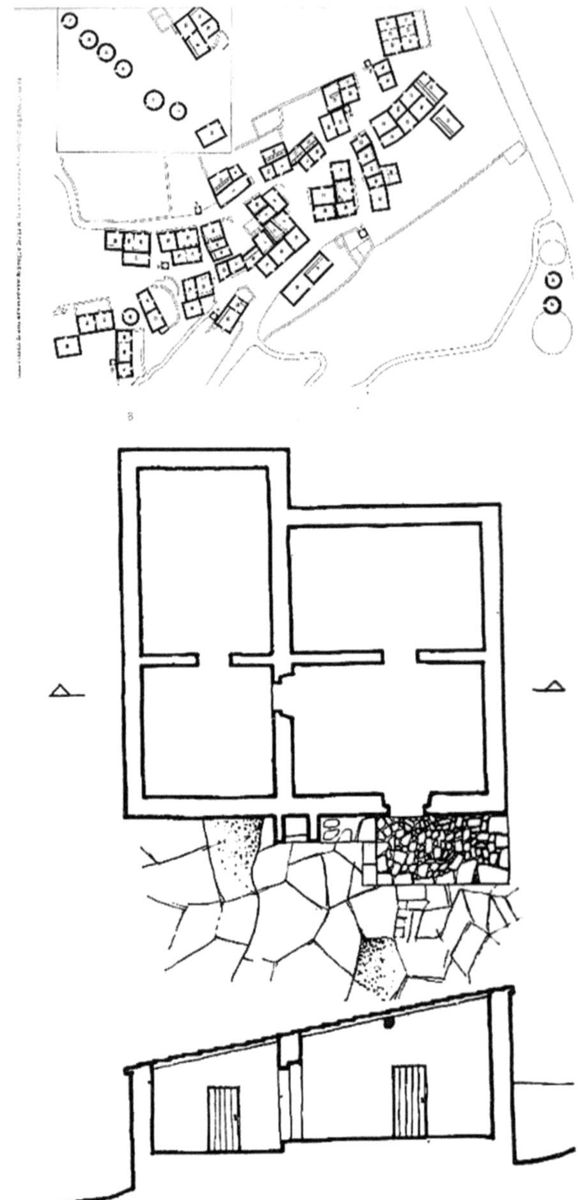

Fig. 4.31 Beliche de Cima, Tavira—plan (Migue Reimão Costa)

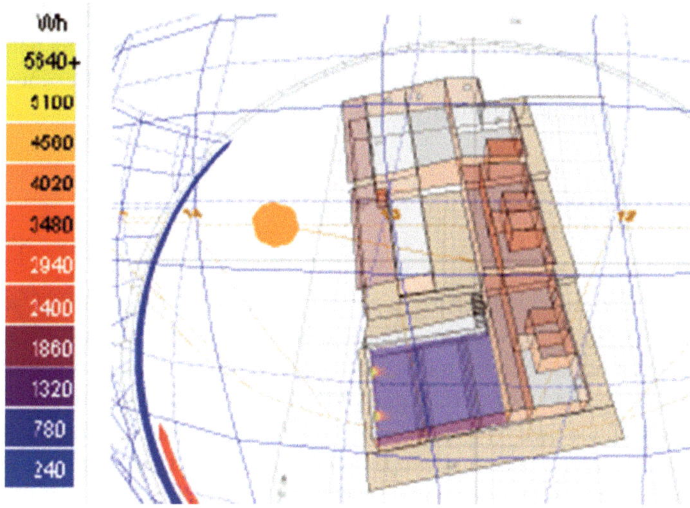

Fig. 4.32 The use of software to assess the environmental performance on a vernacular building in Alentejo (Serpa)

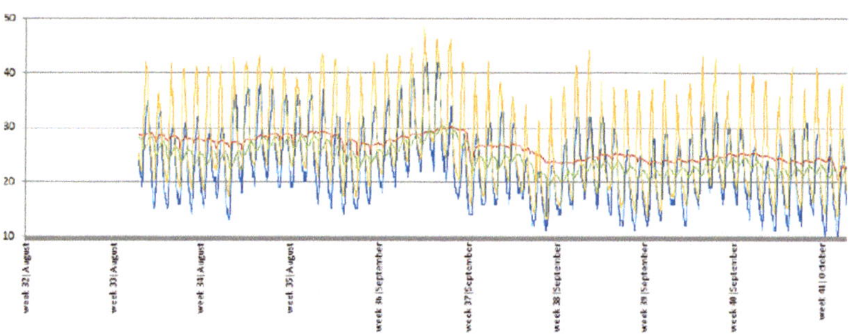

Fig. 4.33 Thermal performance of a vernacular building in Serpa (Alentejo), during the summer heatwave of 2016. Internal temperatures (brown line) are stable, within comfort limits and well below external temperatures (Távora 1947; Viana 1999)

produced by HVAC systems; instead it is the ability to provide natural, variable environments, such as in traditional buildings.

New software tools used by students of architecture show the ability of vernacular to produce comfortable thermal and lighting natural environments through its appropriate bioclimatic design (Fig. 4.32). Figure 4.33 shows the result of measurements made on a vernacular building in Alentejo, during a particularly hot heat wave in 2016: internal temperatures are stable and well below external temperatures, essentially due to thermal inertia.

Furthermore, rural villages across the country that have long been deserted due to migrations to urban areas are now being re-occupied by families that chose eco-friendly, alternative, ways of life. New concepts, such as biological agriculture, are slowly contributing to the revival of the rural space, bringing about a new life to vernacular buildings. Contradicting centralization in mega cities, this movement is a step towards global sustainability and greater quality of life.

Conclusions

This chapter has provided an overview of the regional variations of vernacular architecture in Portugal, resultant from man's dialogue with the local natural context. Vernacular architecture is today slowly being perceived as a core reference for sustainable design and an essential part of Portugal's cultural heritage.

In a world challenged by social conflicts, global warming, and lack of harmony, vernacular stands as a beacon of hope and possibilities. There are clear signs of a renewed interest in rural space and vernacular revival; let us hope that this trend is here to stay.

References

AAP (1961) Arquitetura Popular em Portugal. AAP, Lisboa

Agarez RMC (2013) Regionalism, modernism and vernacular tradition in the architecture of Algarve, Portugal, 1925–1965. Doctoral dissertation, UCL (University College London)

Barros F (2013) Construção do Território e Arquitectura na Serra da Peneda. Padrão (Sistelo) e as suas brandas - um caso de estudo. Município de Arcos de Valdevez, Arcos de Valdevez

Barros F (2015) Sistelo. Entre o Vale e a Montanha in Revista Aldrava, n°18. Aldraba – Associação do Espaço e Património Popular, Lisbon

Barros F, Barroso C, Riveiro B, Oliveira D, Ramos L, Lourenço P (2016) Arquitecturas de Falsa Cúpula na Região Transfronteiriça Gerês/Xurés. Enquadramento e Caracterização Construtiva. In: Proceedings of 2° Congresso Internacional de História da Construção Luso-Brasileira. Culturas Partilhadas, vol II. CEAU, FAUP, Livro de Actas, pp 493–508

Cabral I (2009) Reabilitação ambiental de edifícios em Portugal: o caso da reabilitação da arquitectura vernácula em áreas protegidas. PhD Thesis (architecture), Faculdade de Arquitectura da Universidade de Lisboa

Cabral I, Chalfoun N (1998) Microclimate control and passive cooling in a rural house for tourism in Cova da Beira. In: Proceedings of PLEA'98, June, Lisbon. James & James, London, pp 479–482

Cabral I, Coelho A, Machado G (2013) Assessing energetic self-sufficiency and low environmental impacts in Pontes, Portugal. In: Proceedings of CIAV'13: vernacular heritage and earthen architecture: Vila Nova de Cerveira. CRC/Balkema/Taylor & Francis, London, pp 593–598

Caldas J V, Correia JEH (1999) A casa rural dos arredores de Lisboa no século XVIII, ed. Caleidoscópio, Lisboa

Correia M (2007) Taipa no Alentejo. Rammed Earth in Alentejo. Argumentum, Lisbon

Correia M, Carlos G (eds) (2015) Cultura sísmica local em Portugal. Argumentum, Lisboa

Costa MR (2014) Casas e montes da serra entre as estremas do alentejo e do algarve. Edições Afrontamento, Porto

Fernandes JM, Janeiro A (2008) A casa popular do Algarve: espaço rural e urbano, evolução e actualidade. Comissão de Coordenação e Desenvolvimento Regional do Algarve, Faro
Fernandes S (1963) Arquitectura, nº 75, ed. AA, Lisboa
Leal J (2009) Arquitectos, engenheiros, antropólogos: estudos sobre arquitectura popular no século XX português. Fundação Marques da Silva, Porto
Maia MH, Cardoso A, Leal JC (2013) A arquitectura popular em Portugal: uma leitura critica. Relatório final. Porto, ESAP. Available on https://comum.rcaap.pt/bitstream/10400.26/19898/1/Relato%CC%81rio%20detalhado.pdf. 17 Outubro
Milheiro AV (2012) Nos trópicos sem Le Corbusier: arquitectura [sic, arquitetura] luso-africana no Estado Novo. Relógio d'Água, Lisbon
Moreira C (1987) Populações Marítimas em Portugal. In: Instituto Superior de Ciências Sociais e Políticas, Lisboa
Oliveira EV, Galhano F (2003) Arquitectura Tradicional Portuguesa. Col. Portugal de Perto, 24. Publicações Dom Quixote, Lisboa. (5ª Edição; 1ª Edição de 1992)
Oliveira EV, Galhano F, Pereira BE (1969) Construções primitivas em PortugalPublicações Dom Quixote
Palla J, Gaspar P (2015) Construções palafíticas da bacia do Tejo: Levantamento e diagnóstico do património construído da cultura avieira. ASF, Lisboa
Pereira NT, Freitas A, Dias F (1961) Arquitectura popular em Portugal. Edição da Associação dos Arquitectos Portugueses, Lisboa
Portas N (1958) Arquitectura, nº 63, ed. AA, Lisboa
Ramos RJG (2010) A casa: arquitectura e projecto doméstico na primeira metade do século XX português. FAUP publicações, Porto
Ramos RJG & da Silva (2012) S. D. João Andresen: a casa do pós-guerra e o debate arquitectónico nos anos de 1950. Januário Godinho: Leituras do Movimento Moderno
Ribeiro O (1963/1945) Portugal, o Mediterrâneo e o Atlântico. Sá da Costa, Lisboa
Ribeiro O (1961) Geografia e Civilização. Temas Portugueses. Centro de Estudos Geográficos, Lisboa
Ribeiro O (1968) Mediterrâneo. Ambiente e Tradição, 2nd edn. Fundação Calouste Gulbenkian, Lisboa
Sampaio CG (2009) Povoamento de Ocupação Sazonal em Castro Laboreiro, Brandas e Inverneiras. Master Thesis (architecture), Faculdade de Arquitectura da Universidade do Porto
Santa Rita JM (2012) Building with the climate in popular architecture in Portugal: critical reading of the "Survey on Regional Portuguese Architecture" João Manuel Santa Rita. CEAA I Centro de Estudos Arnaldo Araújo (FCT uID 4041), p 476
Távora F (1947) O Problema da Casa Portuguesa. Cadernos de Arquitectura, Lisboa
Viana P (1999) Território, povoamento, construção: manual para as regiões do Parque. ADERE-PNPG, Ponte de Barca

Chapter 5
Seeking Contemporary Urban Comfort Through Vernacular Architectural Principles in Hot Arid Climate

Mona Azarbayjani

Introduction

It has been more than six million years since our ancestors appeared on Earth. Thus, we can say human beings are one of the most sustainable species that can be found on our planet. Not far from 200,000 years ago, the very first civilizations had been formed and ever since, nations tried to overcome the harshness of their environment by application of various methods. Considering the Primitive Hut, the architecture is formed by the environmental forces in its context.

The environmental forces are important not only to the architecture scale, meaning the scale of a building, but also to the urban scale. Climatic characteristics are part of these environmental forces, including wind patterns, sun path, temperature, and so on.

Looking at sustainability by its definition as longevity, analyzing the architecture and urban patterns of traditional societies which overcame the harshness of environment over the course of time might bring passive solutions to both architectural and urbanistic scales of design. While the architectural scale was more of attention during past years, the purpose of this chapter is to look into urban features of hot arid climate that has been used to bring urban comfort and to abate the harshness of the environment, and then to propose modern design alternatives using the benefit of contemporary advancement in technology and tools in architectural discourse.

The city of Yazd with hot arid climate, the 15th largest city of Iran and a World Heritage Site, is chosen to be studied. Accommodating numerous architectural elements from traditional Persian architecture, Yazd is known for its wind catchers that help the city with adapting to its desert like climate. The elements to be analyzed are wind catchers, courtyards streetscapes, vegetation, water, and street shades.

M. Azarbayjani (✉)
Integrated Design Lab. Energy Performance Laboratory, School of Architecture, College of Arts + Architecture, University of North Carolina at Charlotte, Charlotte, NC, USA
e-mail: mazarbay@uncc.edu

A. Sayigh (ed.), *Sustainable Vernacular Architecture*, Innovative Renewable Energy, https://doi.org/10.1007/978-3-030-06185-2_5

The area of study is 168,121 ft^2. of a neighborhood in Dowlatabad, one of the historical parts of Yazd. To analyze the impact of each architectural/urban element on urban comfort, the context was modeled in Envi-Met (the simulation software) and each environmental category is brought into it in order to study them separately and together to take the lessons for future urban design.

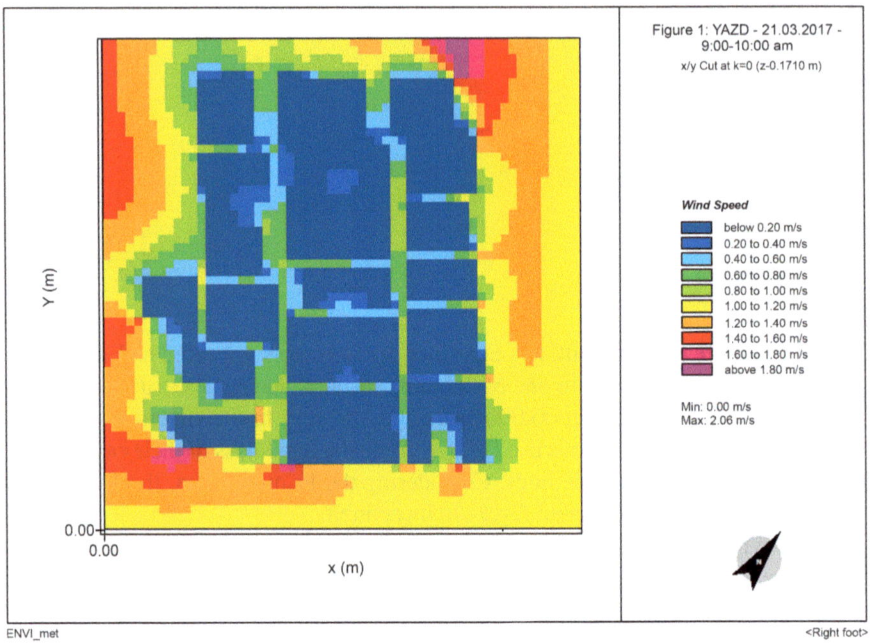

Context (Left). Simulated site (Right)

In hot arid climate, the direct angle of harsh sunlight as well as winds and low precipitation makes these climates severely dry. In addition, especially around noon, dust and sand storms are predictable as a result of heated air layers closer to the ground. Therefore, the architecture is impacted by the effect of its surrounding climate. A closer look to the urban features of these cities reveals some of these factors.

Compact plan design with "low area to volume ratio" can be seen in the architecture of buildings in the hot arid climate. Integration of thermal mass and compact plan results in the reduction of heat loss and heat gain during winter and summers by minimizing the heat transfer between interiors and exteriors. In addition, maximizing the shaded areas and use of light color for the exterior facades, minimizing the number and size of the openings, and orienting the buildings towards south/south east are some of other design strategies to maximize the efficiency. Two other passive strategies studied in this chapter are wind catchers and courtyards with a small pool and vegetation that have cooling effect for the interior space.

In this chapter, several factors that form urban comfort in hot arid climate are discussed and then the design elements that affect those elements are analyzed.

Urban Comfort

Although, in the past, the focus of thermal studies was on the building level, in the most recent studies, there is a shift toward urban microclimate as it affects the livability of urban spaces, strengthening the thermal comfort of the outdoor occupants. Urban microclimate can change the people's lifestyle, also change the use of urban space by affecting the parameters of thermal comfort.

There are several parameters that affect the urban microclimate, such as urban density, form and geometry, vegetation, water bodies, and surface properties as well as urban orientation, which includes the orientation of roads and paths. Each of these parameters can affect one or several outdoor thermal parameters, enhancing the urban comfort.

Reflective Material The use of reflective materials, which can reflect the solar radiation, is an effective approach toward urban comfort. The higher reflectance results in the lower surface temperature, almost about 5 °C decrease in the air temperature under low wind conditions.

Presence of Water Water ponds, fountains, or even artificial waterfalls can help the reduction of air temperature in microclimates up to 11 °C. Water bodies work as a heat sink, dissipate the heat through evaporation. This helps areas with hot arid climate to regain humidity and also to cool down the microclimate. The presence of water with vegetation affects the amount of solar radiation, ambient temperature, and wind velocity, resulting in a better urban comfort.

Vegetation The presence of green spaces and vegetation mediate the heat gain. This can be used to improve the urban comfort at pedestrian level. The green space has various benefits to the urban space. It blocks the noise, provides more shading and probably the most important feature, the cooling effect via evapotranspiration and shading. Maximize the shading is the intention in designing in hot arid climate and planting trees can facilitate this concept.

Shading Elements Shading has a great impact on the outdoor thermal comfort. It reduces Sky View Factor (the percentage of clear sky at specific locations) which results in higher thermal comfort.

Shaded Alleys

Urban Density "The ratio of the buildings' footprint to the total area of the site." This factor affects the amount of solar gain which can be helpful in winters while it can cause discontent in summers. In hot arid climate, the intention is to raise this factor as much as possible to reduce the loss of energy through convection and radiation.

Street Orientation Street network is the structure of every city. Street orientation is important for ventilation and to control the solar radiation. In cities with hot arid climate, because of the low altitude, a north-south street axis provides an equal solar protection for buildings. Each street orientation and pattern, such as Spanish Grid and Jeffersonian, influences the microclimate differently. Vegetation, specifically planting trees, can intensify the impact of street orientation on the microclimate.

Street Aspect Ratio Street aspect ratio, or the height of buildings surrounding the street (H) to the street width (W), is an important factor that affects energy transfer, hence affecting the microclimate. In traditional hot arid climate cities, narrow streets (pathways) can be seen. It means that the higher aspect ratio, higher H and lower W, is beneficial to achieve urban comfort.

Urban Texture

Site Location

The location of this study is a block in Dowlatabad neighborhood, Yazd, Iran. This block is in the historical part of the city, consists of a dense residential fabric as well as the other specifications of traditional hot arid climate architecture with courtyards. The site is 0.72 km (0.45) away from the Dowlatabad garden, an eighteenth century complex with the tallest wind catcher in the country.

Location of the site understudied

Analysis

The analysis process begins with a series of simulations in the ENVI-MET software. Every simulation studies the presence of a specific element by eliminating all other factors.

For each element, the simulation ran three different times, 9 am, 12 pm, and 4 pm for 1-h span on equinox, summer, and winter solstices. Hence, the air temperature, relative humidity, and wind speed were calculated and compared between the presence and the absence of that specific element while the other elements were kept constant.

The elements of study include: courtyards, the orientation, and the width of the streets. To that end, the simulations ran with the presence of the courtyards and once they were filled with mass. To study the impact of the orientation, the model was rotated 180°. And lastly, the width of the streets increased 5.25 m to study the impact of the street width.

Impact of the Presence of Courtyards on Air Temperature

As shown in Fig. 5.1, the absence of courtyard spaces in the neighborhood resulted in a slight increase of air temperature in December. However, the temperature is almost the same for other months.

Impact of the Presence of Courtyards on Relative Humidity

As illustrated in Fig. 5.2, the absence of courtyard spaces in the neighborhood resulted in a slight increase of relative humidity. The largest difference was observed in 21st of December, with a 1% difference.

Impact of the Presence of Courtyards on Wind Speed

The absence of courtyards results in an increase in the wind speed (Fig. 5.3). This helps with better ventilation that can lower the temperature and increase the urban comfort.

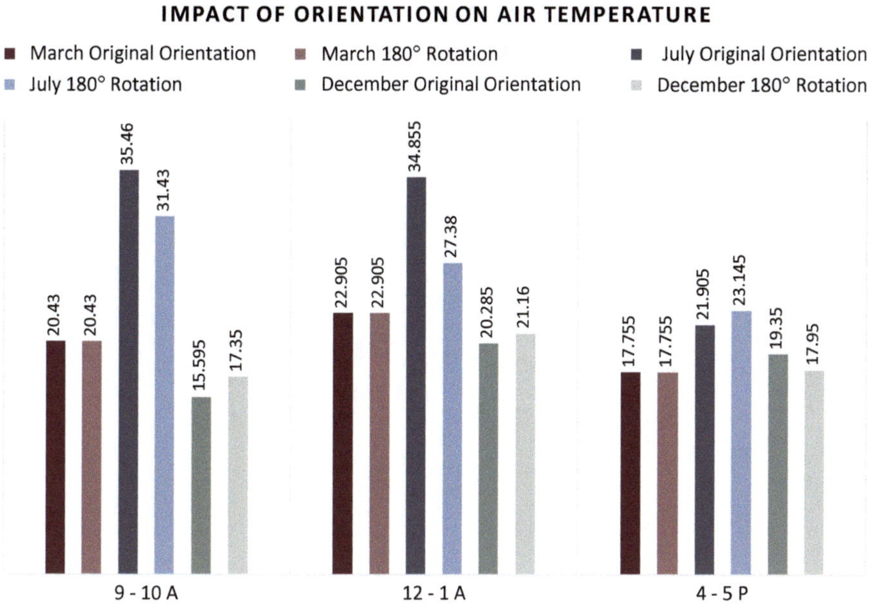

Fig. 5.1 Impact of orientation on air temperature

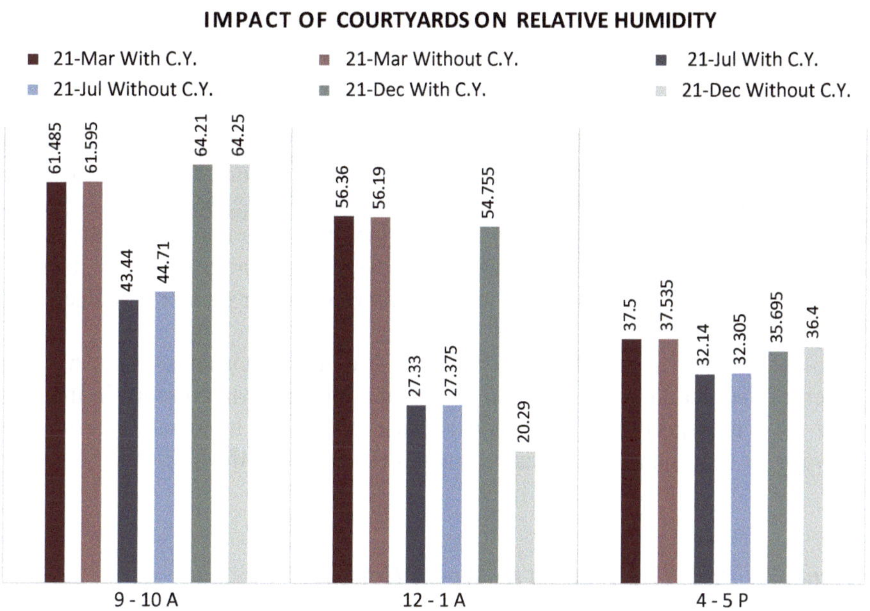

Fig. 5.2 Impact of courtyards on relative humidity

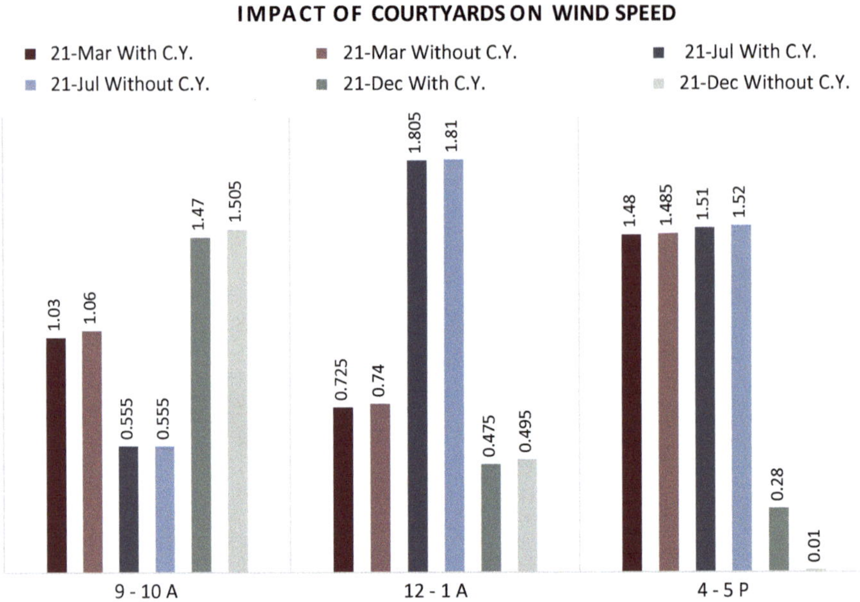

Fig. 5.3 Impact of courtyards on wind speed

Orientation

Orienting the site toward south (180-degree rotation) results in air temperature decrease in summer months and air temperature increase in winter, in the morning and at noon (Fig. 5.4). In the afternoon, however, the results are the opposite, which is ideal. *Hence, by orienting the neighborhood toward south, the air temperature can be cooled down in the hotter hours of the day and in the hot seasons.*

The change in orientation results in reduction of relative humidity in the mornings. At noon the change is not significant. In the evening, there is a rise in the relative humidity, which increases the urban comfort as the air temperature decreases (Fig. 5.5).

In case of wind speed, the results depict a rise, in the mornings (Fig. 5.6). This increases the natural ventilation and results in a higher urban comfort. In the evening the wind speed reduces which because of the air temperature difference between the day and the night that intensifies the urban comfort.

Roads' Width (The Height to Width Ratio)

The results indicated that there is no difference in the air temperature. It can be concluded that the effect of roads' width is insignificant on impacting air temperature (Fig. 5.7).

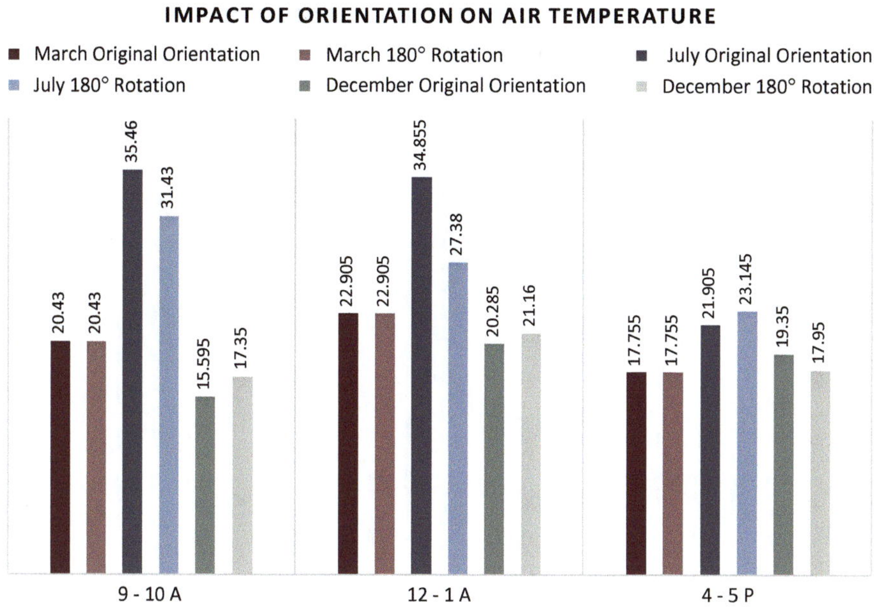

Fig. 5.4 Impact of orientation on air temperature

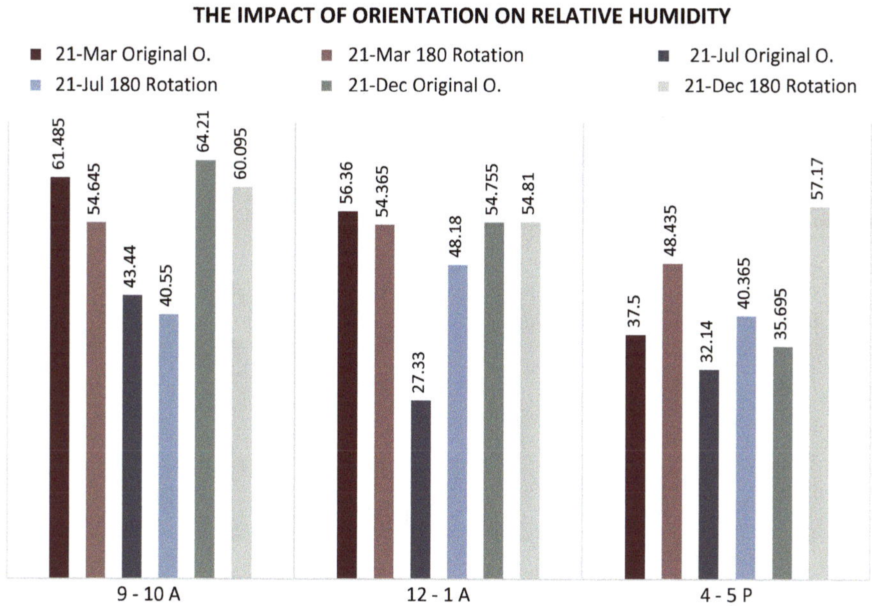

Fig. 5.5 Impact of orientation on relative humidity

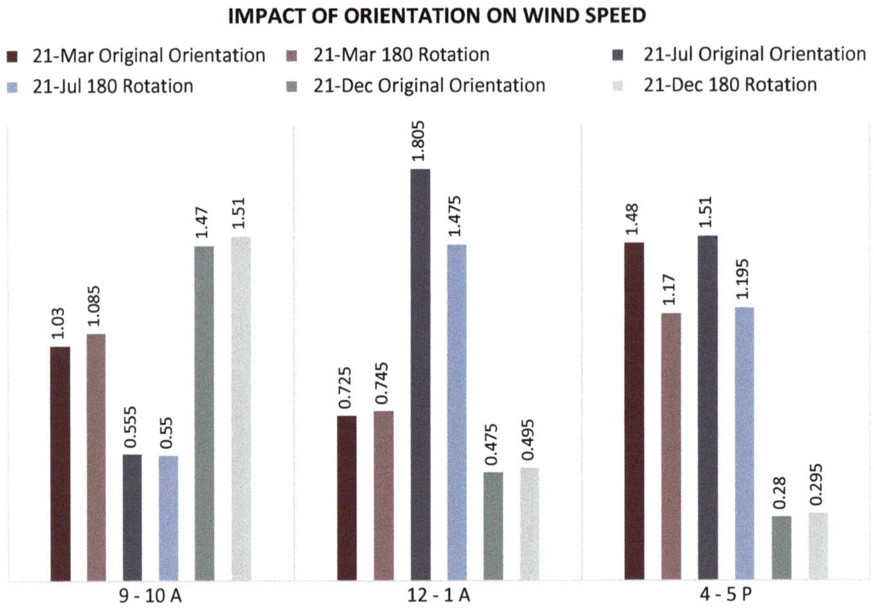

Fig. 5.6 Impact of orientation on wind speed

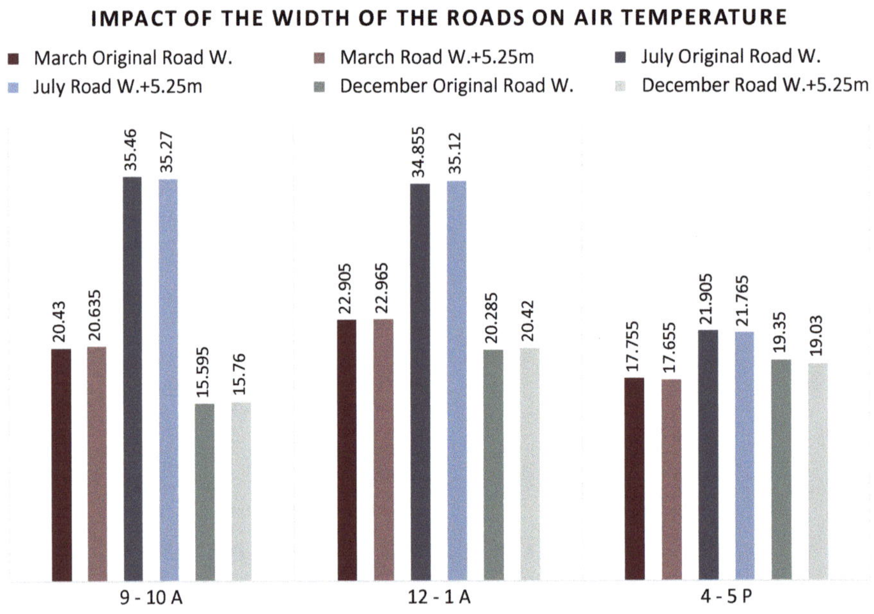

Fig. 5.7 Impact of roads' width on air temperature

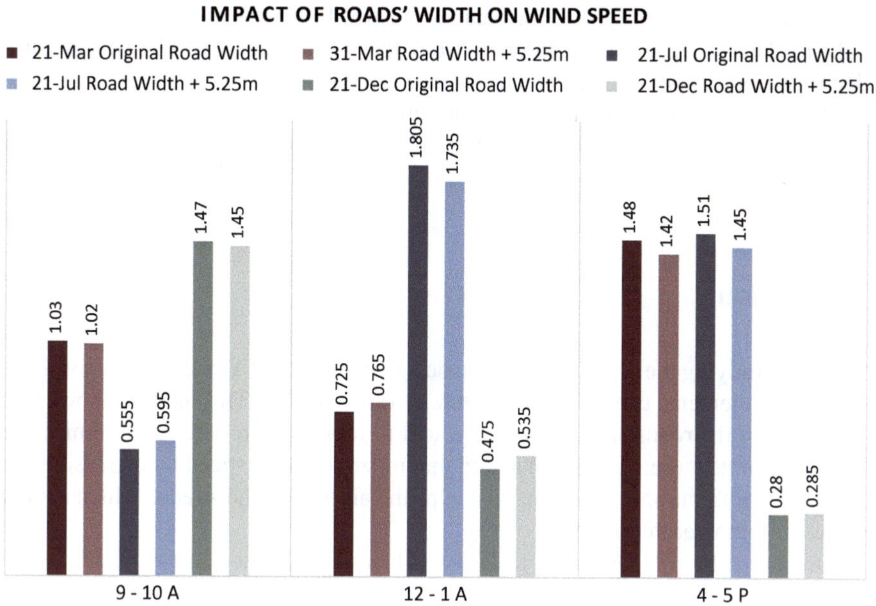

Fig. 5.8 Impact of roads' width on relative humidity

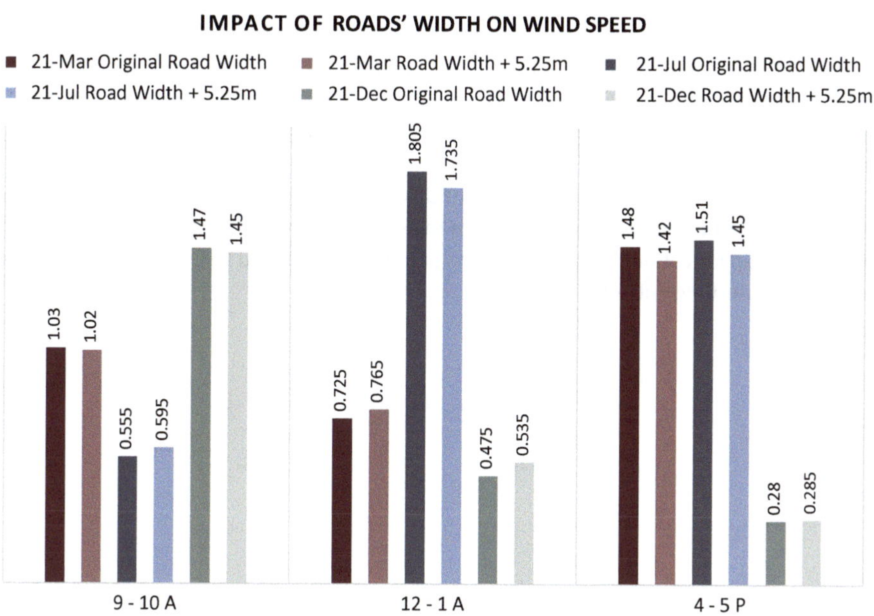

Fig. 5.9 Impact of roads' width on wind speed

The relative humidity did not change drastically after widening the roads by 5.25 m. However, it can be inferred from the results that the relative humidity decreased as it exposed more surface to the sun (Fig. 5.8).

Widening of the roads decreased the wind speed, which adversely affects the urban comfort since it reduces the breeze in the streets and for the pedestrians (Fig. 5.9).

Conclusion

Through studying the elements of urban design in the city of Yazd, one can conclude how urban elements can inform the future of urban design. The analysis shows that the width of the roads/streets should be kept to minimum in order to minimize the reflection of heat (heat radiance) and to maximize the wind speed by increasing the height to width ratio. In addition, this helps the air flow in the streets, which results in the air temperature to drop.

Orienting the neighborhood toward south in a way that courtyards are in the northern part of the house helps to have a higher wind speed. It also reduces the air temperature during summer and increasing during the winter months, resulting in a higher urban comfort.

These results as well as the analysis approach can be used in today's architecture and urban planning through finding the optimized angle for the city grid, to get the best wind speed and air temperature, also to find the correct area ratio of the courtyards. The width of the streets should be kept to its minimum. A multistory street configuration works better as it shades the area beneath it and minimizes the asphalt surface.

Courtyards intensify the relative humidity and lower the air temperature. These spaces can be incorporated in the design to be used as public spaces. The surface of these open spaces should be filled with vegetation and shaded to avoid any heat radiance. Longitudinal form for courtyards.

Chapter 6
The Gödöllő Palace as a Typical Example of Vernacular Architecture in Hungary

Marta Szabo

Introduction

The Royal Palace of Gödöllő is one of the most important examples of Hungarian Palace Architecture of the eighteenth-century period. It is considered a monumental building. The palace was built by Count Antal Grassalkovich I (1694–1771). The builder was a typical figure of the Hungarian aristocracy of the eighteenth century. He was a royal noble and president of the royal Hungarian Chamber as well as a confidant of Empress Maria Theresa (1740–1780). The construction of the palace began around 1733 under the direction of András Mayerhoffer (1690–1771), a famous builder from Salzburg who worked in Baroque and Zopf style (see Fig. 6.1).

The palace has a double U shape design and is surrounded by an enormous park (see Fig. 6.2). The building underwent several enlargements and modifications during the eighteenth century; its present shape being established in the time of the third generation of the Grassalkovich family. By then the building had eight wings, and—besides the residential part—it contained a church, a theatre, a horse riding-quarter, a greenhouse for flowers and an orange orchard.

After the male side of the Grassalkovich family died out in 1841, the palace had several owners, and in 1867 the decision of parliament designated it as a summer residence of the King of Hungary. This state lasted until 1918; thus Francis Joseph (1867–1916) and later Charles IV and the royal family spent several months in Gödöllő every year (see Fig. 6.3).

M. Szabo (✉)
Faculty of Mechanical Engineering, Szent István University, Gödöllő, Hungary
e-mail: Szabo.marta@gek.szie.hu

A. Sayigh (ed.), *Sustainable Vernacular Architecture*, Innovative Renewable Energy, https://doi.org/10.1007/978-3-030-06185-2_6

Fig. 6.1 The palace entrance

Fig. 6.2 Garden of the palace

Fig. 6.3 King Franz Joseph and Queen Elisabeth with the royal family at Gödöllő

During this period the palace became the symbol of independent Hungarian statehood, and, as a residential centre it had a political significance of its own. It was Queen Elisabeth (1837–1898) who specially loved staying in Gödöllő, where the Hungarian personnel and neighbourhood of the palace always warmly welcomed her. She was able to converse fluently in Hungarian. Following her tragic death, a memorial park adjoining the upper garden was built.

The period of the royal decades also brought their enlargements and modifications. The suites were made more comfortable, and a marble stable and coach house were built. The riding-quarter was remodelled.

Between the two world wars the palace served as the residence for Regent Miklós Horthy. No significant renovation took place during this period, apart from an air-raid shelter in the southern front garden. After 1945 the palace, like many other buildings in Hungary, fell into decay.

Soviet and Hungarian troops used the building, some of the beautifully decorated rooms were used for an old people's home, and the park was divided into smaller plots of land.

The protection of the palace as a historical monument started in 1981, when the National Board for Monuments launched its palace project. The most important tasks of preservation began in 1986 and were completed in the end of 1991. During this time the Soviet troops left the southern wing, then the old people's home was closed down.

Within this period the roof of the riding-quarter and the stable-wing was reconstructed, the façade of the building was renovated, as well as the trussing of the central wings and the double cupola. Research was carried out in the archives and in the building, and thus the different building periods of the monument were defined. Painted walls and rooms were uncovered, which revealed the splendour of the eighteenth to nineteenth centuries. Architectural structures were discovered, and so were the different structures of the park (Gál n.d.).

The utilisation of the main front wings of the palace was designed as a clear and well-developed architectural project. The first floor's 23 rooms (nearly 1000 m^2) accommodate the interior exhibition. The emphasis was on the revival of the old atmosphere of the royal period and the introduction of the time of the Grassalkovich family.

Reconstruction is the principle of the interiors completed so far, creating the state as it was around the 1880s. One of the most striking features of the Empress Elisabeth Exhibition is its historical accuracy.

The painted foyer and the Grand Hall on the first floor are also used for various programmes, with a link to the programme organisation and gastronomic activity unit on the ground floor.

Design Aspects of the Building

After the seventeenth to eighteenth century, the new aristocracy became more open to the western life standards which influenced the architecture and the new style of construction of new palaces and civil residential buildings as living spaces for noble families. These palaces have provided (instead of the fortress style and function castles) ample, luminous, and more comfortable living space with aesthetic motives by providing more light.

As a consequence, the design aspect of the building was not anymore being sheltered and having protection but instead become:

- Fulfilling the living standard of the occupants related to their status,
- Adapting itself to the outside environment and climate,
- Fulfilling the comfort requirements of the residents,
- Providing full functionality.

These meant providing

- Right thermal comfort,
- Good visual comfort,
- Relatively good indoor air quality.

These requirements have been different for today buildings.

Adaptation Requirements

Hungary—being at the centre of Europe—has a continental *climate* and is characterised by cold winter and hot summer. For these conditions the buildings have to provide solutions for both: for heating and for cooling in the temperature fluctuation.

Adaptation to Humans

Building residents have to adapt themselves to the outdoor and indoor climate, which was different in the earlier centuries. They managed to adapt easily to the outside temperature, and they got used to the colder indoor environment; however they have to use more clothing while they were in the rooms.

Architecture, Structure, Walls, Windows, Transparent Surfaces

The Hungarian rural castles had, in general, two floors with high slope roof with loft and cellar. The living area has been thermally protected from the ceilings and the floor.

The rooms were built with surrounded thick (60–70 cm) walls (see Fig. 6.4).

Room Division and Their Functions

The rooms were designed in a U shape while the functional rooms were at the front of the Palace (see Fig. 6.5).

Fig. 6.4 The wall thickness and passage in the U-shaped corridor

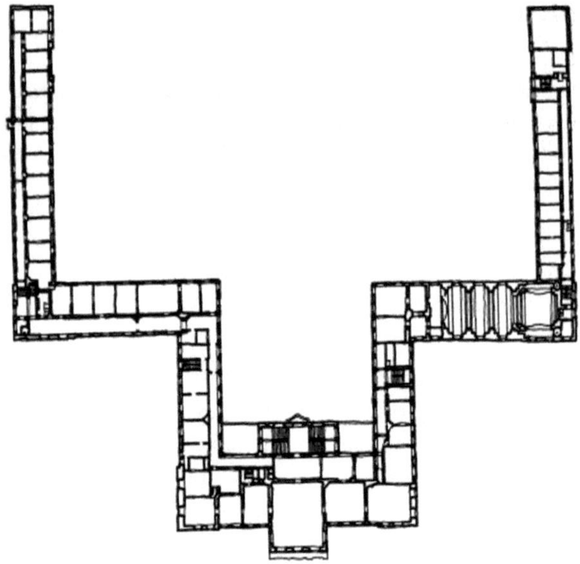

Fig. 6.5 The design of the rooms and their functions at the palace

Comfort Establishments (Cooling, Heating, Ventilation, Lighting and Aesthetic Function)

The heating demand for castles is determined by the size of the heated space and the local climate.

Cooling

The 60–70 cm thick walls had high thermal resistance so they could protect the occupants from the outdoor hot during the summertime. The high thermal inertia is due to the building material used (brick) and the wall thickness. The big thickness primarily provided the secure static of the buildings having the advantage of its thermal characteristics. Because of the high thermal inertia the walls reacted slowly on heat fluctuation, i.e. have heated up very slowly and kept the lower temperature for longer resulting in low indoor temperature as well (Áron n.d.-a).

The heat balance has been influenced by the façade windows as well. Normally the wooden framed windows had one layer of glass with poor insulation; thus the heat insulation capability was low. However, in ceremony rooms and corridors and in the main façade rooms big windows have been used. These structures could not avoid the cold in winter and could not resist the heat fluctuation in summer; however they could provide the ventilation and the light. Using internal shutters could increase the thermal insulation.

Double glass windows have been introduced by the end of the eighteenth century; thus the air between the glass layers increased the thermal insulation. Applying outside jalousie helps to avoid the heating up the indoor air in the rooms. Additionally it could support the ventilation behind the window.

The high thermal resistance and the outside jalousie gave effective protection against overheating the indoor space and supported the ventilation in the night.

Heating

The wooden shutters were not enough protective against thermal loss in winter; furthermore it was not advantageous because it kept out the small amount of light available in winter period. Internal curtains were used as a better solution in providing an additional insulating layer.

The efficiency and effect of heating was very limited because of the oversized wall structure as well as the high ceiling and big area of the rooms which resulted in a large volume of air to be heated up. Heating up this volume is even today with modern heating systems is a big challenge.

In the nineteenth century the most common solutions were to incorporate fireplaces and stoves.

The fireplace is open to the indoor air and by conduction, convection and radiation it makes the rooms warm. The air around the fire is heated up, the hot air raises and cold air flows replacing it providing air movement and a kind of temperature distribution in the room, this is true as long as the fire is there. This heating does not provide homogenous warmth in the room and it means creating uneven thermal comfort. The fire stoves are used for their aesthetic appeal than perfect heating. Another heating concept is called the tile oven, which is widely used.

The tile oven (see Fig. 6.6) has a better advantage compared to the fire places: it keeps the warmth for long period after the fire goes out, by conduction from the surface of the oven which provides the heat gradually to the indoor air. In general this system has good air flow and effective circulation resulting in good heating up of walls and the interior. Although the temperature distribution in the room is not fully homogenous, the heating is more effective because the walls act as good heat storage media from the oven. This is due to the ceramic tiles and the built-in heat ducts carrying the hot air continued from the oven.

For the tile ovens wood or coal was used as combustion material. The oven could be fed from the room or from outside opening. This method was very popular for castles in Hungary so that the heating comfort can be achieved without the ashes being noticed. In another words, the dirt was kept out from the heated room.

The Use of Heating

In the houses with several rooms the heating has been solved from a central room where the firing material was stored as well. From this room the ovens of neighbouring two or three rooms could be fed (Áron n.d.-b).

Fig. 6.6 Heating stove

In bigger castles these rooms could not be inserted within the chain of imposing rooms. In general the rooms in the ground floor were heated from the corridor, i.e. the openings for feeding were going to the corridor. The ovens in the first floor living rooms were fed from the so-called heating or service corridors; these led to the back of the oven. Having this service corridors going to the meeting corner of two rooms could give the opportunity to heat the ovens of more neighbouring rooms at the same time.

Even in using the fireplaces or tile stove the space could not be continuously heated up to a comfortable level. However when the rooms are needed for a function, they still cannot be heated for more than 15 °C.

The bigger room has not been heated at all; a lot of them have been used seasonally. Usually the most representative room has been heated by fireplaces not just for providing heat but for having a representative and esthetical role.

In case the occupants wanted to use a certain room in winter, prior to their use, the heating process must be started a day or two earlier so that a reasonable heating level can be achieved in time for the function to take place. As an advantage this gradual process was very beneficial not only for function to take place but also for the furniture/interiors which were mostly made from expensive wood, to be kept at a moderate temperature to avoid thermal shock.

Conclusion

Vernacular architecture has many iconic features and appeal, which can be used nowadays in modern buildings. Generally speaking, it used to be the public buildings heritage. Discussing the Godollo Palace as an example, one notices the splendour of such buildings, but many features were overdesigned to give the impression that it is grand building. It requires more than can be afforded to run it regarding the energy or electricity supply. Being in the central region of Europe it can function very well during the summer due to its thick walls which reduce the thermal lag from outside to the inside. Also the garden, flowers and shrubs create comfortable environment during most seasons. However it is difficult to keep the place warm in winter.

References

Gál B (ed) (n.d.) Kastélyok, fogadók a 18. Században (Castles and inns in the 18th century). Historia 1993-056

Tóth Áron (n.d.-a) Hőség és fagy: hűtés és fűtés a régi időkben – első rész Időutazás XI. Örökségfigyelő 2015.07.01. http://oroksegfigyelo.blog.hu/api/trackback/id/7586224

Tóth Áron (n.d.-b) Hőség és fagy: hűtés és fűtés a régi időkben – második rész Időutazás XII. Örökségfigyelő 2015.07.16. http://oroksegfigyelo.blog.hu/api/trackback/id/7631432

Chapter 7
Typological Analysis of Vernacular Residential Buildings in Moderate-Humid Climate of North Iran

Seyedehmamak Salavatian and Farzaneh Asadi Malekjahan

Introduction

Through time, climate has greatly influenced building forms in any particular region; additionally culture of people, their religion, availability of materials, and construction technologies are other important factors.

Today with the limited energy resources, vernacular architecture presents us an essential source of architectural knowledge in response to climate and culture without reliance on energy-consuming active systems (Baboli et al. 2015). The study of vernacular architecture is the rediscovery of nonexpensive simple solutions, which are often the best answers to important needs.

The most important aspect of native lifestyle upon creation of vernacular buildings is that this architecture has not been born by the society's individual architects. In this system, people decide to build a house with no preplanned design because they already know what they want based on their experience. Then, the house is constructed by the owner and other local ordinary people, which represents the concept of "dweller as designer" discussed by Pallasma (2005). Thus, no trace of professional architects appears and a natural beauty shines within this architecture (Taleghani 2010a).

The design characteristics of vernacular architecture in the southern part of the Caspian Sea have been investigated by many researchers. Memarian (1992) conducted one of the first studies in this field. Farajollahi (2008) researched Guilan vernacular architecture from construction and structural viewpoints. Another study (Gorji and Yaran 2010) of the vernacular architecture of this region was done in comparison with the similar climate in Japan. The major characteristics of both types accompanied by suggestions for architects were presented. Khoshsima et al. (2009) aimed to evaluate the role of climate in traditional architecture of the Caspian

S. Salavatian (✉) · F. A. Malekjahan
Department of Architecture, Rasht Branch, Islamic Azad University, Rasht, Iran
e-mail: salavatian@iaurasht.ac.ir

A. Sayigh (ed.), *Sustainable Vernacular Architecture*, Innovative Renewable Energy, https://doi.org/10.1007/978-3-030-06185-2_7

shores. Mahani's bioclimatic method was also applied in order to achieve proper design instructions. Their findings were compared to the common vernacular architecture to validate its efficiency. Khakpour (2006a) conducted a comprehensive study on rural houses in the central plain of Guilan and managed to provide new insights for construction of nowadays buildings by taking advantage of present knowledge and systematic direct observation and interviews.

Consequently, this study is an attempt to investigate a number of selected houses remained from the last century as the worthy representatives of vernacular architecture in this moderate humid region. This analysis derives practical design principles usable for designers working in this region and to some extent in similar climates. This research considers morphological aspects of native houses; through the lessons learnt from this analysis, cultural identity of today's architecture could improve. In the second stage, adjustment of the studied buildings with their environment and climatic condition is evaluated to provide climatic techniques for architects.

General Geography and Climate of the Region

There exists four types of climate conditions in various regions of Iran, namely, moderate and humid, cold, hot-dry, and hot-humid climates. Different climatic zones are distinguished in Fig. 7.1. Guilan province is about 14,700 m^2 (9% of whole country) and located in the southern part of the Caspian Sea and northern side of the Alborz mountains which provides it with a specific climate. The coordinates of its capital city—Rasht—is 37.27° N and 49.58° E. Average annual precipitation is between 1 and 2 m and average relative humidity is above 70% throughout the year (Kasmayi 1993). As seen in Fig. 7.2, the entire region is divided into noticeable distinguished zones: plain area along the seaside containing broad fields and highland regions along the mountains covered by massive forests (Gorji and Yaran 2010).

The extensive coverage of vegetation and the intense precipitations, especially in autumn and winter, and high percentage of humidity throughout the year are the distinguishing climatic features of this region (Ghobadian 2003).

The majority of people were involved in agriculture, farming, and fishery in the past. Abundance of greenery influences people's personality. They are mostly calm, good-tempered, philanthropic, and hospitable (Diba and Yaghini 1993). Short fences surrounding individual houses instead of high strong walls are evidences of the constant interaction among people and their inclination to communication and group life. Gender discrimination is subtle in this culture, since women are highly engaged in economic and social activities (Khotbehsara and Nasab 2016). Cultural characteristics of this region differ from rest of the country. Through centuries, people have been engaged in rice cultivation which necessitates the cooperation of all family members and women have been brought to social interactions through this way (Khakpour et al. 2013).

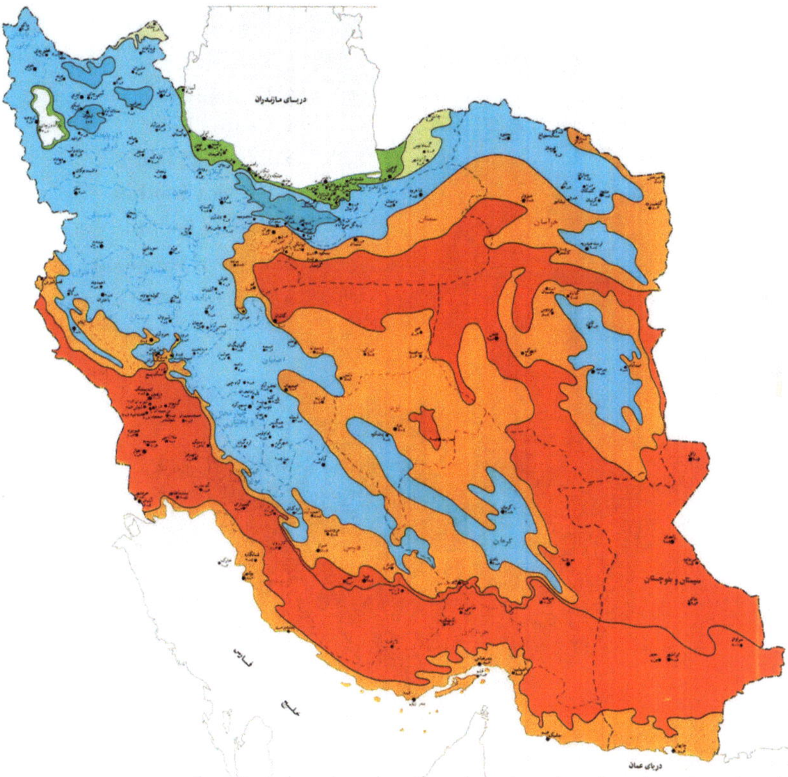

Fig. 7.1 Climatic zones of Iran (Kasmaei 2012)

In this research, the focus of study is upon houses found in plain areas based on two reasons; first, this area has occupied the greatest portion across the province and second, there are rich architectural contexts in these zones which made their general features as the well-known architectural characteristics of this region.

Selection of Dwellings

Among the climatic zones found in this region (mountainous, foothill, plain, and coastal) the plain typology was selected. Although other types are to some extent similar, they are different in some other aspects.

Guilan Rural Heritage Museum located in the forest park of Saravan is an open air museum established to preserve and revive heritage of the region. The valuable houses found in the area were dismantled and reconstructed in the eco-museum (Guilan Rural Heritage Museum n.d.). Six houses among the existed dwellings—belonged to the plain zones—were studied. Table 7.1 presents their general characteristics.

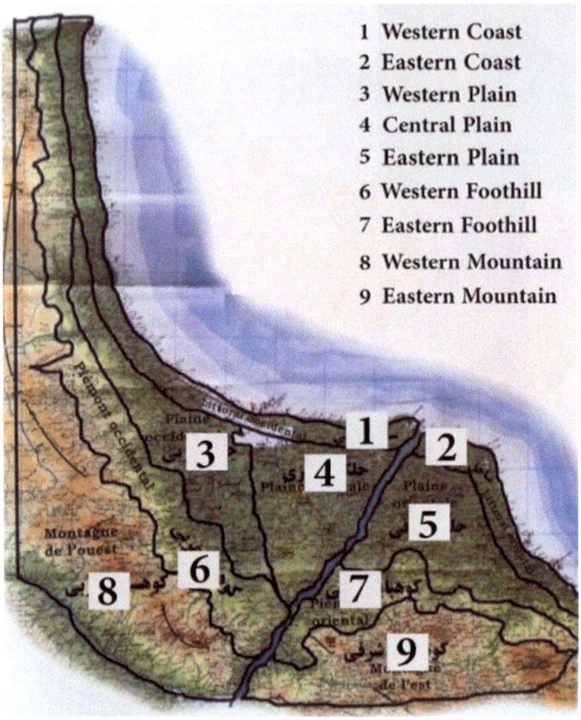

Fig. 7.2 Nine architectural–cultural divisions of Guilan (Taleghani 2010c).

Analysis of the Built Environment

In the following sections, various aspects of residential buildings are discussed. Three main subjects regarding morphological features, construction properties, and climate responsive characteristics are presented.

Morphological Analysis

Morphological analysis of houses is presented in two separate sections: first, the principles of space arrangement in the creation of vernacular houses are reviewed; second, the overall shape of buildings is described.

Spaces Layout

The organization of architectural spaces in four different scales of residential context, site arrangement, open/semi-open spaces, and closed spaces is studied.

Table 7.1 Generic information of selected houses

Amini house • Central plain 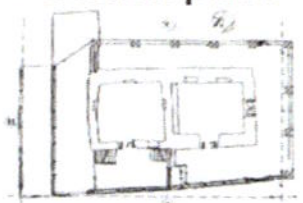 • 8.5 × 15.5 m		Rafiei house • East plain 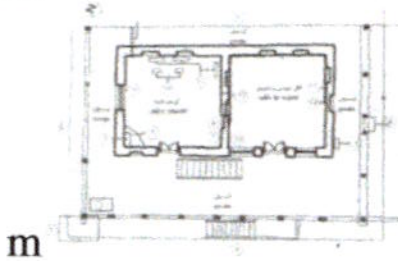 • 10 × 11 m	
Mohtashem talab house • West plain 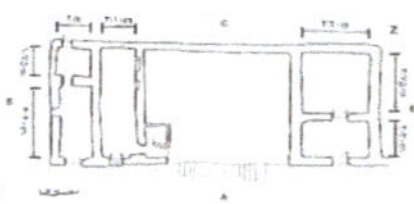 • 7.5 × 14.7 m		Moradi house • Central plain 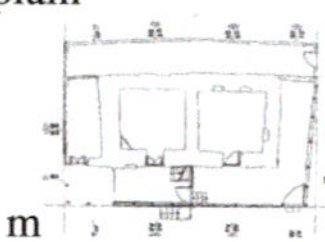 • 9 × 13.5 m	
Mousavi house • West plain 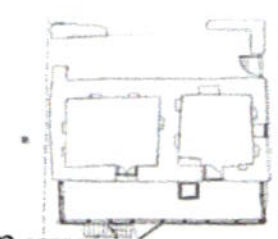 • 10.3 × 11 m		Mousazadeh house • West plain • 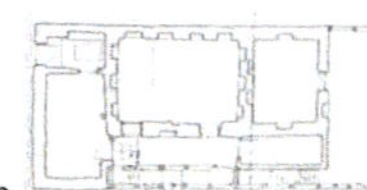 7.5 × 16.5 m	

Residential Rural Context

Houses have a scattered pattern in residential areas. They are formed as home garden or home farm. The first advantage of this organization is that a permanent monitoring of residents is provided and the fields are protected against wildlife. Furthermore, rice crops transportation from fields to storages (*Kandouj*) is facilitated. Above all, the abundance of suitable soil all over the region which eliminates the limitation for land access for cultivation makes it possible to be nearby residences. Residential buildings do not have solid walls; short hedges, water stream, or series of trees are used as buildings division (Khakpour et al. 2013). Short walls or fences which are helpful for ventilation purposes define the boundary of houses; in this way livestock and poultry are also preserved.

Site Organization

Houses in this area are of extroverted types (i.e., without central courtyard). No tangible border appears between buildings and their surroundings. Yards are the practical elements in vernacular houses of this region and could support various activities such as daily affairs, cooking in warm seasons, crops preparation, and producing handicrafts. They are a major linkage between private and public spaces, meanwhile as a place for daily activities. A number of semi-open spaces for purposes such as crops storage, stables, barns, etc. are created (Diba and Yaghini 1993). Some secondary buildings are formed around the house depending on the family's economy. Ranching activities lead to barns, rice cultivation needs "*Korouj*"—a shelter whose walls are covered by thin tree branches and roofs by rice stalk (*kolush*)—and sericulturism develops "*Talambar*" (Figs. 7.3 and 7.4) (Moradchelleh and Rashid 2011).

Houses are generally located in the central part and away from the perimeter boundaries of the site. Quite large areas are devoted to landscaping. There is a noticeable distance between the entrance and the building which makes residents cross a path to reach houses (Khakpour et al. 2013). Other auxiliary buildings are influenced by the location of the main building and are gathered around it which increases accessibility and views.

Forms of yards vary according to the site condition but are usually in a nongeometric polygon. In plain zones, house areas are quite flat and no considerable slope is appeared; thus, service buildings could be constructed at any position of the area. In some cases subtle slope of the land is from center (house location) toward edges (farm sides). Figure 7.5 depicts organization of the main and service buildings around their site. For example, in Mohtashem Talab (Table 7.2a) house, spaces organization is as follows: the entrance is located in the south side. *Kuti* (18.5 × 5 m) at the north side works as the rice crops storage and in 18 m distance from the house. *Talambar* (6.5 × 5 m) is a shelter made by clay and reed for sericulturism in the east side in 9 m distance from the house. The west side is allocated to truck farming. Fruit garden, paddy, and forestry are other adjacent lands. Trees (especially deciduous) are planted all around the buildings and in particular in the southern parts.

Fig. 7.3 An example of *Korouj* (Taleghani 2011a, b)

Semi-Open Spaces

Since women and men are not divided in their daily life, open and semi-open spaces could be quite useful in this area. There is no tendency to hide women in enclosed spaces; therefore, porches and balconies have become a key element of houses carrying a lot of activities. They are also a supreme spot to provide extensive views of surrounding green nature (Khotbehsara and Nasab 2016).

Porches are usable about 9 months a year (mid-March to mid-November). Their width varies between 2 and 2.5 m and provides an effective linkage among various rooms. Activities like cooking, dining, daily affairs, and even sleeping might be happened in porches. Southern porches are larger and higher than other sides and work as the main sitting area. It possesses more wooden ornaments and denser plants design (Memarian 1992).

Porches formation might vary depending on the building situation. One of the porches is certainly in front of the main façade. They have a critical connecting role in multiroom houses and also connect different floors in multilevel houses. For example, the key role of porches and balconies in the circulation in Amini house is shown (Figs. 7.6 and 7.7); pathways in the building are designed such that they provide a complete movement around the building in addition to access to the rooms (Fig. 7.8). The porch of main façade connecting to the lateral porch created a semi-open space (Fig. 7.9). The hierarchical connection between outside and inside is as shown in Fig. 7.10.

Fig. 7.4 An example of *Kuti* (Taleghani 2011a)

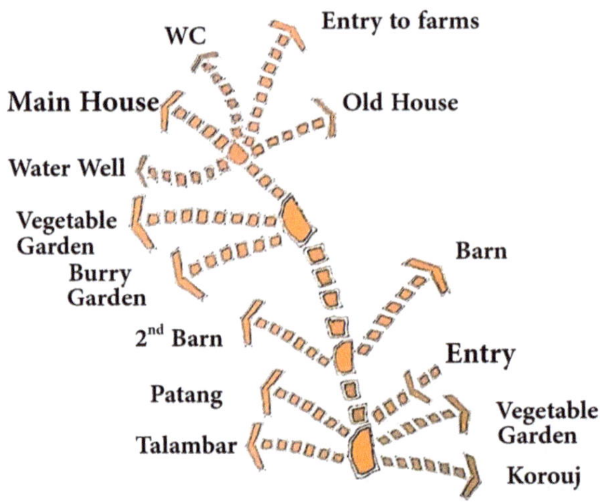

Fig. 7.5 Access to the main building and service buildings in site plan (Taleghani 2011a, b)

Table 7.2 Examples of site layout

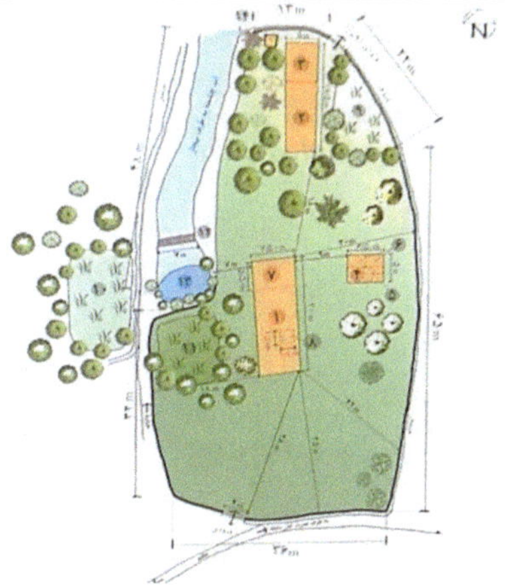	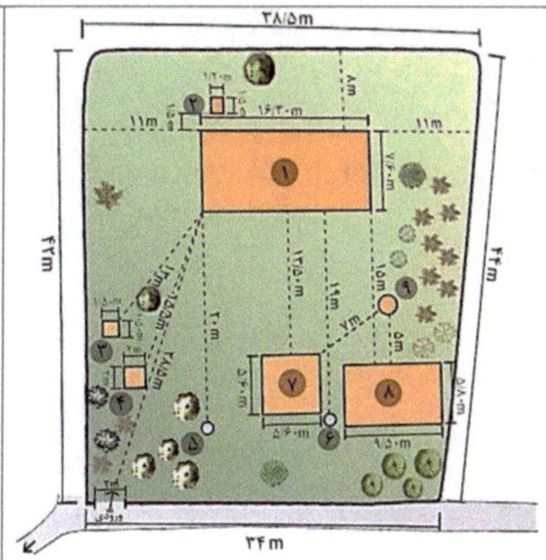
a. Mohtashem Talab house (Taleghani 2010b)	b. Mousazadeh house (Moradchelleh and Rashid 2011)
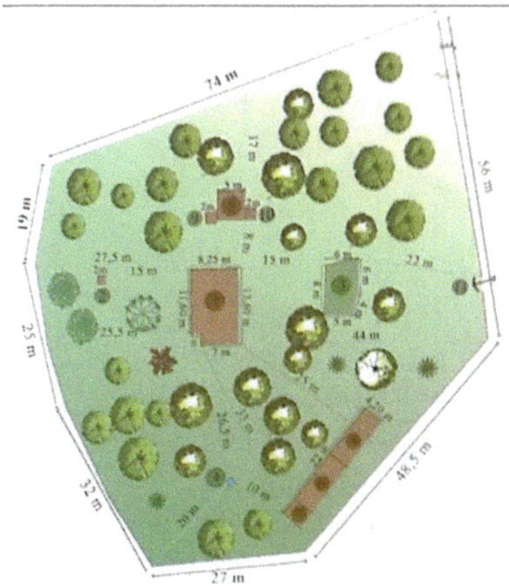	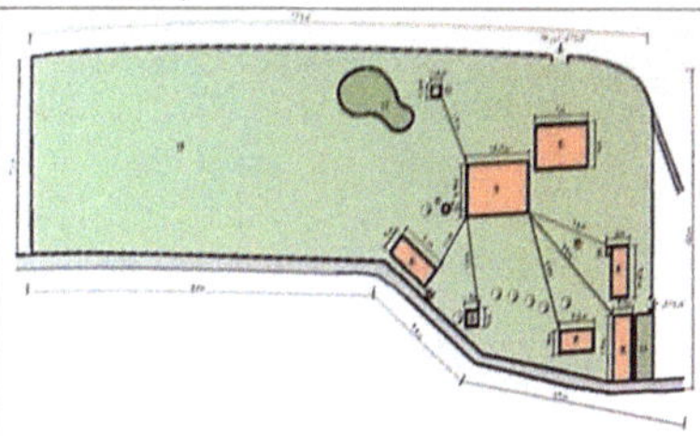
c. Amini house (Taleghani 2010c)	d. Moradi house (Taleghani 2011b)

Closed Spaces

Multifunctional spaces have an important role in this architecture. Each room is not just devoted to a fixed purpose, but it also takes different functions depending on the situation (Khakpour et al. 2013). The overall form of houses in Guilan vernacular architecture is longitudinal and rooms are set together in a linear shape. The geometry of houses is based on the square-shaped module. Ordinary rooms are square-shaped (about 3 × 3.5 m) and are used in winters. A few openings are considered and cross-ventilation is provided in the south-north direction. In two-story buildings, activities are divided; sitting area is organized at the upper level and service area at the lower level. Another arrangement in levels divides spaces seasonally; e.g., in Mohtashem Talab house, closed spaces are located in three levels: first, ground level (+1.30 m

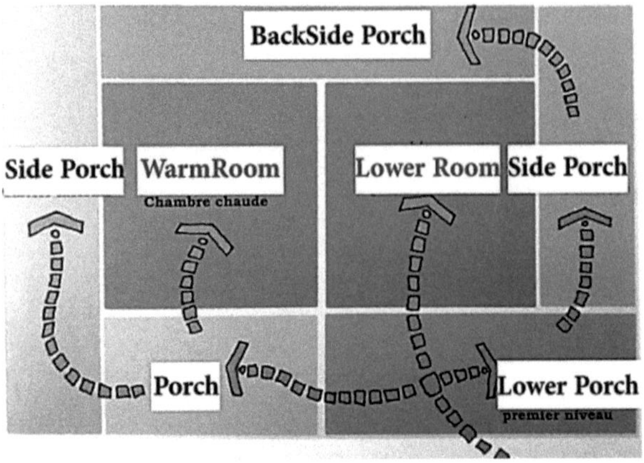

Fig. 7.6 Circulation diagram in the ground floor, Amini house (Taleghani 2010c)

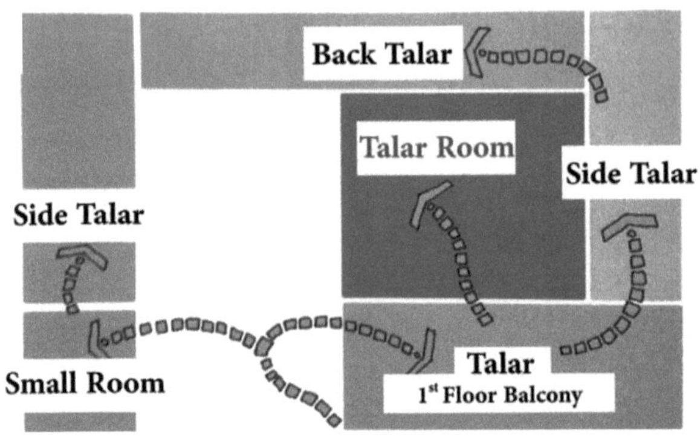

Fig. 7.7 Circulation diagram in the first floor, Amini house (Taleghani 2010c)

height) allocated to service functions as barn and storage, second is the mezzanine (+1.90 m height) for winter rooms, and third, first floor (+2.30 m height) which includes summer and guest rooms (Figs. 7.11 and 7.12) (Taleghani 2010b).

In the same way, Figs. 7.13 and 7.14 show rooms organization in Moradi house. The whole house consists of four rooms: two rooms in the first floor, one room in mezzanine, and the forth one in the second floor. There is no inner connection between the rooms and the circulation is through the porch and balcony.

Central core of houses extends longitudinally as family develops or economic condition improves. The development process from one-room version is as follows: (a) one-room house, (b) extension at the same level with one to four porches, (c) extension at the same level by adding extra rooms with porches, (d) extension in levels as a mezzanine, (e) extension in levels as an entire level (Memarian 1992).

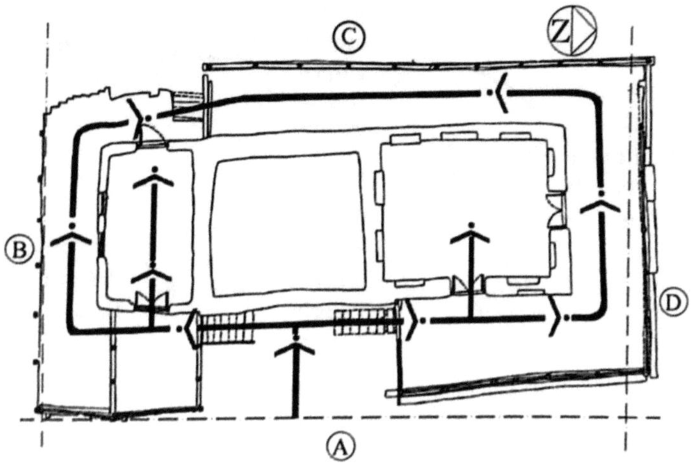

Fig. 7.8 Circulated paths in the plan of Amini house; ground floor (Taleghani 2010c)

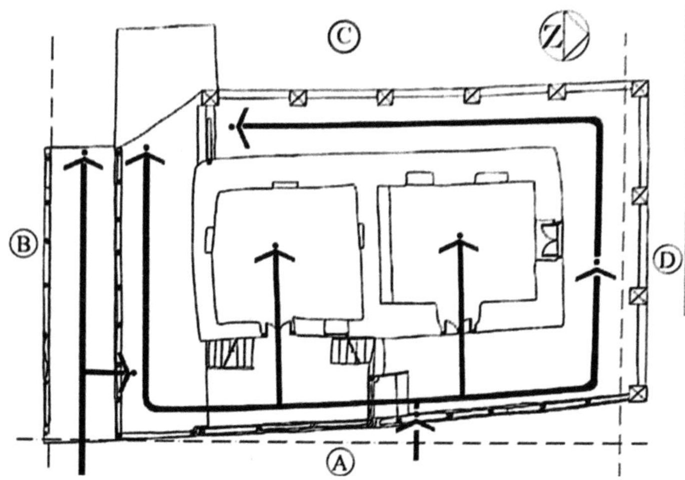

Fig. 7.9 Circulated paths in the plan of Amini house; first floor (Taleghani 2010c)

Buildings Form

Since the perception of simple shapes and volumes is quicker and easier, vernacular building forms are generally simple and pure. Buildings mostly consist of rectangular forms which are among the most stable geometric volumes. They are usually constructed in two-story levels that are oriented to the north-south direction along the Caspian Sea (Khoshsima et al. 2009) and are elevated from the ground floor by a pilot or crawl space. The basement is usually prohibited due to the high amount of humidity and moisture (Moradchelleh and Rashid 2011). Hipped roof is a common

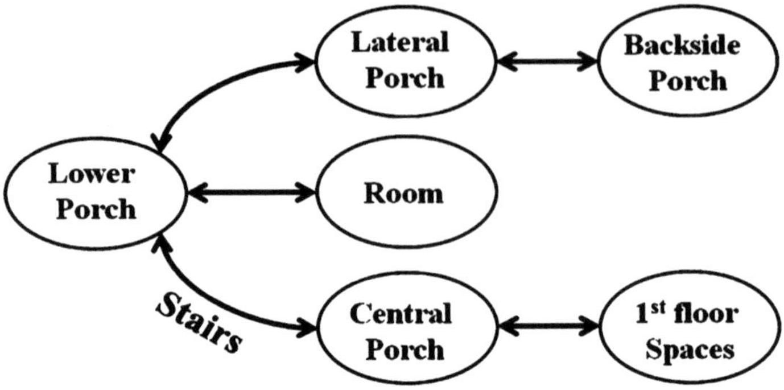

Fig. 7.10 Hierarchical connections between semi-open and open spaces

Fig. 7.11 Mohtashemi Talab house, closed spaces layout in ground floor (Taleghani 2010b)

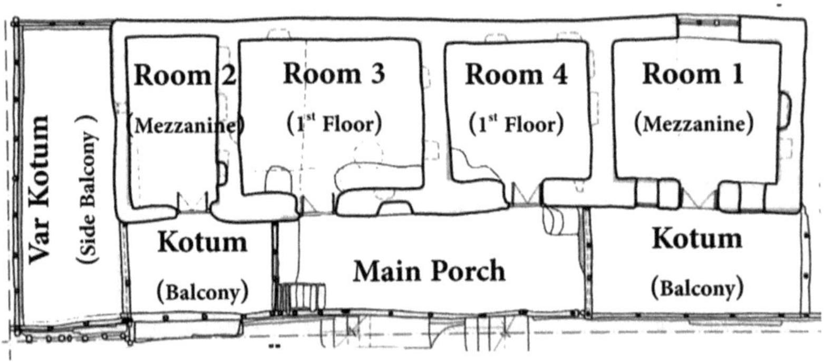

Fig. 7.12 Mohtashemi Talab house, closed spaces layout in first floor (Taleghani 2010b)

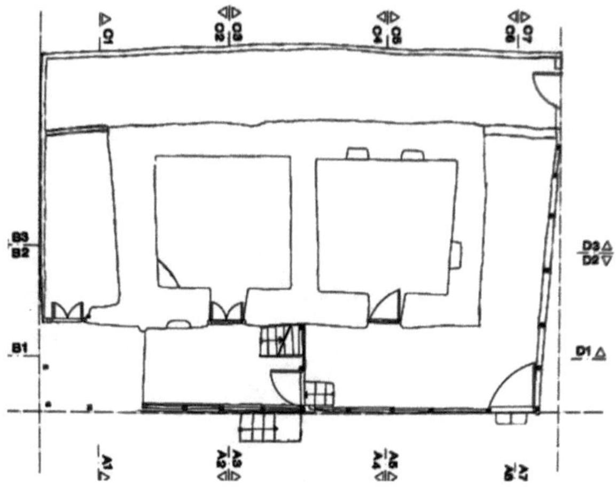

Fig. 7.13 Moradi house, ground floor (Taleghani 2011a, b)

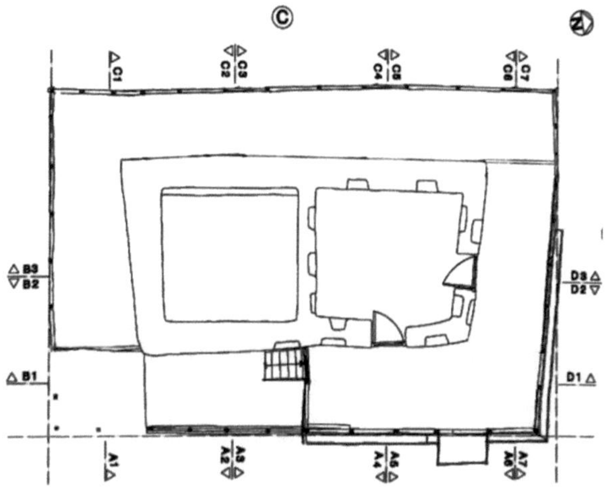

Fig. 7.14 Moradi house, first floor (Taleghani 2011a, b)

type in this area, which is usually very steep. Figure 7.15 illustrates a typical vernacular house in Guilan province.

A large balcony at the upper level is considered around the buildings, called "*Talar,*" to protect interior spaces from heavy rain. There is usually a large overhang roof to cover balconies especially in the west side, called "*Facon*" (Fig. 7.16) (Kasmaei 2012).

Facades are generally symmetric; however secondary parts are asymmetric. The lack of absolute symmetry is due to the requirements of space organization coming from different space needs. This feature does not interfere with the overall balance and is an important principle in vernacular buildings.

Fig. 7.15 Typical rural houses in Gilan province—Rafiei house (Taleghani 2007)

Fig. 7.16 Integrated balconies and Facon (west side)—Moradi house (Taleghani 2011a, b)

Although local builders were not aware of geometric rules, a number of well-known ratios as golden ratio and $\sqrt{2}$ taken from classic architecture have been intelligently applied in building forms. Consideration of these aesthetical ratios creates pleasant proportions among buildings length, width, and height (Figs. 7.17 and 7.18) (Taleghani 2011a, b).

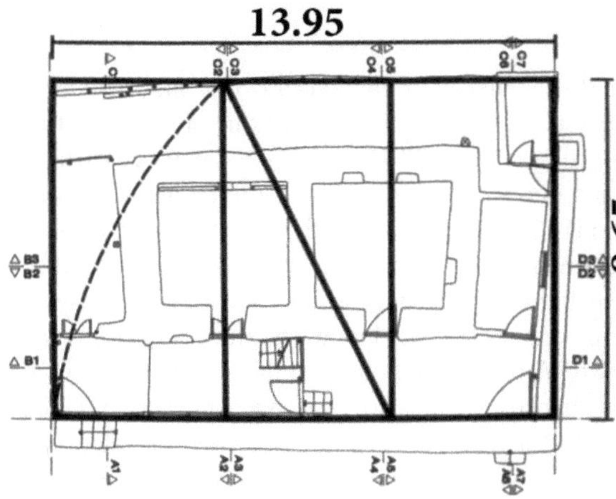

Fig. 7.17 Golden proportions in ground floor, Moradi house (Taleghani 2011a, b)

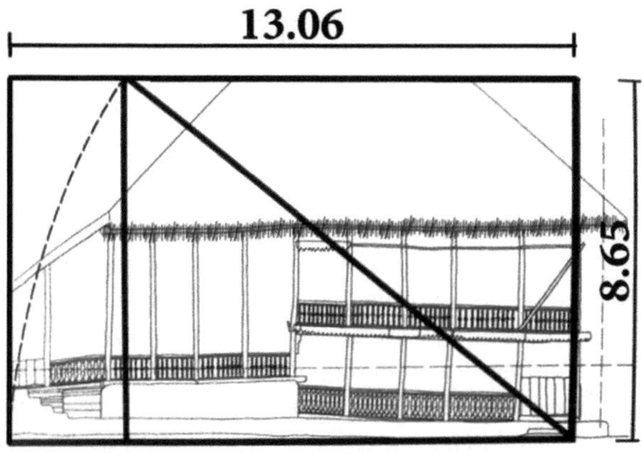

Fig. 7.18 $\sqrt{2}$ proportions in main façade, Moradi house (Taleghani 2011a, b)

Rhythmic columns as vertical elements and balustrade as horizontal elements provide a visual dynamism. Ornaments are in two main types: clay (wall and shelf carvings), and wooden (capitals, columns, balustrade, joists).

Constructional Techniques and Materials

In this section, a detailed study of construction in vernacular buildings is neglected; such an accurate study requires mechanical and structural analysis which is out of the scope of this study. Besides, in order to derive effective lessons to be applied in nowadays architecture, vernacular construction systems—which are made due to lack of facilities and materials—are less practical.

Materials

Although clay and stone masonry methods are also observed, wood-based systems are the most frequent ones. According to the principle of self-sufficiency in Iranian traditional architecture (Pirnia 2004), wood is the main construction material used in structures, walls, floors, and also roofs due to its abundance and availability in this region. Wood is a lightweight material, resistant to earthquakes, with high thermal resistance to encounter the environmental thermal discomfort. There are vernacular houses built completely with wood known as *Zegali* systems whose walls are made up of timber frames coated with a mixture of clay, straw, and water. Wood shingled roofs were also commonly used in this region. In general, hardwood such as acacia, oak, and mulberry which are resistant against water, moisture, and insects' penetration is useful for foundations while softwood like alder, maple, and pine is used mostly for long joists and beams. Wood pieces were cut in autumn and winter and kept in a closed place far from moisture because wood in spring is juicy and absorbs fungi. Then, a year later they will be used (Taleghani 2011a, b). Besides, the extensive amount of plants made them a key building material in traditional architecture (Ghobadian 2003). Rice straw thatched roofs "*kolush*" and "*Gali*" which is a kind of plant growing next to lagoons are the typical roof coverings.

Foundation

Plain regions are accumulated by alluvial sediments and the level of underground water is high; moreover, the bedrock must be strengthened. In the traditional practice, the ground is dug about 1.5 m and filled by wood ash, charcoal, and river gravel layer by layer. Then it is rammed till it reaches about 20 cm above the ground level. The flowing water will not touch the foundation surface in this way. In the other common practice, the ground is dug about 1 m and filled by rocks; therefore, a uniform loading is achieved. The dimensions are the same as the house perimeter (Khakpour 2006a).

The layering of wooden foundations is as follows: six cylindrical timbers (about 35 cm diameter and 100 cm length) are the lowest layer; next layer with 25 cm diameter timber, and the third layer in a perpendicular direction with thicker timbers (35 cm) are set; and the final layer is a stout trapezoid-shaped timber (Fig. 7.19). Number and rows of this type of foundation depend on the room numbers (house area) and number of stories. No mortar is used to connect elements and rigid connections are avoided. Buildings stability is due to the compressive loads and incomplete pyramid-shaped foundations. Overall wooden beams (equal to building length) are set on the ends of foundations; secondary beams are put over the longitudinal timbers (equal to building width) and this way the sequential load distribution occurs (Khakpour 2006a).

Walls

In clay system, the raw clay pieces which are flattened by hand in 4 cm thick are cut in dimensions of 20 × 20 cm and these strata are set all over each other when they are still wet (Fig. 7.20). To cover the clay strata, rice stalk (*Kolush*) is mixed up with clay and water and is applied to walls in three layers. As the final step, when the wall completely dried, white clay provided from nearby mountains is mixed with water and is rubbed to it in a thin layer.

In *zegali* system which is based on wood material, one or two sides of columns are covered by thin tree branches (or reeds) and the empty spaces are filled by clay (Taleghani 2011a, b).

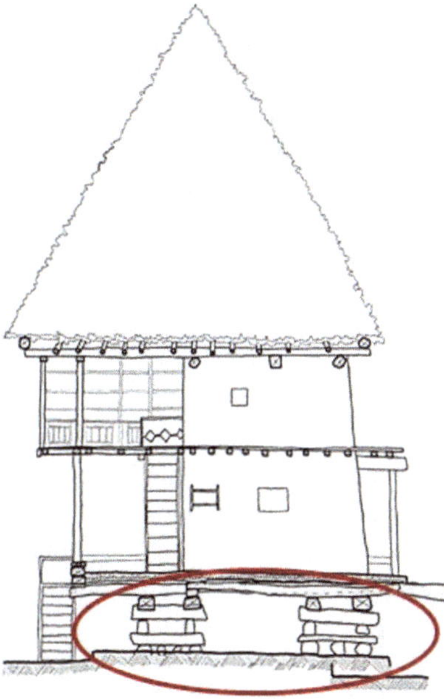

Fig. 7.19 Schema of the foundation, Rafeyi house (Khakpour 2006b)

(a) (b)

Fig. 7.20 (a) clay stratification (b) clay wall construction method (Taleghani 2010a)

Floors

Floor construction is accomplished through installation of small beams (*Vashan*); in the first step, a setting composed of long beams over the clay strata wall is located. Then, *Vashan* is mounted over them and as the final layer, boarding is done and a clay plaster is applied (Taleghani 2010a).

Roofs

Four-sided pitched roofs with the slope of 100–150% are the most frequent shape of roofs in this region which conducts rainwater and preserves buildings outer surfaces as an appropriate solution for this prevailing rainy climate. A variety of materials were used to cover roofs in this region such as wooden laths, clay roof tiles, and galvanized iron sheets, but the most commonly used material was straw and *Kolush* (rice stalk) due to its cheap price, availability, abundance, and easiness of installation (Khoshsima et al. 2009) (Fig. 7.21).

The structure of the pitched roof includes a small frame (as shown in Fig. 7.22) connected with several pillars which are built all around the building at 2 m distance from the building perimeter (Taleghani 2010a).

Climatic Analysis

Traditional architecture has always been a good example of climatic design representing efficient techniques which our precedents utilized to improve their environment. In the following sections, house features are analyzed in particular from this viewpoint.

Fig. 7.21 Floor structural elements (Taleghani 2010a)

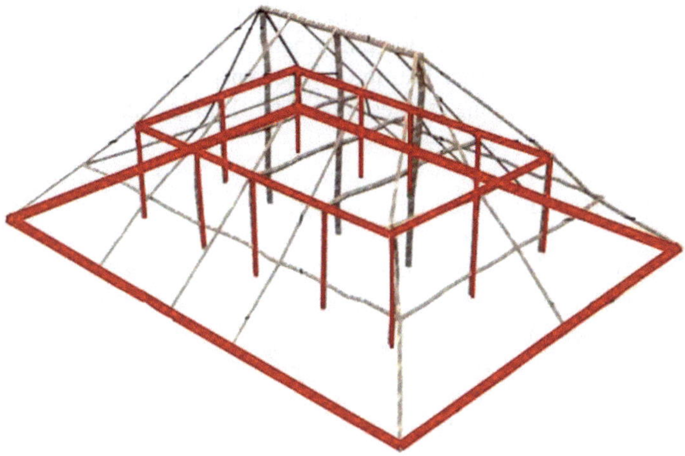

Fig. 7.22 Pitched roof structural elements (Taleghani 2010a)

Volumetric Configuration

Dispersion of houses in the residential area facilitates air to flow and access the openings. The main building that consists of occupants' living space is located in the center to northern part of the site with shallow-depth plan. Their shallow plans work ideally to capture the desirable wind flow. It helps to ventilate and capture the solar heat to dehumidify the interior spaces. The rectangular form of buildings provides larger interior areas with open air and natural ventilation potential.

Building form in this climate must be such that an air inhale for windward and an air exhale for leeward side are provided (at opposite or adjacent walls). Thus, it is desired that each room be connected to outdoor from two sides. Therefore, the linear development of buildings is a suitable solution to this requirement (Kasmayi 1993).

In the west side of buildings, the sloped roof descends to the ground level like an umbrella and no windows are considered in this façade. The rooms in the triangular space under the roof—called *Facon*—are allocated to utilities as storage and cooking area (Miryousefi 2006).

Regarding space priorities, it could be stated that generally upper rooms are in a better situation compared to lower ones due to better natural ventilation, less accumulated humidity, and more comfort. Besides, outer rooms, enclosed by porches and buffer zones, are preferred to the spaces in the central parts of the building due to their openness toward outside (Gorji and Yaran 2010).

Buildings are detached from the land and constructed above the ground level: first of all, keeps the building away from the moisture of the soil and high level of underground water; secondly, ventilation takes place in the compartment under the building meanwhile the rooms located at upper levels benefit more efficient breezes (Moradchelleh and Rashid 2011).

Barns in the ground floor provide extra heat to the upper floor and prevent heat loss from the main rooms. Floor to ceiling height in buildings is considered adequately high to allow more air into the interior spaces and to facilitate humidity elimination in enclosed rooms (Zarkesh et al. 2012) (Figs. 7.23 and 7.24).

Open and Semi-Open Spaces

Backyard is an architectural feature, considered to provide the possibility of cross-ventilation for the interior spaces. It is a narrow yard at the back of the house just for ventilation purposes (Baboli et al. 2015).

Semi-open spaces in this climate are common and useful since air flow is allowed while protection against rainfall and sun radiation is provided. Balconies also protect wall plasters (mainly made of a mixture of mud and straw) against intensive rainwater. Porches determine the main features of outward architecture of Guilan. Apart from social, cultural, and economic benefits, they also have a critical climatic role. Proper analysis of their function is a key factor to gain modern solutions in order to take its multi-aspect advantages.

Porches of upper stories are called *Talar* (balcony). Balconies are located at a higher level than porches—at the arrival step—and the stairs extend to the balconies (Khotbehsara and Nasab 2016). Porches particularly located in the east and south sides can prevent the rain reaching walls surfaces and in summer protect the walls from intense sun radiation (Miryousefi 2006).

Orientation

Building orientation in this climate is expected to supply required solar gain in cold seasons and avoid direct sun radiation in hot season while natural ventilation is also provided in almost two third of the year. Building orientation benefits from east/

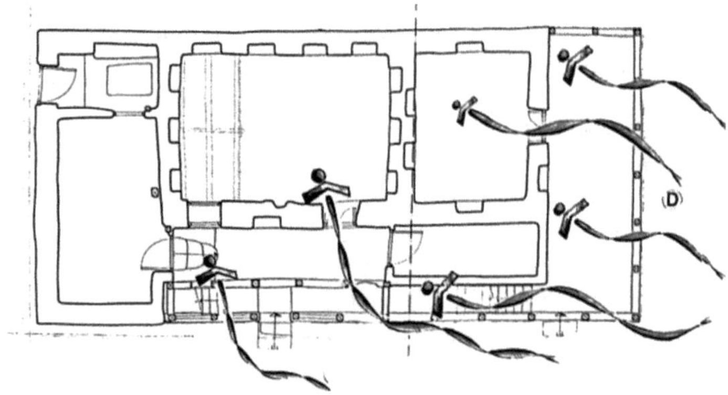

Fig. 7.23 Natural ventilation in plan, Mousazadeh house (Taleghani 2011a)

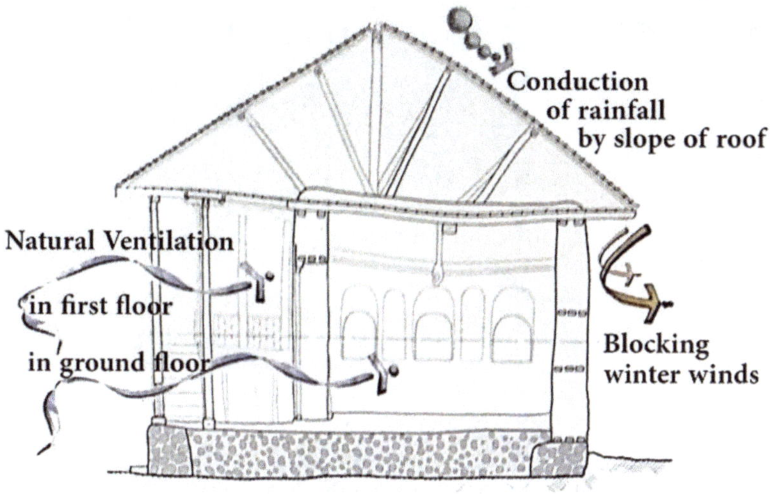

Fig. 7.24 Natural ventilation in section, Mousazadeh house (Taleghani 2011a)

north-east prevailing winds and avoids unwanted north-west winter winds. The frequent orientation observed in this area is north-south with an inclination toward east. The recommended best direction of a building in this region is south to south-east (30° skewness from south) if the building has one open front and for two open front buildings; direction to south is the best decision (Kasmayi 1993).

Openings

Openings are mostly placed according to the sun orientation and wind patterns. In this climate, too large windows are not used since they are not very effective and tend to increase wintertime heat loss. Instead, if small windows are positioned properly, the air movement amplifies through the building. In general, openings of ground-floor rooms—cold season spaces—are smaller in order to avoid heat loss while openings of first-floor rooms—warm season spaces—are larger to facilitate natural ventilation. In most cases, the windows and doors are installed in the northern and southern sides of the house. The south walls have more windows that promote the introduction of winter sun.

Studying various elevations reveals that the warmer sides (south, south-east, and east) have more window surfaces than colder sides. No windows are considered for the west facades so that severe winter winds and also inclined rains are blocked completely. In contrast, winds from east side (appropriate summer winds) are allowed to penetrate through wide windows; therefore, eastward facades own more windows than southward ones (Figs. 7.25 and 7.26). In addition, windows in the opposite sides (north and south) are considered to intensify the north wind being circulated through the building. As reported in a study by Nia (2013), the ratio of window to the story area is about 5%.

Fig. 7.25 Amini house, south façade (Taleghani 2010c)

Fig. 7.26 Amini house, east façade (Taleghani 2010c)

In order not to interfere natural ventilation, proper distance between openings and surrounding barriers/walls/buildings must be considered. This distance in north-east and south direction should be more than four times the building height (Moradchelleh and Rashid 2011).

Building Envelope

As already discussed in the section "Materials," materials in the majority of buildings have low thermal capacity. Apart from wood and clay (mostly in mortars and plasters), limestone, river stone, and sand—found in the coasts—are used in exterior wall construction. Thick external walls are considered in order to prevent high amount of heat loss in winters. Since the daily temperature is quite consistent, saving heat is not needed and buildings are made up of the least thermal capacity (Habib 2012).

Transparent facades are the result of double-façade configurations. The colonnade in front of balconies forms the outer layer of envelope while room walls are the inner layers. Porous envelopes embrace the central core of buildings in all—or at least—two sides. Façade colors are mostly white because it absorbs less heat from the sun, so it prevents inner space of the building to get so hot during warm seasons (Habib 2012).

Lessons Learnt to Be Adapted in Urban Mid-Rise Apartment Buildings

According to the studies carried out in three sections of morphological analysis, construction methodologies, and climatic analysis for the investigated cases of vernacular architecture in this region, a number of guidelines for the design of residential buildings are suggested. Currently, the majority of houses in this region are devoted to multistory apartment buildings (averagely four-story) and these guidelines must be adjusted to tenants needs of nowadays as well as being transferrable from one-story type to multistory building type.

In this respect, the solutions for sustainable architecture to be applied in recent buildings in Guilan are as follows.

Overall Design

- Locating the main building in a proper distance from the perimeter barriers quite in the center of the site.
- Orienting buildings toward the prevailing summer winds.
- Designing pilot as an assistance to intensify air movement around the building.
- Organizing rooms in thin layers with permanent opposition to air.
- Flexibility in rooms design and taking advantage of partitions and lightweight walls as the space dividers to facilitate natural ventilation as required.
- Using semi-open spaces as the circulatory connectors for closed spaces.
- Organizing spaces with natural ventilation needs in the outer layer of plan and service spaces (heat producers as kitchen) in the center of plan.
- Attaching spaces like terraces in front of windows to control depth and amount of sun shine into the building during all seasons based on residents' demand for sun heat.
- Using the recognized proportional ratios in traditional architecture.
- Average surface of openings, 25–40%, protecting them where located opposed to heavy rain, and positioning them at body height.
- Storing rain water to be utilized for nondrinking water use.

Natural Ventilation

- Causing a considerable air pressure difference between various wall sides of interior spaces.
- Designing wind catchers and atriums to enter the air flow inside.

Natural Lighting

- Considering a proper position and angle of roof elements in order to penetrate indirect natural lighting.
- Predicting suitable sunshades to reduce the intense direct sunshine.

Materials and Construction

- Using local materials such as wood accompanied by thermal insulation.
- Using materials with low thermal capacity and short time lag.
- Double-layered inclined roof which works as a buffer zone for the building reducing the heat gain by the sun as well as reduction the building weight.
- Double-layered floor which provides natural ventilation and setting for the mechanical ventilation equipment.

Conclusion

All in all, vernacular architecture all across the world including Guilan province hosts a great deal of sustainable solutions; taking them into consideration would improve space quality and energy efficiency in today's buildings. For this purpose, selected houses of the plain zone were investigated and their main features were identified to be transformed to usable recommendations in architecture of current residential buildings.

A summary of main architectural characteristics in this region discussed in this study is as follows:

- A close linkage between inside and outside spaces.
- Porches/balconies in one to four sides of the building as an important usable space.
- A noticeable height of main spaces from the ground level.
- Four-sided roofs with steep slopes.
- High-porosity spaces at the outer layer of buildings.
- Porous parapet to let air penetrate inside.
- Transparent south façade.
- Depth of buildings not more than one room.

The above-mentioned properties along with other detailed features led to a list of instructions of design in this area which could be taken into consideration in various phases of design. Vernacular architecture of other geographical regions in Guilan (i.e., coastal, foothill, and mountainous) also indicates worthy lessons which need to be studied in separate researches.

References

Baboli FBM, Ibrahimb N, Sharifc DM (2015) Design characteristics and adaptive role of the traditional Courtyard Houses in the moderate climate of Iran. Proc Soc Behav Sci 201:213–223
Diba D, Yaghini S (1993) Analysis of Guilan vernacular architecture. J Archit Urbanism 24:6–16
Farajollahi A (2008) Structural analysis of Guilan vernacular architecture (Eastern coast). Art Archit 8:112–122
Ghobadian V (2003) Climatic analysis of the Iranian traditional buildings. Tehran University, Tehran
Gorji Y, Yaran A (2010) Solutions for sustainable architecture in Guilan in comparison with Japanese architecture. honar-ha-ye ziba 41:43–54
Guilan Rural Heritage Museum. http://www.gecomuseum.com/LandingPage.aspx
Habib F (2012) Evaluation of a sustainable Lahijan Vernacular Settlement (N Iran) and its landscape: architectural typology and building. World Appl Sci J 20(4):527–531
Kasmaei M (2012) Climate and architecture. Khak, Esfahan
Kasmayi M (1993) Climate and architecture. Khak Publisher, Tehran
Khakpour M (2006a) Construction of Shikili houses in Guilan. Honarhaye Ziba 25:45–54
Khakpour M (2006b) Construction of Shekili houses in Guilan. HONAR-HA-YE-ZIBA 25:45–54
Khakpour M, Ansari M, Sheikhmehdi A, Tavoosi M (2013) Elements of rural residential premises of Northern Iran and the regulations governing their layout. World Appl Sci J 24(1):89–95
Khoshsima E, Mahdavi A, Inangda N, Rao SP (2009) Evaluation of traditional architecture of Southern Shores of Caspian Sea Region in Iran. In: Conference of the international journal of arts and sciences, pp 164–177
Khotbehsara EM, Nasab SN (2016) Porch and balcony as sustainable architecture factors in vernacular houses of west of Guilan: case studies in Khotbehara, Iran. World Rural Obs 8(3):48–56
Memarian G (1992) An introduction to Iranian residential buildings (extrovert typologies). Elm O Sanat University, Tehran
Miryousefi P (2006) Guilan rural heritage museum. Memar 39:10–13
Moradchelleh A, Rashid H (2011) The study of the form of the building in climates of Iran. Middle-East J Sci Res 8(2):306–313
Nia EG (2013) Architectural structure and environmental performance of sustainable Lahijan Vernacular Settlements, N Iran. Int J Environ Prot 3(1):15–23
Pallasma J (2005) The eyes of the skin: architecture and the scenes. Academy Press, Cambridge
Pirnia M (2004) Styles of Iranian architecture. Soroush Danesh, Tehran
Taleghani M (2007) Rafiyi house, Guilan rural architecture heritage. Moein Publisher, Tehran
Taleghani M (2010a) Mousavi House, Guilan rural architecture heritage. Matn Artwork Publisher, Tehran
Taleghani M (2010b) Mohtashem Talab House, Guilan rural architecture heritage. Matn Artwork Publisher, Tehran
Taleghani M (2010c) Amini House, Guilan rural architecture heritage. Rozaneh, Tehran
Taleghani M (2011a) Mousazadeh House, Guilan rural architecture heritage. Matn Artwork Publisher, Tehran
Taleghani M (2011b) Moradi House, Guilan rural architecture heritage. Matn Artwork Publisher, Tehran
Zarkesh A, Moradchelleh A, khnlari E (2012) Region based classification of design and construction consistent with climate. Middle-East J Sci Res 11:66–76

Chapter 8
From Genius Loci to Sustainability: Conciliating Between the Spirit of Place and the Spirit of Time—A Case Study on the Old City of Al-Salt

Shaden Abusafieh

Sustainability concept has become the common interest of numerous disciplines in the last two decades. This popularity was in the wake of escalation in energy prices, shortages, war, pollution concerns, and resource depletion, along with heightened environmental degradation and climate change. This issue dramatically increased our awareness of environmental impact, as architects and building design professionals. Buildings of the future have no longer just simply created to be functional and esthetically pleasing, but it should be environmentally responsive as well. Designing sustainable building is a great challenge that architects face nowadays.

Sustainability "The Spirit of Time"

> How do you build things that will help us thrive in the present as well as in the future? How do you build structures that are as relevant today as they will be in a hundred years? (Green 2015).

The most widely used definition of sustainability is meeting "the needs of the present without compromising the ability of future generations to meet their own needs" (Brundtland Report 1987). For a complete definition of the sustainability, the economic, social, and environmental pillars were used as three main powerful dimensions of sustainability. If any one pillar is weak, then the system as a whole is unsustainable. When all three pillars are strong, people live in a system of high quality life (Fig. 8.1).

S. Abusafieh (✉)
College of Architecture and Design, Department of Architecture Engineering, Al-Ahliyya Amman University, Amman, Jordan

A. Sayigh (ed.), *Sustainable Vernacular Architecture*, Innovative Renewable Energy, https://doi.org/10.1007/978-3-030-06185-2_8

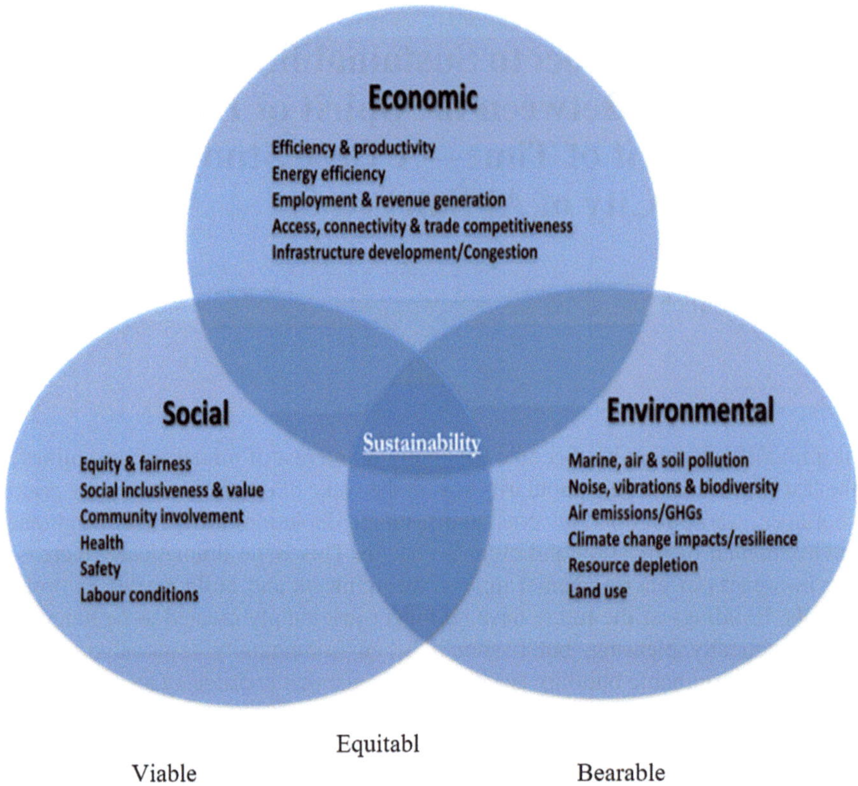

Fig. 8.1 The three main dimensions of sustainability. Source: United Nations Conference on Trade and Development (UNCTAD XIII), 2012, Doha, Qatar

Sustainable

The 2030 Agenda for Sustainable Development underlines a global commitment to "achieving sustainable development in its three dimensions—economic, social and environmental—in a balanced and integrated manner" (Ludwig et al. 1997).

The economic dimension refers to the ability of an economy to support a defined level of economic production for an unlimited period of time. This can be achieved by using various strategies to utilize natural and human resources optimally. Economic sustainability should satisfy present consumption levels without compromising future needs. Therefore, financial condition of the organization is not the main focus of economic sustainability. It has many aspects such as economic performance, market presence, indirect economic impacts, and procurement practices (Meadows et al. 1972).

The environmental dimension of sustainability concerns about the renewable resource harvest, pollution creation, and non-renewable resource depletion that can be continued indefinitely. Environmental sustainability is the responsible interaction with the environment and the maintenance of the factors and practices to avoid depletion or degradation of natural resources and allow for quality of environment on a long-term basis (Ludwig et al. 1997).

Social sustainability is an often neglected aspect of sustainability. Predominantly sustainable development discussion focuses on the environmental or economic dimensions of sustainability. Social sustainability is "a process for creating sustainable successful places that promote wellbeing, by understanding what people need from the places they live and work. Social sustainability combines design of the physical realm with design of the social world – infrastructure to support social and cultural life, social amenities, systems for citizen engagement, and space for people and places to evolve." (Social Life 2012)

Principles of Sustainable Architecture

To achieve sustainable architecture, the above-mentioned three main dimensions of sustainability can be interpreted through six fundamental principles (Wheeler and Beatley 2014).

1. *Optimize site potential*: the proper site selection including the existing building's rehabilitation, orientation, and landscaping of a building are aspects that affect local ecosystems, transportation methods, and energy use.
2. *Optimize energy use*: use renewable energy resources such as solar energy, wind, and biomass. It is important to increase our energy independence by reducing our dependence on fossil fuel-derived energy. This could be achieved by operating net zero-energy buildings.
3. *Protect and conserve water*: reducing water consumption and protecting water quality are key objectives of sustainable design. Buildings should decrease their need for water by increasing efficiency and by maximizing the use of water that is collected, used, purified, and reused on-site.
4. *Optimize building space and material use*: across its entire life cycle, a sustainable building is designed and operated to use and reuse materials and natural resources in the most productive and sustainable way.
5. *Optimize operational and maintenance practices*: designers should use materials and systems that reduce maintenance requirements and life-cycle costs.
6. *Enhance respect for user*: this can be obtained by enhancing the indoor environmental quality of a building that has a significant impact on occupant health, comfort, and productivity. Sustainable building should maximize daylighting, have appropriate ventilation and moisture control, optimize acoustic performance, and avoid the use of materials with high-VOC emissions. Moreover, sustainable architecture should recognize the importance of respecting the human needs and the human dimension in its broad meaning.

Culture and Sustainability

Culture represents a mental map which guides us in our relations with our surroundings and with other people (Downs 1973). It is the predominating attitudes and behaviors that characterize the functioning of a group or organization. It includes all creations, material and non-material achievements, the inherited expectations, and the past and present gains as a result of living together (Toffler 1973). Culture encloses meanings, beliefs, values, customs, norms, and symbols relative to society. Beliefs are the means by which people make sense of their personal and social experiences. Values direct people on what should and should not be done, what is good or bad, and what, why, and how to choose. Norms are shared patterns of behavior in a particular culture that informs its members what they should or should not do in a given situation (Downs 1973).

There is a significant link between culture and sustainability. Culture with its rules, values, beliefs, and norms carries the sustainability of vitality of the community (Matsumato and Juang 2003). Based on the main logic and methods, Guy and Farmer classify sustainable architecture under six different categories as eco-techno, eco-centric, eco-esthetic, eco-cultural, eco-medical, and eco-social (Guy and Farmer 2001).

The eco-cultural logic leads to transformation and reuse of traditional construction techniques, building typologies, and settlement patterns as an expression of the cultural sustainability. It denies universal and technologically based design methodologies that often fail to coincide with the cultural values of a particular place and people (Guy and Farmer 2001). At the same time, eco-cultural method emphasizes that the design in architecture should spread sustainability of the culture.

Cultural and environmental sustainability could be achieved by adopting a regional design approach. This approach intends to respect climate and sustain the culture of the region through considering the existing pattern of the region, the existing architectural features of the buildings, the existing lifestyles of the inhabitants, and the existing cultural issues (Kultur 2012).

Sustainability in Arabic Cities

Since the beginning of the eighteenth century, Arab cities have suffered major setbacks. Early in the last century and due to the occupation by other cultures, these cities have been denied their own identity. So, Arab cities witnessed a historical and cultural detachment that leads to the current crisis in Arab architecture.

Many scholars have dealt with the current situation in Arab architecture. They discussed the impact of inappropriate imported trends. Khalid Asfour claims that a process of "cutting and pasting" was introduced as a cultural mechanism. The process involves cutting ideas from western cultural field and pasting them with their logic in the new Arabian field (Asfour 1998). Unfortunately, the majority of new buildings in the Arab States remain within the vocabulary of western architecture.

In architectural thought and in sustainable design, not all of the good ideas in a specific region are appropriate for another, especially if these ideas are related to social, cultural, and climatic considerations.

Due to socio-cultural and economical transformations and the cultural and intellectual openness to the western thought in Arab region, sustainability became an inevitable trend. Sustainable design principles in housing took different expressions. This produced architectural trends that are different in their forms and characters but alike in their content. Accordingly, multiplicity of sustainability trends has a great impact on the architectural identity and image of the Arab city. The main trends were the modern technology, the NEO-traditional, and the contemporary interpretation trend.

The modern technology trend calls for the universality of sustainable architecture. This trend indicates the application of the advanced western technology. In the Arab world, followers adopt and copy the most advanced technological western solutions in their designs without giving appropriate attention to the local conditions. After examining and filtering, taking what is applicable in terms of its appropriateness for the social and cultural context, is considered the appropriate way to deal with western technologies.

The NEO-traditional trend consists of two approaches, the absolute adoption of the traditional model without any intervention and studying the architectural legacy and learning from its experiments. Both approaches relay on the architectural legacy, and this emphasizes the importance of this legacy as a source of inspiration. While choosing between both approaches, we should seek authenticity. Authenticity in architecture means "investigating and studying the architectural values of the past and then take advantage of their lessons without copying the models" (Olfat 1987). The best way to achieve the authenticity is to understand the values of the architectural past rather than copying its elements. It is not the absolute adoption of the traditional model, it's the implications and lessons from the present and the past. This trend would ride out of the cultural crisis and the loss of identity.

Contemporary interpretation trend is the trend that attempts by utilizing the appropriate modern technology *to* express the traditional approach in a contemporary manner. It compromises between the modern technology and the NEO-traditional trends. In this trend, followers understand, recruit, and fuse the heritable values in traditional architecture as well as the aspects of the modern age to create the appropriate regional identity. The contemporary interpretation trend has an important challenge in forming this identity. This challenge appears in the attempt to respond to local climate and in the process of blending and obtaining the balance between the advanced generalities in the universal and the local or traditional aspects from the ancestors.

Genius Loci "The Spirit of Place"

Genius locus, "the spirit of place," is a term worth noting for its eminence in the development of architecture thinking. In ancient times it referred to the guardian divinity of that place more than physical place itself (Turner 1996). In the eighteenth century, the Latin phrase was usually translated as "the genius of a place," meaning its influence (Jackson 1994). Norberg-Schulz re-fetched this concept to architecture in the 1980s, curiously after the environmental failing of global architecture.

Turner claimed the first law of design and landscape planning "Consult the genius of the place" (Turner, 1996), while McHarg's came with the second law "Any land use in the care of a specialist profession tends towards a selfish disregard for other land users" (McHarg, 1971). These two laws go along with the place-based design trend in architecture. It possesses an intimate relationship with the genius of the place. Therefore, situational characteristics must be highlighted in regional design guidelines that account for environmental sustainability at the same time.

According to the dominant qualitative and quantitative frameworks of sustainable architecture principles, the existing criteria of sustainable architecture do not prioritize place-based design (Dinep and Schwab 2010). A question here should rise about the possibilities of overlap with place-based and sustainable design principles.

The current Leadership in Energy and Environmental Design (LEED) rating system is made up of a combination of 11 credit categories: integrative processes, location and transportation, materials and resources, water efficiency, energy and atmosphere, sustainable sites, indoor environmental quality, innovation, regional priority credits, smart location and linkage, neighborhood pattern and design, and green infrastructure and buildings (USGBC 2009). Regional priority has been categorized with a too low credit value, and it lacks sufficient explanation as well.

Architecture design should be more situational, rather than universal. This makes the site's characteristics essential sources of design, not only circumstances to be accommodated from the environmental features to the cultural characteristic (Meyer 2004). Therefore, how might the principles of architecture sustainability be incorporated into situational characteristics of place-based design?

A place-based sustainable design guidelines for architecture should be provided and tested in order to create a connection between place-based design and sustainable design criteria. This would lead to produce a new place-based sustainable design guidelines that conciliate between the spirit of time "sustainability" and the spirit of place "genius loci." This conciliating will be an interdisciplinary effort to bring local identities back into the realms of sustainable architecture design.

In the past, vernacular architecture reflects the immaterial variables of a place: meaning, identity, and history. It defines the spatial character that called genius loci. This ancient concept should always be present, as the fourth conceptual vertex of our vernacular architectural conscience, along with firmitas, utilitas, and venustas.

Vernacular Architecture

Genius loci was transversal to all vernacular architectural poles. They were able to accomplish a genius loci for specific human settlements by using local materials and resources to create a shelter that is adequate to its use, to its environmental surroundings, and to its occupants.

In order to preserve the continuity of their existence, mankind has throughout history tended to strength entities with the world in order to preserve the continuity of their existence. He was always trying to meet his basic needs from natural sources. Eating, drinking, covering, sleeping, breathing, and security are the basic tangible requirements which is ranked first in the Maslow's hierarchy of requirements (Maslow 1943). Culture of the human life has been clarified in the local and traditional construction products that came from the connection between human and the environment. These products went along with topography, local climate, local materials, and their genius loci "the spirit of place."

According to Rudofsky, some concepts that gain popularity in architectural perspectives such as sustainability seem to have taken place in past examples of vernacular architecture (Rudofsky 1964). Buildings that are designed in accordance with a community's culture, lifestyle, and climatic conditions are called vernacular architecture. A harmonious relationship between climate, architecture, and people has been always established in vernacular architecture. Because of the common goal between sustainable design and vernacular architecture, vernacular architecture products could be associated with sustainable architectural principles. Both are aspiring to produce environmentally friendly constructions that are compatible with the surrounding conditions and that can last for many years.

Although sustainability seems to be a new concept in today's architecture but it is not. Many architectural examples found in different parts of the world showed that sustainable architectural design principles refer to vernacular architecture. Vernacular architecture has a knowledge and features that can be studied and classified from a sustainability point of view. *Designed for the Future* book is a proof that there were many smart sustainable moves we made in the past, and how we would be remiss in not learning from them today. Eva Franch Gilabert mentioned that "Many of the deep concerns about architecture are timeless." Green (2015)

We have to take into consideration that the new strategies, materials, and means of sustainable design might put to risk the vernacular way of thinking in architecture, which made us to achieve the sustainable character of the buildings overwhelming on region's genius loci. So, since we have a successful vernacular architecture that has accomplished all of the sustainable design principles "the spirit of time," at the same time, it has respected and preserved the genius loci as a "the spirit of place," *why not to take lessons from the past to build out sustainable future?*

Vernacular architecture represents a morphological response to environmental constraints and to the socio-economic and cultural characteristics of societies. Moreover, the materials and architectural components used in it are climate responsive and tailored according to distinct locations and topographical features. Vernacular architecture is a cost-effective architecture with a low environmental impact. Therefore, it represents a great resource that has significant potential to define sustainable design principles (Correia et al. 2014).

The following analysis used some of the buildings of the historic center of Al-Salt as a case study. These vernacular houses are analyzed in its architectural components and potential retrofit strategies. There should be a balance and conciliating between sustainability and genius loci in order to mainly respect the social relevance, cultural identity, and local ethics of our historic city.

Al-Salt: A Case Study Analysis

Al-Salt, where you can find architecture that sheltered the body and nourished the soul. It is an ancient agricultural town and administrative center located in west central at Jordan. The town is built in the crook of three hills, close to the Jordan Valley, and about 790–1100 m above sea level. It has many architecturally elegant buildings, many built in the Nabulsi style from the attractive honey-colored local stone. A large number of buildings from this era survive as of recent time. These buildings are considered examples of architecture with inherent sustainable characteristics. By analyzing these buildings and the strategies that have been used through various urban regeneration architectures, it is possible to recognize sustainable lessons that were established in the past and that are crucial to be respected and to be used in the present.

At Al-Salt old city, there are many healthy buildings constructed to suit the physical and climatic conditions of the area. They are considered symbolic of the cities heritage and legacy. Local materials have been expertly used in constructing the structures that have been shaped in respect to nature. All of the principles of sustainable architecture can be found in Al-Salt old houses. The geographical structure of the region as mountains enabled the formation of the great yellow stones that motivated people to use it as building materials. The architecture design relies on simple, renewable, and naturally insulating materials and passive strategies like courtyards, thick walls, and natural ventilation to keep houses cool in summer and retain heat in winter.

Environmental and Social Sustainability Dimensions in Traditional Houses in Al-Salt

The house needs three objectives to be successful: it has to fulfill social needs (social sustainability), affordable to all people (economical sustainability), and provide livable indoor environment along building life cycle (environmental sustainability).

An analysis has been made of the traditional house according to sustainability dimensions. This analysis is essential to identify the environmental and social sustainability potential in traditional houses at Al-Salt.

Environmental Sustainability Dimensions

There are many environmental factors that play an important role in creating architecture responding to its environment. Wind movements, its directions, humidity, rain, sun path, and resulting heat or cold are some of these factors. In addition, there are physical features and natural resources such as the site, topography, soil type, indoor air quality, water treatment, waste management, and building materials available in the ambient environment.

Site Selection

Choosing the site is an essential step for sustainable design for its effect on building environmental performance along the building life cycle. Selecting places for settlements was based on geographical location, topography, and resources availability. Al-Salt houses were built on the mountains leaving the valleys with its fertile soil for agriculture as Salt known in Jordan with its high quality of its fruits and vegetables (Fig. 8.2).

Urban Fabric

Compact urban fabric, attached houses, narrow passageway, and adaptable house design were solutions proved their success in Al-Salt architecture in a trial to adapt to the prevailing climate (Fig. 8.3). The spatial organization of residential districts at Al-Salt was a clustered organization that made up of a dense composition of dwellings. Houses often shared walls to promote a lesser surface-to-volume ratio that considered to be a possible environmental solution to the common problem of heat gain in buildings (Khawaja 2002). Moreover, this closeness made the walls act as sun-shading devices. During the daytime, walls create shadow on each other and this leads to reduce sun heat effect and create shaded alleys for pedestrians (Fig. 8.4).

Fig. 8.2 Al-Salt old city, Jordan. Source: (The author)

Fig. 8.3 Narrow passageway at Al-Salt old city to protect users from sunlight. Source: (The author)

Fig. 8.4 Attached and compact urban fabric as climate treatment, Al-Salt old city. Source: (The author)

Orientation

Protection from the direct sun radiation and natural ventilation was the most important thing in planning features at Arabic regions. This used to ensure minimizing the thermal load and at the same time allowing the prevailing winds to pass easily in order to eliminate high temperature during summer time. The main objective was to establish the optimum orientation of the building with regard to wind and sun heat.

As in most Arabic cities, buildings in Al-Salt old city were constructed close to each other. This type of compact structure created narrow alleys, known as *Harah*, which were shaded for most of the day (Fig. 8.5). The houses were planned in a way that let air movement pass easily through urban fabric. *Harah* ran from northwest to southeast, permitting the prevailing south winds to pass through. The alleys were started with a wide width then were started to narrow deep between houses. This caused increase in wind velocity as it breezed through, creating a comfortable pedestrian zone (Fig. 8.6).

Courtyards

Traditional urban fabric in Al-Salt was an introverted design for different environmental, social, and economical reasons. The courtyards that were the focal point design element and the main open space in traditional houses, create inward design. Even though plot areas vary according to owner's economic level and social status, all houses were designed within the same design principles and planning features.

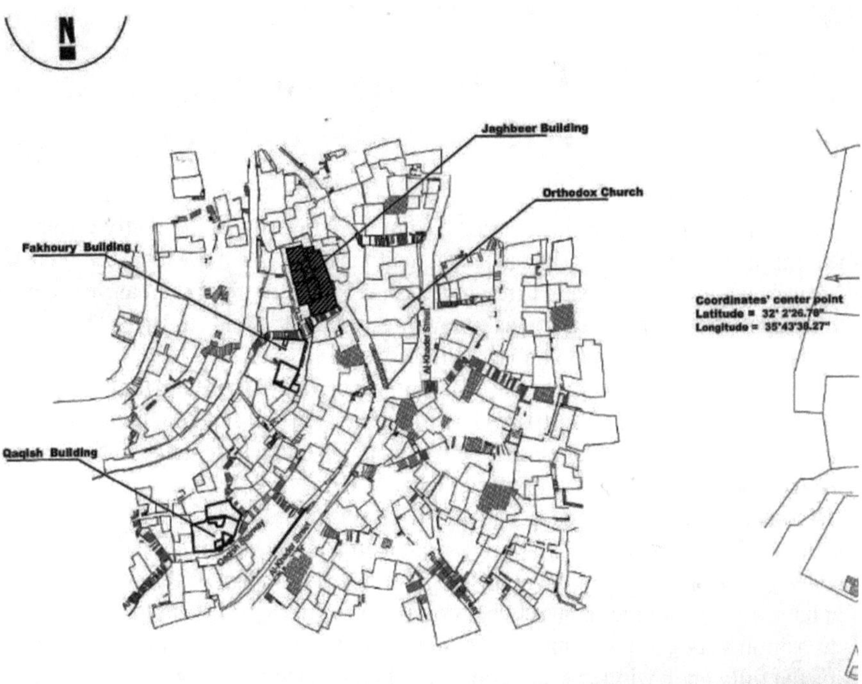

Fig. 8.5 Stairs "Harah" ran from northwest to southeast, permitting the prevailing south winds to pass through. Source: Architecture heritage in As-Salt city, Jordan (As-Salt City Development Projects Unit 2016)

Fig. 8.6 The alleys were started with a wide width then were started to narrow deep between houses. Source: (the author)

Most of the openings in introverted design were directly or indirectly opened to the central courtyard. The shaded zones and surfaces that have been created by courtyard walls help to create different pressure zones. Therefore, an air movement will start from high-pressure zones to low-pressure ones. This enhances the natural ventilation around and inside houses (Fig. 8.7).

Socially, courtyards were the best solution for privacy issues, and they meet most of the family uses. Some courtyards had trees that kept environment pleasant inside the houses and preserve the relation between human and nature that create always a positive energy inside the living places.

Another architectural space called *Iwan* considered as a sustainable feature inside Al-Salt houses (Fig. 8.8). It's a transitional space covered outdoor area between the front door and the courtyard. The *Iwan* was ideal so that a person does not have to step into the relatively hot courtyard to get to the other spaces. It makes a transition between the sunny courtyards and the semi-dark rooms. At the same time the fully open windows and doors are often opened to the *Iwan*. While outside windows were covered with *Mashrabiya*, traditional wooden treatments for outside windows that protect from direct sunlight and preserve privacy for indoor (Fig. 8.9). Courtyard and *Iwan* achieve one of the main sustainable dimensions which is the

Fig. 8.7 Courtyard walls help to create shaded areas and enhance the natural ventilation and lighting (Touqan house). Source: Architecture heritage in As-Salt city, Jordan (As-Salt City Development Projects Unit 2016)

social pillar since they sustain people needs and traditions. It respects the privacy and the dramatical transaction between public, semi-public, semi-private, and private spaces inside houses.

Energy Consumption Renewable Energy Sources

Energy is an essential issue in building life cycle during the construction process or the building's occupancy. Embodied energy is the energy that required to produce building materials including collection of raw materials, processing transportation, construction, and maintenance during building life cycle (Kim and Rigdon 1998).

Traditional houses in the Al-Salt, like traditional houses in the region, depended on natural resources in building materials. Traditional building materials and construction systems are based on low use of energy and sources, and work in harmony with the natural environment. Therefore, traditional houses did not release carbon dioxide or pollutants. They are considered sustainable and friendly to the environment.

Fig. 8.8 The *Iwan* in Khatib house. Source: Architecture heritage in As-Salt city, Jordan (As-Salt City Development Projects Unit 2016)

Fig. 8.9 Windows treatments in Abu Jaber house. Source: Architecture heritage in As-Salt city, Jordan (As-Salt City Development Projects Unit 2016)

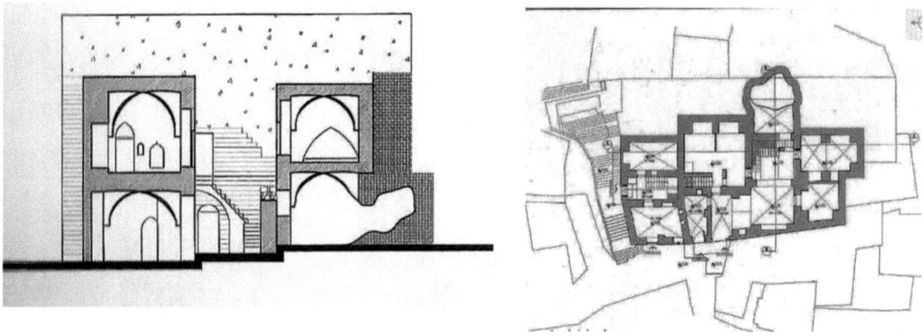

Fig. 8.10 Food storages at Jaghbeer house. Source: Architecture

Moreover, a great passive food and grains storages were founded in the traditional houses of Al-Salt. Large cavities were created in the interior walls of the house that been attached to the mountain. These cavities located inside the mountains with a fixed 12° C (Fig. 8.10).

Traditional house represents a sustainable architecture for its passive energy building that corresponded to the ambient environment. They used passive cooling and heating systems that utilized natural renewable energy resources.

Building Materials

Houses envelope at Al-Salt old city consists of different building materials which forms effective barrier against the unpleasant external climate. Building materials from available resources in region were used in traditional houses. Walls were built with stones, mud, and lime. Indoor walls were used to cover with plasters. These materials are ideal for the regional climate. Therefore, they consider sustainable and environmentally friendly materials with low embodied energy and less in toxicity than manufactured materials (Fig. 8.11).

Walls were thick (from 80 to 100 cm) to provide thermal masses that delay the heat transference and acoustic insulator for interior spaces (Fig. 8.12).

Construction Systems

Traditional houses used simple construction methods, easy to implement and manipulate. Thick masonry walls were used to support roofs made of stone vaults.

Vaulted and flat stone roofs were finished with clay on gypsum from the top. In large spaces, masonry cross-vaulted ceiling was used. The construction system was suitable with the building materials that were available in the ambient environment (Fig. 8.13).

Fig. 8.11 Building materials (yellow stone). Source: Architecture heritage in As-Salt city, Jordan (As-Salt City Development Projects Unit 2016)

Fig. 8.12 Building materials (yellow stone). Source: Architecture heritage in As-Salt city, Jordan (As-Salt City Development Projects Unit 2016)

Fig. 8.13 Construction system (vault and cross-vaulted ceiling). Source: Architecture heritage in As-Salt city, Jordan (As-Salt City Development Projects Unit 2016)

Water Harvesting

People of Al-Salt depended on wells to get water by digging a vertical shaft underground inside the houses. They used to collect rainwater inside their own wells. Their technique relied on harvesting the rain on top of the roof. Then, it will reach the well through a spout within the thick walls (Fig. 8.14). Rainwater collected for general drinking, washing, and watering. Drinking water was stored in big water jars "*Jarah*" that were placed in on hollowed-out niches inside walls. Small water jars were placed on alcoves near the windows to be cooled by the cold breeze.

Social Sustainability Dimensions

As mentioned before, one of the most important dimensions of sustainability is social dimension though it was almost neglected most of the time. It includes several aspects as quality of life, satiability, empowerment, safety, equity, accessibility, and cultural identity. Social dimension can be discussed within the place-based design trend in architecture that possesses an intimate relationship with the spirit of place "genius loci."

Houses of Al-Salt old city will be portrayed as a symbol of cultural identity in this region society. These houses were shaped upon the principles of the society values and traditions derived from Islam. The dogma of Islam is not only concerned with the spiritual life, but it also embraced man's entire life.

The word "house" in Arabic is "*Sakan,*" which means rest and quietness. It has been mentioned in several Quran verses as "It is Allah Who made your habitations homes of rest and quiet (16 Al Nahal: 80)." So, house architectural design was the reflection of Islam conventions, traditions, and social values that revealed society and cultural identity and fulfilled human physical and social needs.

Fig. 8.14 Water harvesting system at Touqan house. Source: Architecture heritage in As-Salt city, Jordan (As-Salt City Development Projects Unit 2016)

Privacy

Individuals have the right to be free from public intercourse especially in their private life. This system would protect the family life and preserve the stability of the society by keeping social health and well-being. The traditional houses arrangement took privacy as an important factor in design. Privacy in traditional houses was achieved through city planning, neighborhood unit, street width, entrances, house design, almost solid walls, openings form, location, and details of louvers and screens (Fig. 8.15).

The acoustical privacy was achieved through house design that consists of three spatial zones: public, semi-public, and private that were connected through the courtyards and lobbies. Building envelope was acoustically insulating by its massive materials as stone and brick.

Urban Planning and Design Philosophy

Cluster special organization was used in urban fabric of residential areas. The main entrances to the housing areas were located on the mail streets of the *Souq*. Within shops a main stairs would lead into a district located on the mountains of Al-Salt.

Fig. 8.15 Introverted design and windows treatment to achieve privacy. Source: (The author)

These districts would have a secondary stairs that define main gates to another residential clusters. The residential clusters were made up of a maze of winding alleys moving vertically through a sequence of spaces. Doors ensure privacy by not be placed opposite or near each other. Moreover, they will never open directly into the house, and so *Iwan* was created.

Courtyards and introverted design were the best strategy to maintain privacy inside houses. Family then can perform their social activities without any visual intrusion. The roof also provided a vital connection to the street and social life. It was also used as a place to play when the courtyard may be unavailable.

In traditional houses, courtyards as multipurpose family space maintained social sustainability within the family, whereas stairs and streets in front of houses as a social gathering space maintained social sustainability within neighborhoods.

External Elevations

Traditional houses in Al-Salt have plain elevations without any projecting surfaces, with some longitude openings at the ground and first levels of the building. The exterior facades contain decorative elements such as round arches, blind arches, columns, and round small openings.

Windows in the ground floor were high enough for privacy issues. Windows should be placed above eye level on the street side, about 1.70 m above the ground level. Upper windows and balconies were often decorated and carved by wooden or masonry louvers (Fig. 8.16).

Fig. 8.16 Exterior elevation at Mouasher house. Source: Architecture

Social Relationships and Neighborhood

Islam strongly encourages social life on a wide scale. It recognizes the main need of social intercourse within humankind. Muslims and Christians have been living in Al-Salt since long time ago. They have been sharing the same culture, lifestyle, and traditions in strong coherent and homogeneous ties. The city urban fabric was set in a way that greatly contributes to the enhancement of the social interaction and strong relationships. Attached and nearby houses were demonstrated this issue where they were linked together, separated only by narrow *Harahs*.

Neighborhood planning and design was the product of social relationship and cultural progress. The organic pattern of residential districts provided different spaces and environments for social interactions. While the clustering of the houses reflects the strength of the social fabric, streets and stairs functioned as public spaces and provided a convenient space for people to walk, to gather, and to interact (Fig. 8.17).

Conclusion

In order to think of architecture in a way that combines place-based design trend and sustainability trend, it is necessary for the two design realms to overlap and merge. To establish a theoretical framework for place-based sustainable design that

Fig. 8.17 Social interaction spaces were created between neighborhoods and within streets (*Harahs*). Source: Architecture heritage in As-Salt city, Jordan (As-Salt City Development Projects Unit 2016)

connects the spirit of time "sustainability" with the spirit of place "genius loci," a merge between sustainability and the four paradigms of place-based design should be achieved. The four paradigms are site structure, design decision, memorable experience, and culture.

The site structure is where the historical and cultural events took place. It shapes the spaces of the site and builds the microclimate. The structure of the site is composed of the plants, topography, architecture, and water. Design decision-making process is the cross-disciplinary work between designers from many fields. It responds to local issues and trends to bring a better future for the architectural design.

Memorable experience is about the outcome and the process at the same time. It is the deep collaboration between designers and user's experiences. *The culture* is the integrated pattern of human knowledge, belief, and behavior that depends upon the human capacity for learning and transmitting knowledge to the succeeding generations (Merriam-Webster 2015).

This combination of the ideas on place-based design as well as sustainable thinking can serve as a framework or a guide that conciliate between the spirit of time and the spirit of place. Sustainable lessons learned from vernacular architecture such as Al-Salt old city district might be the best subsistence and the strongest foundation for this authentic framework.

References

Asfour K (1998) Cultural crisis: cut and paste leads to disaster in the Middle East. Archit Rev 1213
As-Salt City Development Projects Unit (2016) Architectural heritage in As-Salt city, A-Salt Greater Municipality
Brundtland Report- World Commission on Environment and Development (1987) Our common future. Oxford University Press, London
Correia M, Dipasquales L, Mecca S (2014) VerSus. Heritage for tomorrow. Vernacular knowledge for sustainable architecture. Firenze University Press, Firenze
Dinep C, Schwab K (2010) Sustainable site design: criteria, process, and case studies for integrating site and region in landscape design. Wiley, Hoboken
Downs RB (1973) Famous American books. McGraw-Hill, New York
Green J (2015) Designing for the future: 80 practical ideas for sustainable world, New York
Guy S, Farmer G (2001) Reinterpreting sustainable architecture: the place of technology. J Archit Educ 54(3):140–148
Jackson JB (1994) A sense of place, a sense of time. Yale University Press, New Haven
Khawaja M (2002) DAR AS-SAKINA: reinterpreting the traditional emirate, house in the design of a modern apartment building in Sharjah, Final project research & design, American University of Sharjah, Sharjah

Kim J-J, Rigdon B (1998) Sustainable architecture module: introduction to sustainable design. National Pollution Prevention Centre for Higher Education, Michigan

Kultur S (2012) Role of culture in sustainable architecture. In: Archi-cultural translations through the Silk Road, proceedings of 2nd international conference Mukogawa Women's University, Japan, July 14–16, 2012

Ludwig D, Walker B, Holling CS (1997) Sustainability, stability & resilience. Ecol Soc 1(1):7

Maslow AH (1943) A theory of human motivation. Psychological Review 50(4)

Matsumato DR, Juang L (2003) Culture and psychology. Wadsworth Publishing, San Francisco

McHarg I (1971) Design with Nature, Falcon Press, Philadelphia

Meadows DH, Meadows G, Randers GJ, Behrens WW (1972) The limits to growth. Universe Books, New York

Merriam-Webster (2015) Definition of culture. Accessed 3 Aug 2015

Meyer-Arendt K (2004) Tourism and the natural environment. In AA Lew, CM Hall and AM Williams, eds. A Companion to Tourism. London: Blackwell

Olfat (1987) Hamuda, Architectural character between authenticity and modernism. Al-Dar Al-Masriah Al-Lubnaniah, Cairo

Rudofsky B (1964) Architecture without architects-an introduction to non-pedigreed architecture. The Museum of Modern Art, New York

Social Life (2012) Design for social sustainability: a practical framework for building communities

Toffler A (1973) The culture consumers. Random House, New York

Turner T (1996) City as landscape: a post-modern view of design and planning. E & FN Spon, London

USGBC (2009) USGBC LEED green associate study guide. U.S. Green Building Council, Washington

Wheeler SM, Beatley T (2014) Sustainable urban development reader. Routledge, London

Chapter 9
Urban Planning Enriched by Its Representations, from Perspective to Thermography

Benoit Beckers and Elena Garcia-Nevado

Introduction

The human population has exceeded seven billion, and more than half of this population is now living in cities. This trend is growing rapidly: more than two billion people are expected in cities around the world by 2030, and they will probably join the billion people already living in slums. In addition, today's cities grow mostly spreading, often on the best farmland, threatening the food self-sufficiency of many parts of the world: the central valley around Santiago de Chile, the south of Montreal (one of the few fertile lands in Canada), England as a whole, or the Beijing region are well-known examples. Therefore, the major challenge of our century is to create denser cities, which nevertheless guarantee acceptable living conditions for all of their numerous inhabitants (Beckers 2016).

In this perspective, past architectures that can enrich the future—to paraphrase the title of this book— are primarily urban architectures, conditioned by densely built environments. Numerous studies have shown that Haussmann Paris (late nineteenth century) and central New York (early twentieth century), compared to more recent cities of their respective continents, are more economical both in terms of transport (from their higher density), but also regarding the thermal behavior of their buildings. By betting on lighter buildings, at the expense of thermal inertia, but also by offering more generous spaces for housing and work (and therefore larger air volumes to heat or cool), contemporaneous cities of the richest countries have been less successful than their elders a hundred or a hundred and 50 years ago.

B. Beckers (✉)
Urban Physics Joint Laboratory, Université de Pau et des Pays de l'Adour, Anglet, France
e-mail: benoit.beckers@univ-pau.fr

E. Garcia-Nevado
Architecture & Energy, Universidad Politécnica de Cataluña, Barcelona, Spain

A. Sayigh (ed.), *Sustainable Vernacular Architecture*, Innovative Renewable Energy, https://doi.org/10.1007/978-3-030-06185-2_9

In 1800, at the dawn of the Industrial Revolution, the human population had reached its first billion, but less than 5% of this population was living in cities. With a little more than a million inhabitants, Beijing was then, and for a few more years, the largest city in the world, the last of these pre-industrial megacities, from imperial Rome, through Constantinople and Xi'an, which had reached this imposing figure, a clear limit which could only be reached in very centralized empires, capable of making the resources of very vast territories available to their capital. In 1800, London had just reached the million inhabitants, but in very different circumstances, which would soon allow it to grow much more and to enter, the first, in a new urban world, practically beyond measure with the former one. In the mid-nineteenth century, half of the English were already urban.

Is it therefore worthwhile to go back in time in cities, and is it possible to draw lessons for our today megacities, so different in size, in operation, but also in location? Today the urban growth bulk takes place between the tropics, in areas of the world that were virtually uninhabited only two centuries ago. Aristotle considered that a city should not extend beyond what can be perceived at one glance, but there is not a high enough tower in Pudong from where one can only make an idea of the Shanghai urban area, with its 60 million inhabitants.

And yet, the skyscrapers that mark the skyline of Shanghai are reminiscent of the medieval towers of Bologna and San Gimignano. Since ancient times, large urban structures have changed little: a network of streets that meet to form squares and which delimit built plots, which have window-pierced facades. Sometimes the buildings get too tall for too narrow streets, and the windows do not get enough sun. It happened in Ostia, the port of imperial Rome and, almost 2000 years later, in Manhattan and Hong Kong. In all three cases, it was necessary to issue rules on the height of the buildings or their size, according to the width of the streets, to guarantee everyone a minimum access to the Sun and the sky.

In this chapter, we will recall the three great historical revolutions concerning the cities, the Urban Revolution, about 5000 years ago, with the simultaneous appearance of the first cities and the writing, the Perspective Revolution, which offered the Modern Times the means of designing three-dimensional streets and squares, and, finally, the Industrial Revolution, and the disruption of urban scales, until their loss. We will finally propose a new way of thinking and designing the city, and we will show how this, at the heart of urban physics, could participate in the progress of the cities of our time.

The Urban Revolution

Since the Neolithic Revolution, there are human grouped habitats, but they are not fully considered as cities, even when they gather a few thousand inhabitants, as Çatalhöyük (Turkey, nearly 10,000 years ago). This habitat is homogeneous, without monuments and without specialized buildings, because all the inhabitants have the same activity (agriculture or breeding) and therefore the same status.

A city, *stricto* sensu, includes various activities, with specialized neighborhoods (craftsmen, military, etc.) and monuments (palace, temple, market, manufacture, etc.). In this sense, the first of all the cities was perhaps Uruk, in Sumerian country, where archeologists have found temples, remains of craft activities, and the first traces of hieroglyphic writing as well as another major invention: the wheel. But there is nothing left of the original habitat. The oldest city we know was founded at the end of the fourth millennium BC by Uruk settlers on the site of Habuba Kabira, today in Syria. At the beginning of the third millennium BC, still in present day Syria, the city of Mari was founded (Margueron 2003).

These first two cities were created *ex nihilo*, the first on an orthogonal plane, and the second on a perfectly circular plane. This is easily explained by the fact that Mari was endowed with a double enclosure: a wall to protect it from enemy incursions, and an embankment to save it from floods. These walls were expensive, and the circular shape is the one that guarantees the best relationship between area and perimeter. From the beginning, the cities had to face three challenges that remain valid, which is why we can express them with the today words:

- A city must be *sustainable*, that means it must continuously have the resources to maintain its population and infrastructure. This is a condition of existence. Cities could only appear when the local populations were able to release surplus food. These extras have generated the wish to trade over longer distances, and cities have established themselves along waterways, or at the end of long channels, as interfaces between river transport and land transport. Their specialized inhabitants consumed surplus food and produced handicrafts. Uruk ceramics are now found in Syria, and Sumerian craftsmanship has spread throughout the Fertile Crescent, traded, for example, for timber from Phoenicia or the Zagros Mountains, which was singularly lacking in the Mesopotamian plain.
- A city must be *safe*, in the face of the attacks of enemies, floods, and flames, and especially epidemics. It is therefore not only a question of surrounding it with a wall, damming it or building it with stones, it is still necessary that the summer breezes circulate there and that the winter sun penetrates it. It will be a constant preoccupation, especially when the cities will go up toward the north of the Mediterranean, and the Hippodamian plan will then show all its advantages (Hippodamus of Miletus, contemporary of Pericles, Herodotus, and Socrates, was considered by the Greeks as the inventor of the checkered town plan, which is none other than the orthogonal plan already apparent in Habuba Kabira, and which he will apply to Millet, his hometown on the Turkish coast, in Piraeus, the port of Athens, and probably in Thurii, Panhellenic colony inspired by Pericles in southern Italy).
- A city must be *smart*, because, in spite of all the efforts, it remains full of dangers, and, until the contemporary time, the urban life expectancy was much lower than in the countryside. To get on, a city must be attractive, and it must offer forms of development that do not exist elsewhere. Babylon, Athens, Alexandria, Rome, and Constantinople, but also Xi'An, Kyoto, Cordoba, Baghdad and, later, Bologna, Paris, Naples, or Florence, are, first, places where one can learn and

study, where everyone can hope for himself or for his descendants a life of high official, wise tradesman, respected prelate, victorious general, and renowned professor.

The checkerboard city has a very notable feature: it must be oriented, and its founding choice depends on its relationship to solar paths and prevailing winds. When the Roman legions conquer a new province, they set up camps on carefully chosen sites, so that everyone benefits from the best climate possible (Vitruvius heavily insists), and that they mesh the territory better. Two perpendicular axes are traced: the *decumanus,* ideally from east to west (but variations are possible to better adapt the resulting checkerboard to the terrain and the prevailing winds), and the *cardo* (whose name is found in the *cardinal* points), from north to south. These town seeds are connected by paved roads, and then grow more or less quickly, according to the military needs, the commercial possibilities, and then the prestige that their monuments and their cultural life are worth to them. By their individual qualities, but also by the mesh of the territory that constitutes their whole, these Roman cities are what could be done at the time of smarter, safer, and more sustainable. Thus, 18 of the 21 largest French cities today were designed in Roman times, and we can still guess their first outline. Marseille was founded much earlier (in 600 BCE) by Greeks from Phocaea, Montpellier is merely the heiress of ancient Maguelone, and only Lille is a true creation of the Middle Ages, at the entrance of the Flemish territory that had not seduced the Romans.

The Hippodamian plan subsequently had great successes, for example, during the construction of "bastides" in southwestern France (thirteenth century) and, later, during the European colonization of the Americas, while older cities (like Paris) or newly built ones, but in particular conditions (like Amsterdam, with its channels), developed more naturally in a radial or spiral way. This decisively two-dimensional urban planning, starting from a cadastre possibly completed by the information of the number of floors of each building, is still dominant today, and it is legitimate in order to manage intrinsically two-dimensional problems, like those of road traffic, or requiring only the additional consideration of relief (e.g., water supply and drainage networks). On the other hand, a problem as simple but as fundamental as the distribution of solar energy in an urban environment cannot be quantified correctly on a simple extruded 2D model (Beckers et al. 2010), or on urban forms reduced to typologies (Garcia-Nevado et al. 2016). To do this, one must get a precise and detailed three-dimensional vision of the city, and it was created only late, in Tuscany, at the beginning of the fifteenth century.

The Perspective Revolution

The great cultural contribution of the European Middle Ages was the polyphonic music, which allowed structuring the aural space on very rigorous arithmetic bases. For the Tuscan discoverers of the central perspective, it first allowed painters to finally emulate music, structuring the visual space on just as rigorous geometric bases (Beckers and Beckers 2014). But to show the perspective, and in particular its harmonic law, it was necessary to have very regular architectural or urban scenes. On the other hand, the success of the first perspectives led to the fact that the scenes—often invented, such as the "città ideale"—served as models for real constructions: buildings, squares, and entire cities soon submitted to the emphasized effects of the perspective view.

Central perspective was a first step, followed by the development of projective and descriptive geometries, rendering theory, photography and cinema, and finally computer graphics, which in turn modified and enriched our visual perception, but only it. The current acquisition techniques, from satellites, drones, or simply from the street, give very precise point clouds from which one can extract varied views of the city.

In a picture, the perspective effect is given by a deformation (simple obliquity in the axonometries, convergent parallels in the linear perspective, gradation in the hot–cold contrast of the color in the aerial perspective). When the deformation is weak, as in the front view of a façade, the image collects little information. But if the camera turns and takes the whole street in a row, the information increases substantially, at the cost of a strong deformation (and therefore an emphasized perspective effect).

The Industrial Revolution

From the early nineteenth century, the direct effects of the Industrial Revolution on cities are experienced in London and Paris, the two major cities of that time. Nighttime temperatures are significantly higher in the city center than in the surrounding countryside, which will be called "Urban Heat Island" (UHI). As factories move into cities, air and water pollutions become serious problems (the famous London smog). There is also the problem of refueling wood and coal for domestic heating, which is expensive and also very contaminating. Everything already indicates that the time to come will be inscribed, for the cities, under the sign of heat transfer.

Thermal science is indeed a recent discipline. In 1800, William Herschel discovers infrared radiation, and, the following year, Johann Wilhelm Ritter for the first time describes the ultraviolet rays. The equation of heat is proposed in 1811 by Jean Baptiste Joseph Fourier, but it deals only with conduction, and it is not until Max Planck's publication in 1900 of his famous article on the radiation of the blackbody to dispose of a complete heat theory.

In the second half of the nineteenth century, the first urban planners (Georges Eugène Haussmann in Paris, Ildefons Cerdà i Sunyer in Barcelona) find effective solutions to the problems of that time, but based on basic physical concepts (prevailing winds, Sun paths) that would hardly have surprised Hippodamus. Their most innovative ideas about mobility and the combination of modes of transport are further reinforced by two major advances: the elevator, which allows overcoming the limit of six or seven floors, and motorized transport, which accompanies the development of the garden city, a concept theorized by the British town planner Ebenezer Howard in 1898. Starting with New york and Chicago, the big American cities will then adopt a now familiar morphology: a very dense downtown consisting of skyscrapers, surrounded, as far as the eye can see, of very sparsely populated residential neighborhoods. In the twentieth century, however, the very stimulating thoughts developed at the Bauhaus or in the CIAM congresses lead, in the post-war years, to the relative failures of Chandigarh and Brasilia, generally attributed to the loss of scale following the substitution of the reference to the urban pedestrian by that of the automobile (excessive zoning, too broad avenues, urban highways, urban sprawl).

In the twenty-first century, the "smart grid" (electricity grid made "smart" by the contributions of computing) raises the idea of a "smart city," a city using information and communication technologies for improving the quality of urban services and reducing their costs. However, the expected savings, even if they are significant (of the order of 15–20% for electricity consumption), do not provide any answer to the hard urban issues mentioned in the introduction of this chapter. By barely touching the structures and infrastructure of the city, these techniques give the illusion of acting in the right direction, but without having to make difficult decisions. The concept of "sustainable city," on the other hand, probably faces the heart of the problem, but, because it is impossible to precisely quantify the flows of energy, material, and momentum that take place in the city, present solutions are limited to "eco-districts," a sort of technological showcase where it is impossible today to analyze the physical impact or the real cost, and it is therefore misleading to imagine a generalization to the entire city.

However, recent advances in measurement, particularly short-wave and long-wave radiation (thermography) and computer science (in particular, climate-based simulations) point to changes in our understanding of phenomena. To achieve this, it will probably be necessary to adapt standard simulation methods to the complex geometries of cities (for example, the finite element method for the thermal aspect (Beckers and Beckers 2012)) and to propose new representations for the measured or simulated data of urban physics. By taking up again the reasoning of the first "perspectivistas," we propose here to carry out urban perspectives, by seeking the most distorted view, and therefore the most informed one, but from a thermal camera, in order to obtain infrared perspectives of the urban environment, gathering a maximum of physical information.

The Bayonne Test Case

Thermography Campaign in the Rue des Tonneliers

To explore the possibilities of perspective thermography within the urban environment, a first experimental campaign has been developed in the Petit Bayonne district of Bayonne. This urban tissue is considered one of the densest ones in France and constitutes an iconic example of a highly occluded morphology. Within such urban geometries, radiative exchanges between built surfaces are specially challenging, due to the effect of multiple inter-reflections (Harman 2004). In this vein, this location is considered to be an interesting case study to assess radiative processes, both in short and long wave.

The Rue des Tonneliers, a typical narrow street of the Petit Bayonne, was selected for the measurement campaign (Fig. 9.1). A time-lapse of photographs and

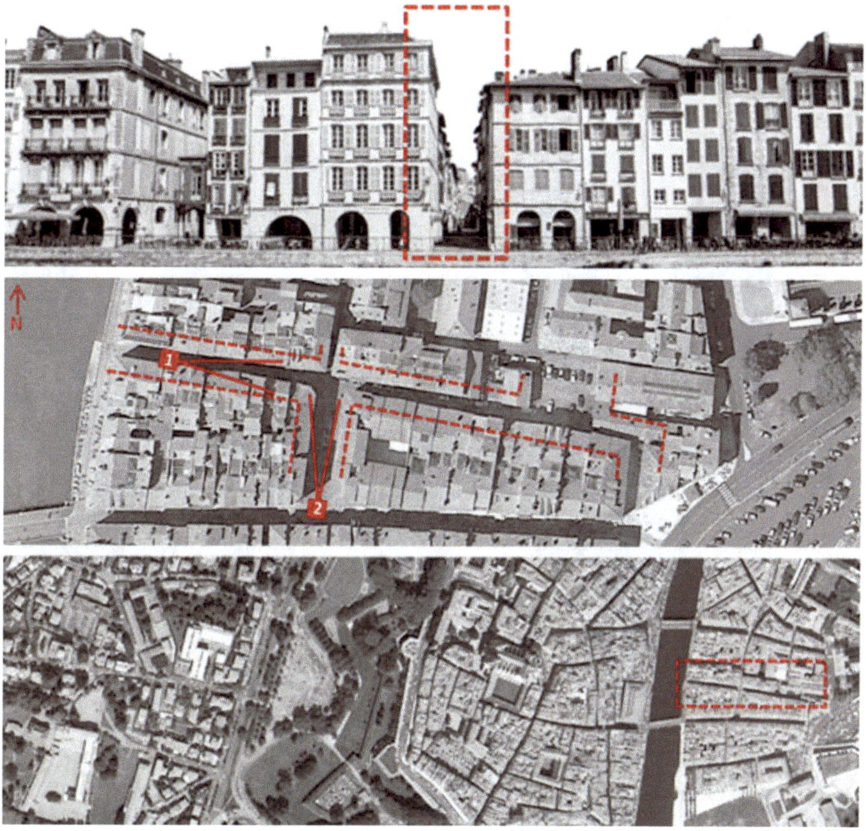

Fig. 9.1 Frontal view of Rue des Tonneliers from the riverside (up); aerial view of Rue des Tonneliers with IR camera positions (middle); aerial view of the old city of Bayonne and its surroundings (down)

thermographies was carried out on 23rd April 2017, a sunny day with soft temperatures and virtually no wind. A FLIR B200 IR camera was placed aligned to the E-W axis of the street (position 1 in Fig. 9.1—middle), shooting pictures every half an hour between 4:00 h and 23:00 h (solar time). Thermographies were calibrated according to the environmental conditions recorded during the test (air temperature and relative humidity), laying down the emissivity value in 1 in order to obtain apparent temperatures of the scene during the day. Time and spatial analyses of the thermal performance of urban surfaces were performed.

A detail from the image sequence obtained is shown in Fig. 9.1. By comparing photographs and thermographs to solar radiation and SVF (sky view factor) simulations on a 3D model of the street (Beckers et al. 2018), conclusions were drawn about the three main principles governing surface temperature: solar energy absorption, radiative cooling toward the sky, and thermal inertia. Results highlighted the key role of urban geometry (orientation, sun and sky obstruction), material properties (reflectivity and emissivity), and boundary conditions (sky and air temperatures, humidity, solar radiation, etc.) on the evolution of surface temperatures during a sunny day (Fig. 9.2).

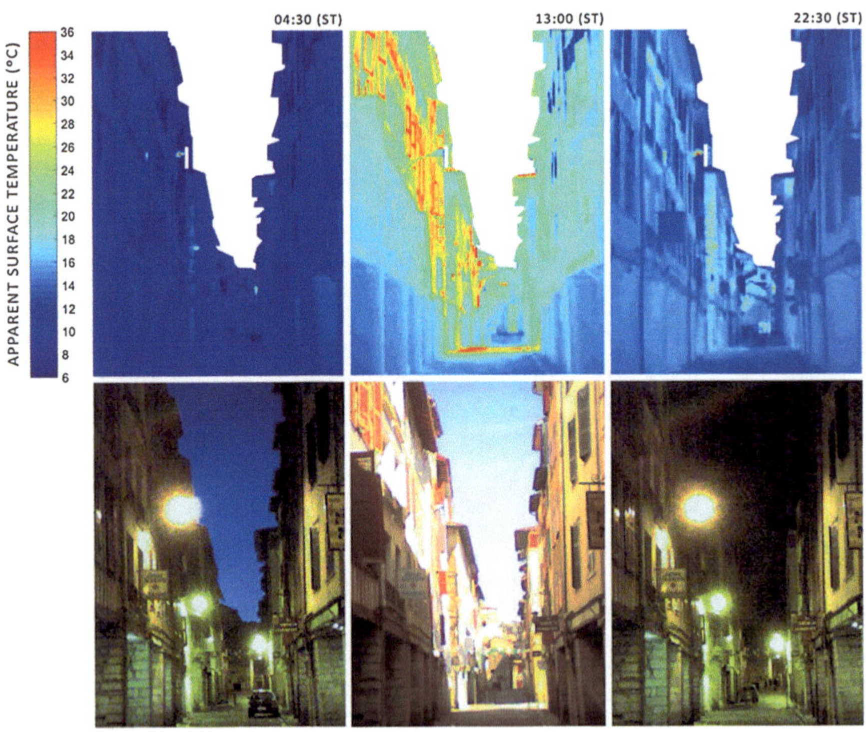

Fig. 9.2 Extract from the photographic and thermographic time-lapse obtained from the measurement campaign carried out in Rue des Tonneliers (Bayonne, 23rd April, 2017)

Perspective Thermography in Urban Environments

For construction-related applications, references agree on suggesting the development of thermographic measurements under the simplest conditions: focusing on a single target, viewed perpendicularly and from a short distance, and at dawn on a cloudy day (Vollmer and Möllmann 2010; Balaras and Argiriou 2002; Kylili et al. 2014). This rule has undoubtedly proven to facilitate the interpretations of thermographs, allowing the users to draw conclusions on the behavior of an individual element or even at a building level (Grinzato et al. 1998; Fokaides and Kalogirou 2011). However, concerning thermographic studies on an urban scale, achieving these wanted conditions is not always possible. Moreover, a simplicity-oriented approach may be in contradiction with the actual material and geometrical heterogeneity of urban environments and the variability of its boundary conditions.

Here, we aim to explore the use of perspective in thermographic studies at a street scale. By changing the observation direction, the area of built surfaces surveyed in the same picture is maximized, providing thermal information of a certain built environment from a wider point of view. Interactions between elements are more evident to the eye on a single image, allowing us to assess energy processes taking place in a more comprehensive way. This is the reason of the attractiveness of this approach and, at the same time, the source of significant challenges regarding results interpretation. Some of these new issues arising with the use of perspective are linked to the inherent complexity of urban environments, which is now present in a single scene: irregular geometries, material heterogeneity, moving elements, uneven boundary conditions, etc. However, some of the interpretation problems are merely due to the changes on the relative geometry between camera and target, as explained below using the example of the Rue des Tonneliers (Fig. 9.3).

Within a street canyon, changing the camera focus from a frontal view of the façade to an oblique one has two geometrical consequences. Each one of these aspects has particular physical-related effects that interfere with the interpretation of thermography in different ways, causing a bias in the observed temperature.

The first geometrical change linked to perspective views is an increase in the distance between the camera and the target (Fig. 9.3 middle line). This enlargement of the observation distance may be significant, varying from distances of several meters for a frontal focus (distance <8 m in Fig. 9.3 middle_left) to hundreds of meters for an oblique one (distance >150 m in Fig. 9.3 middle_right). The increase in the camera-to-object distance may result in an underestimation of the observed temperatures by means of thermography. This fact is explained by two reasons (Faye et al. 2016; Minkina and Dudzik 2009). First, because the higher the distance, the more atmospheric absorption of the thermal radiation emitted by the objects of the scene along its path to the camera sensor. Second, the increase in the pixel size affects the thermal image resolution. According to the literature (Vollmer and Möllmann 2010; Faye et al. 2016), the relationship between distance and surface temperature is non-linear and strongly dependent on the boundary conditions of the measurements (relative humidity, solar radiation impinging, air temperature, etc.).

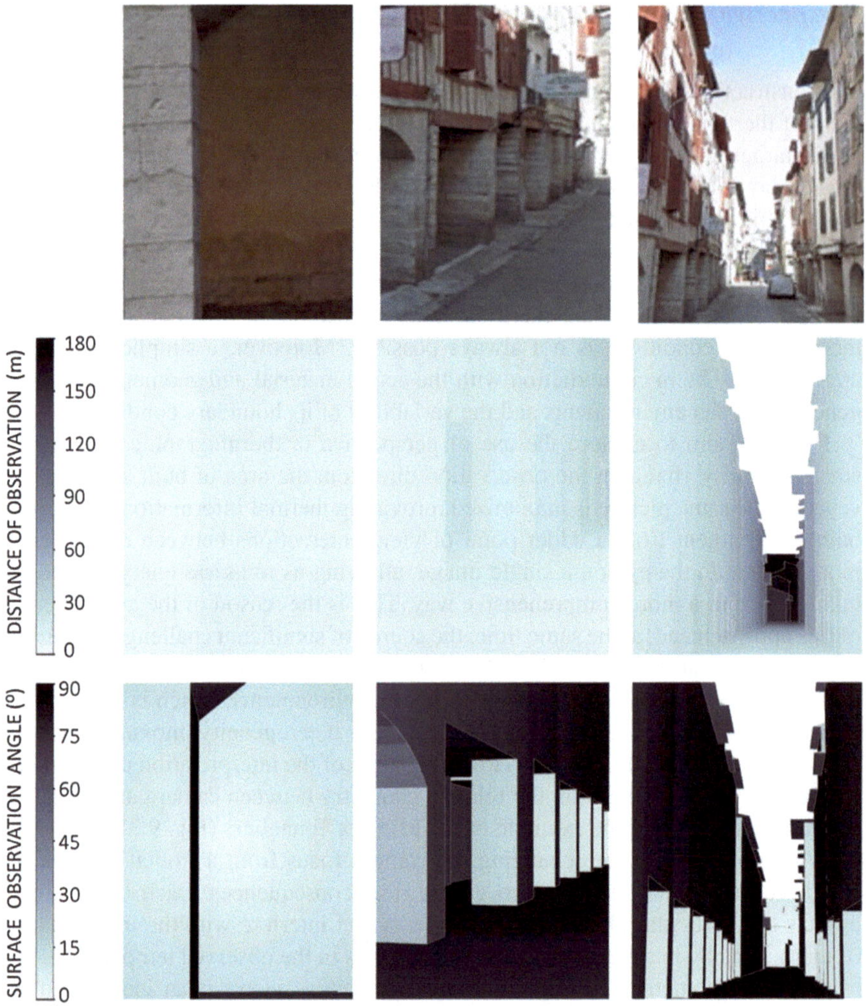

Fig. 9.3 Images of the Rue des Tonneliers (up) for a camera placed in postion 1 rotating from a frontal direction to a perspective one (from left to right). Distance (middle) and angle of observation (down) between the IR camera and the target calculated over a 3D model of the street for the same camera positions

Further investigation would be suitable regarding the interaction between all these factors by means of on-site measurements within the built environment.

Another aspect to be noted regarding the distance in thermography is that, when perspective is accentuated, the observation distances present in a single scene vary to a greater extent. In the Rue des Tonneliers example, this effect is obvious from the color gradient displayed between the foreground and the background of the image in Fig. 9.3 middle_right. Since commercial software allows users to fix a unique distance for the whole image, this aspect may be troublesome for thermography analyses.

The second geometrical consequence of casting pictures from a non-frontal focus to the façade is the increase in the viewing angle (Fig. 9.3 down line). In the case of the Rue des Tonneliers, it can be seen that, for normal observation directions, surfaces framed in the scene are mainly viewed under angles close to 0° whereas the ones tangentially observed constitute a minority (Fig. 9.3 down_left). As the camera rotates, seeking to enlarge the framed area, the viewing angle increases and the differences between the observation angles in the scene are further magnified (see the color gradient on Fig. 9.3 down_center). In the extreme case, when the camera is aligned with the street axis (Fig. 9.3 down_right), a major part of surfaces in the scene is now observed under large angles (>75°) and only the west-oriented elements are perpendicularly viewed.

Under large observation angles, the accuracy of surface temperatures measured by IR techniques can be questioned due to the unevenness of the radiative behavior of bodies depending on direction. Directional effects in the determination of surface temperatures in urban environments have proven to be significant from aerial measurements (Lagouarde et al. 2010), where roofs and pavements are the dominant surfaces in the thermographic scene. In the following section, this phenomenon will be addressed related to the use of perspective in thermography at a street level, where façades constitute the main focus.

Directional Effects of Emissivity on Surface Temperature

All bodies with a temperature above the absolute zero emit thermal radiation. IR cameras are able to measure this radiation and derive from it surface temperatures by means of an algorithm based on Planck's law. In order to obtain accurate temperature results, at least five parameters should be specified by the user in order to calibrate the thermal image: relative humidity, air temperature, reflected temperature, distance to the target, and emissivity of the object. Among them, emissivity is one of the most significant regarding the accuracy of the final temperature results. And, paradoxically, it is the one presenting more difficulties to be accurately measured, since its value fluctuates depending on several factors: wavelength, temperature of the body, time, and also viewing angle (Minkina and Dudzik 2009).

The emissivity is a thermal property which represents the ratio between the radiation emitted by a body and the one emitted by a blackbody at the same temperature. This parameter is therefore defined referred to a theoretical model which is assumed to be a body able to absorb and, consequently to re-emit, 100% of the radiation impinging on it (Kirchhoff's Law). Additionally, radiance distribution for blackbodies is considered to be isotropic, that is, constant regardless direction. Unfortunately, behavior of real materials may differ from this assumption, introducing a bias in results beyond a certain angle of observation. Most usual building materials are non-conductors (i.e., plaster, stone, concrete, etc.) and have high emissivity values, usually higher than 0.8 (Avdelidis and Moropoulou 2003). This is from a normal direction to observation angles until 45°–50°. Beyond that, emissivity values are dramatically dropping for angles larger than 85° (Vollmer and Möllmann 2010).

To illustrate this effect, measures were performed over a concrete tile. For the test, the sample was warmed up to 60 °C within a climatic chamber. Once out of the device, it was placed on a rotating platform inside a room at an ambient temperature (21 °C). Starting from a frontal position (0°), thermographs were shot under increasing observation angles (10°, 20° … 85°) for a fixed emissivity of 1. During the measure process, two K-type thermocouples were recording the surface temperature. The apparent temperature of the sample for three representative viewing angles (0°, 60°, and 85°) is shown in Fig. 9.4. It can be observed that, the more tangential is the view of the sample, the colder it "seems" to be.

According to the thermocouples measurements (T_K), the temperature decreased 3.7 °C between the first and the last image (Table 9.1). However, in terms of the apparent temperature (T_{IR}) showed in thermographs, this decrease was remarkably higher (8 °C). This behavior is due to the fact that the radiation emitted by the concrete sample for tangential direction is lower, and the IR camera associates that to lower temperatures. The emissivity of the sample for the three viewing angles was deduced by reducing the emissivity value until T_{IR} matches T_K. For the sample under study, the emissivity drops from 0.92 to 0.75 between a frontal view and the most tangential one.

Urban thermographs under oblique viewing angles will be affected by the same phenomena. To assess this issue, a second thermography campaign was developed in the Petit Bayonne on 8th May, 2017 (Fig. 9.5). This time, a single building was focused by two infrared cameras placed at different positions. Camera 1 viewed the target building tangentially since it was located in the same spot of the first thermography campaign. Camera 2, placed in Rue Pontrique, framed the building from a frontal direction (Fig. 9.1). Both cameras were switched on at the exact same time,

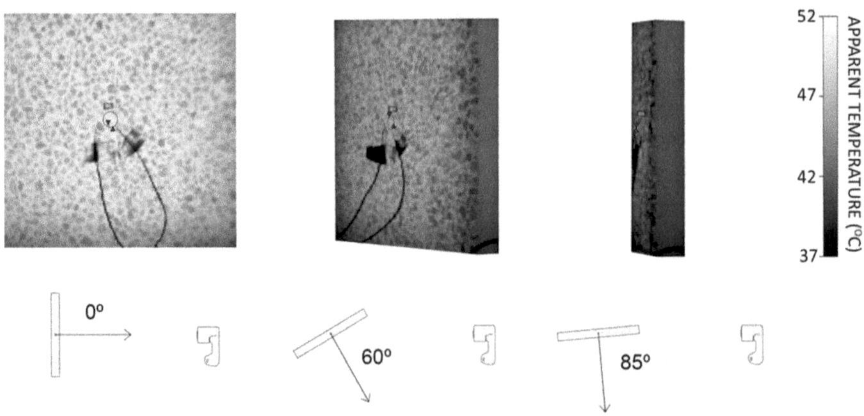

Fig. 9.4 Infrared images of a concrete sample viewed under observation angles of 0°, 60°, and 85°

Table 9.1 Average surface temperatures measured by thermocouples (T_k), apparent temperature recorded by the IR camera FLIR B200 (T_{IR}), and calculated emissivity (ε) for a concrete sample under observation angles of 0°, 60°, and 85°

Viewing angles	0°	60°	85°
T_K	52.0	49.4	48.3
T_{IR}	49.5	46.1	41.5
ε	0.92	0.89	0.75

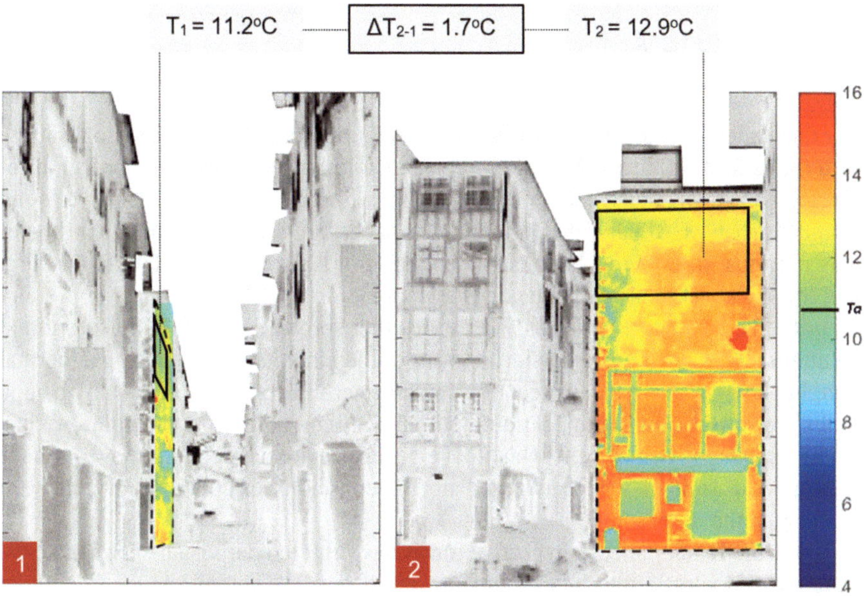

Fig. 9.5 Extract from the second thermographic campaign in Rue des Tonneliers (Bayonne, 8th May, 2017). Thermographs from cameras 1 and 2 focusing the target building (dashed line) at 4:00 h (ST) and average apparent temperature of the region of interest (black line)

every half an hour between 4:00 h and 22:00 h (ST). Thermographs from both cameras were calibrated for an emissivity equal to 1 and using on-site environmental conditions during the test.

The impact of angular effects on emissivity was assessed by comparing the apparent temperature of a particular area of the wall obtained by each camera. To ensure comparability of results, it was important to verify that the analysis was performed over the same part of the wall façade. For this purpose, the availability of a 3D CAD model of the urban area resulted to be a key tool. First, a region of interest (3 × 5.5 m) was defined in the aforementioned model. Then two virtual cameras were placed in the model at the same position than the thermal ones. Finally, images of the region of interest obtained from the virtual cameras were used to select the equivalent pixel area on thermographies obtained from IR cameras 1 and 2.

Results in Fig. 9.5 correspond to the snapshot of 4:00 h. At that time before dawn, there was virtually no wind, the air temperature was 10.6 °C with a relative humidity of 75%. Under these circumstances, the bias in apparent temperature for the region of interest (black line) was of 1.7 °C. As shown, the differences in the observation angle result in a non-negligible temperature deviation for the perspective view adopted in the case of Rue des Tonneliers. The magnitude of this bias is expected to vary during the day, depending on the boundary conditions.

Discussion and Conclusions

For thousands of years, men have been building their cities by adapting to a variety of terrains and climates. They often repeated solutions that had been proven elsewhere. The cities copied each other a lot. The white facades of Carthage are found in the Kasbah of Algiers, the Roman colonnades in Paris, the Haussmann boulevards in most major French cities, the channels of Amsterdam in Saint Petersburg, and the skyscrapers of Manhattan in Singapore.

In the best cases, it is not only lifestyles that spread like this, but also optimal solutions for the inside and outside comfort of the inhabitants. By placing a thermal imaging camera in a typical street of a southwestern Europe historic town, we hope to discover, in the long term, certain optimal configurations from the point of view of the exchanged heat fluxes, which have never been explained until now, because they are the fruit of empiricism and not of conscious calculations, which were unlikely at the time.

Since the Renaissance, perspective forms have entered our visual experience; we have no difficulty in interpreting them, even in much distorted images. It is quite different for the infrared perspectives, which easily disturb us, and of which we struggle to identify the biases, because we ignore the variations of the emissivity of one material to another, from one wavelength to another, from one angle to another, and from one distance to another.

Figure 9.6 summarizes two of these aspects. In the image on the left, all darkened surfaces are at an angle greater than 60° with the direction of the camera. The temperatures indicated by this one will therefore be underestimated, because of the anisotropy of the emissivity. In the image on the right, the darkened surfaces are further away from the camera, and their temperature will also be underestimated, because a greater thickness of atmosphere separates them from the camera, absorbing a part of the infrared radiation, and thus distorting the measurement, in a proportion which depends, partly, on the humidity of the air.

How to overcome these difficulties? A first solution would be to use the previous images as filters, making it possible to correct the temperature of each pixel as a function of the angle that the camera makes with the surface to which it corresponds, the distance to the camera, but also the material of this surface with its own

Fig. 9.6 Angle and distance of observation between the IR camera and the target calculated over a 3D model of the Rue des Tonneliers from a perspective point of view

emissivity. This involves building each time a 3D model that corresponds to the photographed scene, detailed enough to read all the necessary information.

Another solution would be to produce a series of frontal thermographs in the studied street, which is not impossible, since such series are already made in visible light ("Google street"), but obtaining frontal images of each floor of the buildings would require a drone and, in any case, it would be difficult to repeat such an operation regularly in order to obtain both spatial and temporal sequences. However, it is not sure that a single series of frontal thermographs corresponding to a specific moment can correct the infrared perspectives made at other times: the solar spot moving on the facades, the hottest points of the image change during the day, and the atmospheric absorption of the infrared seems to depend not only on the humidity of the air, but also on the temperature of the emitting surface.

To all these complicated problems, we add the fact that it is very difficult, in practice, to make measurements in a city, and even more if the flight of a drone implies to close the studied street to the traffic and the pedestrians. So, the only truly practical solution is to become able to simulate thermography, thanks to a fairly precise numerical mockup, which remains to be realized, but which is no longer impossible today. The thermography would not be used anymore but for the calibration of the model, which seems much more realistic with the currently available hardware.

The main advantage of the digital model is that it allows simulating modified urban scenes, and even scenes that do not exist yet, and thus, ultimately, it can provide a tool for supporting urban planning.

Acknowledgements The work developed by Elena Garcia-Nevado was supported by a PhD fellowship granted by the Spanish Ministry of Education and Science (FPU-13 0388).

References

Avdelidis NP, Moropoulou A (2003) Emissivity considerations in building thermography. Energ Buildings 35:663–667

Balaras CA, Argiriou AA (2002) Infrared thermography for building diagnostics. Energ Buildings 34:171–183

Beckers B (2016) Multiscale analysis as a central component of urban physics modeling. In: Ibrahimbegovic A (ed) Computational methods for solids and fluids, multiscale analysis, probability aspects and model reduction. Springer, Cham, pp 1–27

Beckers B, Beckers P (2014) Reconciliation of geometry and perception in radiation physics. Wiley, Hoboken, p 192

Beckers P, Beckers B (2012) Chapter 10: radiative simulation methods. In: Beckers B (ed) Solar energy at urban scale. Wiley, New York

Beckers B, Rodríguez D, Antaluca E, Batoz, JL (2010) About solar energy simulation in the urban framework: the model of Compiègne. In: 3rd International Congress Bauhaus SOLAR, November 10 & 11, 2010, Erfurt, Germany

Beckers B, Aguerre J, Besuievsky G, Fernández E, García Nevado E, La Borderie C, Nahon R (2018) Visualizing the infrared response of an urban canyon throughout a sunny day. In: Sayigh A (ed) Sustainable energy for all. Springer, Cham

Faye E, Dangles O, Pincebourde S (2016) Distance makes the difference in thermography for ecological studies. J Therm Biol 56:1–9

Fokaides PA, Kalogirou SA (2011) Application of infrared thermography for the determination of the overall heat transfer coefficient (U-Value) in building envelopes. Appl Energy 88(12):4358–4365

Garcia-Nevado E, Pages-Ramon A, Coch H (2016) Solar access assessment in dense urban environments: The effect of intersections in an urban canyon. Energies 9(10):796

Grinzato E, Vavilov V, Kauppinen T (1998) Quantitative infrared thermography in buildings. Energ Buildings 29:1–9

Harman IANN (2004) Radiative exchange in an urban street canyon. Bound Layer Meteorol 110:301–316

Kylili A, Fokaides PA, Christou P, Kalogirou SA (2014) Infrared thermography (IRT) applications for building diagnostics: a review. Appl Energy 134:531–549

Lagouarde JP, Henon A, Kurz B, Moreau P, Irvine M, Voogt J, Mestayer P (2010) Modelling daytime thermal infrared directional anisotropy over Toulouse city centre. Remote Sens. Environ. 114:87–105

Margueron J-C (2003) L'apparition des villes au Proche-Orient. In: Ville et pouvoir: origines et développements. Editions L'Harmattan, Paris

Minkina W, Dudzik S (2009) Infrared thermography. Errors and uncertainties. Wiley, Oxford

Vollmer M, Möllmann KP (2010) Infrared thermal imaging. Fundamentals, research and application. Wiley, Weinheim

Chapter 10
Vernacular Architecture as Model to Design a Prototype for Affordable Housing in Mosul

Antonella Trombadore and Filomena Visone

The Challenge of the International Competition to Explore Vernacular Architecture

The project idea was born during the experience of international competition *The Rifat Chadirji Prize 2017. An open international prize focusing on ideas responding to local challenges in Iraq.*

The Rifat Chadirji Prize is the newest prize to be established as part of the Tamayouz Excellence Award program of championing and celebrating the best of Iraqi architecture.

The prize is named after Dr. Rifat Chadirji, a great Iraqi architect whose influence and importance is far beyond built work. The award will officially be launched in January 2017, celebrating the 90th birthday of the brilliant architect. The Rifat Chadirji Prize is a thematic open-ideas international prize focusing on design proposals responding to local challenges throughout Iraq. (The Rifat Chadirji Prize n.d.)

The primary objective is to establish an accessible source of ideas to combat the country's social challenges through design. The competition's aim was to develop projects for the reconstruction of Mosul in Iraq, after the severe destruction and damaging caused by the ongoing war and the occupation by Daesh, which was terminated in July, 2017.

As underlined in the *Official Announcement of the Competition* (The Rifat Chadirji Prize n.d.), the Mosul city suffers from a chronic housing shortage. The deficit in housing units in Nineveh is estimated to have reached 172,000 units in mid-2016, with a 53,000 units' deficit in Mosul alone. The major contributing

A. Trombadore (✉)
DIDA Department of Architecture, University of Florence, Firenze, Italy
e-mail: antonella.trombadore@unifi.it

F. Visone
Visone Studio Architettura, Firenze, Italy

A. Sayigh (ed.), *Sustainable Vernacular Architecture*, Innovative Renewable Energy, https://doi.org/10.1007/978-3-030-06185-2_10

factors to this shortage can be defined as: (1) the scarcity of tracts of land for new housing projects; (2) the failure to update the city's 1973 master plan and create formal urban expansion zones for housing development.

New housing provision was limited to the private sector. The housing demands of poorer members of society were mainly met in the old city of Mosul where existing buildings became cramped with families living in shared accommodation. The United Nations and the International Organization of Migrants warned that the current number of internally displaced people from Mosul is estimated at over 500,000 (January 2017) and could reach 1.2 Million as the military operations continue. Some formal IDP camps have been established, but they will not have the capacity to accommodate the majority of new displacements. The 53.000 units' deficit is predicted to significantly rise due to the current military operation to retake the city of Mosul from Daesh. The conditions for returning refugees and internally displaced are extremely challenging. The question of how to support those who wish to return to their homeland will become extremely pressing. Limited resources in terms of finance and land mean that carefully considered material and spatial responses are needed.

Participants are asked to propose a solution for the Mosul's upcoming housing crisis, which will affect the city as more neighborhoods will be freed and internally displaced persons and refugees will start to return. Design a prototype for affordable housing for the post-Daesh Mosul, which can be easily replicated with the objective of increasing the capacity of housing in the city and providing a practical and inspiring solution for returnees (The Rifat Chadirji Prize n.d.). The prototype should be flexible enough to adapt to various sizes with different inhabitant capacity requirements. The designs should also be adaptable, allowing adjustments to be made in order to suit different residential capacity requirements (Fig. 10.1).

Fig. 10.1 Map of Iraq and general view of Mosul

The Design for People: Social Sustainability from the Past Experience

The aftermath of an armed conflict is always complex: an immediate solution for dwellings is needed, but attention to further developments is also an inescapable necessity. Given the brief and the topic, and in respect to the traditional architectural characteristics of privacy and modesty (being privacy, modesty, and hospitality milestones of Islamic architecture), the focus is on the organization and layout, therefore the proposed project is sober, functional, and realistic, with little room for frills and fancies, yet considering features taken and reinterpreted from local tradition, but most of all, the call of the project is for social sustainability, in an attempt to involve people granting them the possibility of active participation in the process of rebuilding their own lives.

"What is the city without the people," is a quote that best describes the intentions and the approach to the project: to give a house to people and not to give people to a house. The city is the stage where human stories unravel, the background where the lives of people are taking place and should be considered accordingly. One of the assumptions of the project was to find viable solutions aligned with local traditions, society, habits, climate, rather than importing them from some foreign environment. In this light, people are the focus of the project. The dichotomy between the needs of planners (planning, control, and organization to avoid the sprawling growth of informal settlements) and those of residents (customizing their dwellings) has overcome, thanks to the application of a very simple concept: the unfinished construction. No visionary esthetic utopias but the search of a reasonable balance between the need of the planner and those of the real people living in real places.

Another fundamental aspect is the consideration in many ways of the rich cultural heritage of the area: urban fabric, architectural features, traditional layouts, social expectations, environmental passive strategies, yet implemented with contemporary technological devices and systems.

The approach is therefore articulated in a few domains that are the guidelines of the project: social sustainability, cultural heritage, flexibility, *non finito (not jet finished—customizable)*, and sustainable architectural features (Fig. 10.2).

The Ordinary Shapes of the Cities A subtle network of social and human relationships is made of informal work, mutual solidarity, daily habits, and ordinary living that are often invisible to the planner. It's a hidden layer that gives a place to its atmosphere and helps people getting a sense of social identity and belonging. The well-being of the people just coming out of an armed conflict goes through the regaining possess of their lives. In this light, the project advocates the active participation of dwellers in the decision-making process (Figs. 10.3, 10.4).

Cultural Heritage and Iraqi Vernacular Architectural Features

Iraqi cultural heritage has been known since its dawn and goes beyond the intents of this dissertation.

Fig. 10.2 View of typical settlement of Mosul

Fig. 10.3 The Nineveh gate

Fig. 10.4 The general view and the door of Great Mosque of Samarra

The project idea starts from the observation of some architectural features that are crucial for the formal definition of the project. The definition of the project inescapably derives from the analysis of the rich cultural heritage and its architectural vocabulary, techniques and social values aiming at a solution that can be feasible but also in line with the local traditions, habits, economics, and society.

The Courtyard House

The courtyard house existed in the Mesopotamian region before the Islam appearance, but its massive use and application is due to the perfect compliance with the principles of the Quran. A house in Islam is presented as a shelter where family members can find their privacy and comfort within the realm of Islamic values (Warren and Fethi 1982), i.e., privacy, modesty, and hospitality. From this approach an introverted design takes shape with traditional courtyard house, generating an inner space where women are not exposed to the outside and street vision, ensuring the privacy of Muslim family. Daylight and natural ventilation are allowed by the courtyard that also separates easily the different functions of the house and the genders.

A predominant aspect is the connection between architectural features and social/climate restrictions. The vast use of solid wall masses with small openings grant thermal mass; similarly, the predominant adoption of courtyard houses' types is a good strategy for climate control. Both allow a high degree of privacy as requested by religious laws and social customs.

The *iwan* is a direct implementation of the shaded space created by the Arabian tent. It becomes an important part of the courtyard house creating a transitional area (separating outside from inside, public from private, glare from darkness) and a

Fig. 10.5 The reconstructed Ziggurat of *Ur*

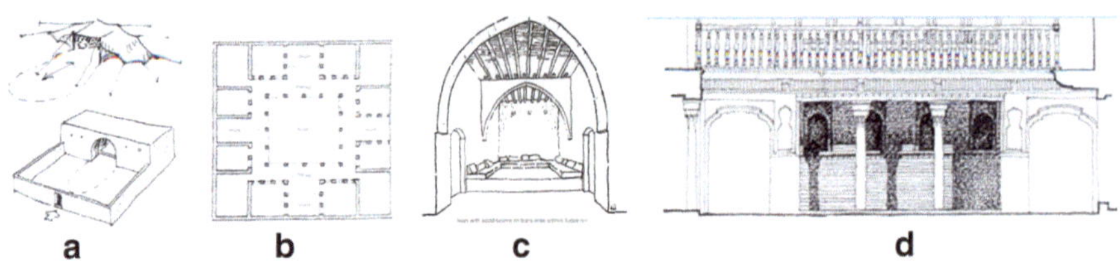

Fig. 10.6 (**a–d**) From yard to house: the structure of courtyard house with typical *iwan* and *tarma*

social place with shaped seats. In the opposite site, since the roof is usually overheating, we can found the *tarma*, as an open gallery, or the *ursi*, a room closed from three sides. In an extremely unfriendly environment passive methods of climate control such as wind scoops, the presence of *iwan* and *tarma*, courtyard fountains, and accessible roofs (*satah*) as extensions of the living spaces are widely spread and characterize the urban fabric of Mosul, and in general Arab cities (Figs. 10.5, 10.6, and 10.7).

A more complex type is the courtyard house, with solid walls that encircle the yard and protect from the sand and the dust driven from the desert and at the same time offering a high degree of privacy, in line with the precepts and values of Islamic tradition. This closed compound is surprisingly open in the inside: first the yard is defined and then, all around it and as needed, closed and semi-closed spaces are located, like *talar* and *tarma* that beyond being pleasant and welcoming spaces with a high degree of decorative sophistication, provide additional shading and

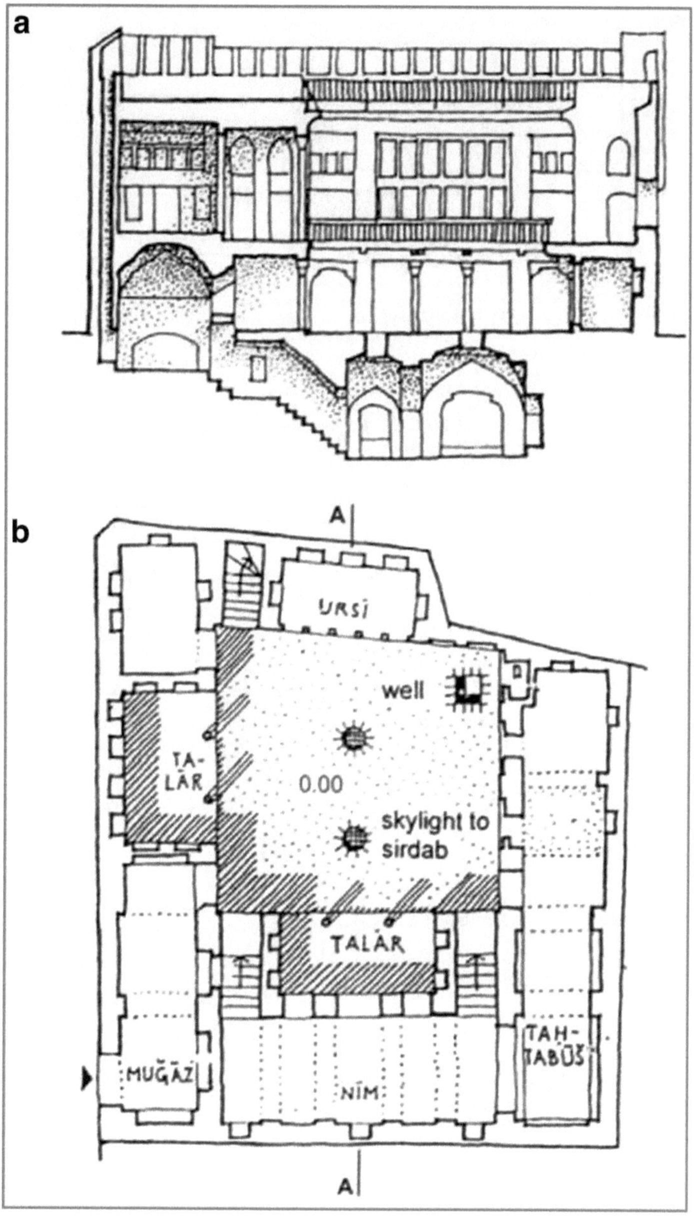

Fig. 10.7 (**a**, **b**) House Shashur, Bagdad: from courtyard to *neem* to *sirdab* and on to a well *bir-at-tabila*

transitional spaces. The courtyard house has a very introvert character, since outward openings are avoided. Light and air are provided by the interior yard, and a sequence of spaces defines a sort of shelter and an oasis in a hostile environment. Here the basic values of privacy, modesty, and hospitality are fully met.

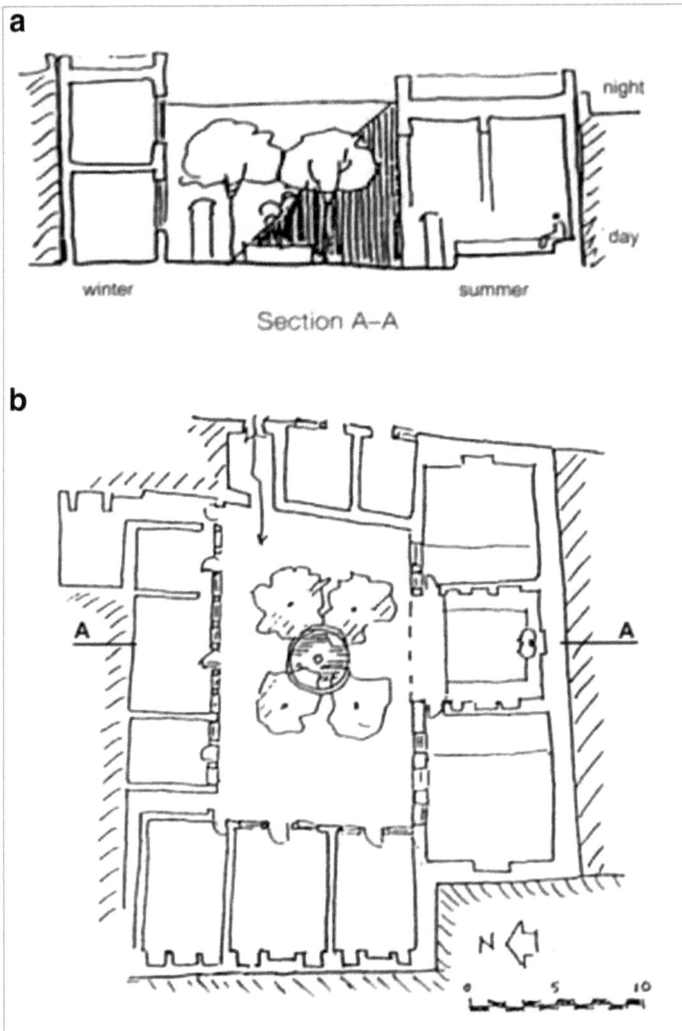

Fig. 10.8 (**a**, **b**) Typical courtyard house in Damascus. Low, glazed two floor construction facing south, high open *iwan* opening north

The architectural result is *a house hidden in the anonymous fabric of the town, with a single exterior elevation facing a narrow alley. Thus the house is not perceived as an object to be seen, but experienced as a succession of spaces* (Ragette 2003) (Fig. 10.8).

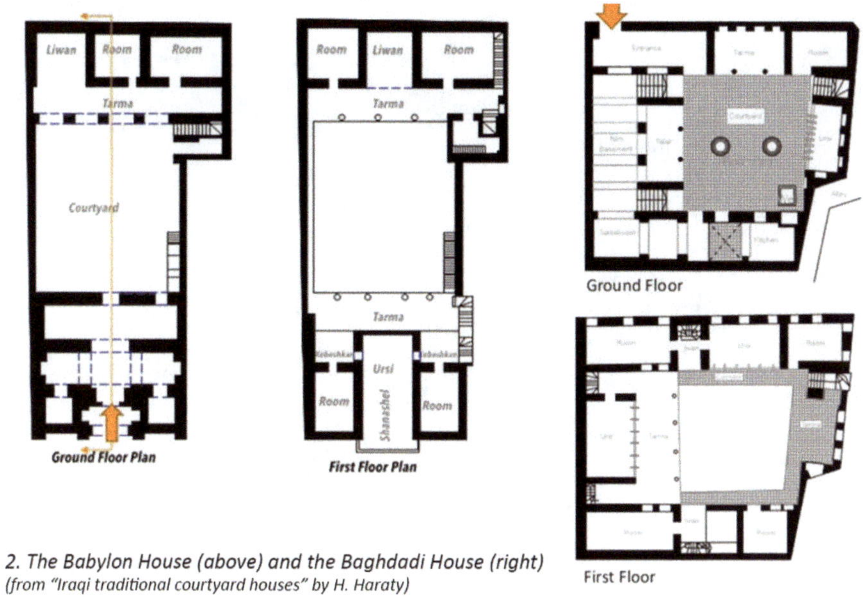

Fig. 10.9 The difference between plan and typology of a typical Babylon house (left) and Baghdadi house (right)

The Flat Roof

The flat roofs are usually an extension of the living space. In a hot dry climate, they are often used in summer for sleeping outside, but also, especially in rural areas, for drying grain and fruit or other food processing. Walls and screens protect privacy from neighbor's views.

The most elementary type for a flat roofed dwelling is made of a single cell (*bait*), with solid walls, small windows (*shubbak*), and a low door (*bab*). In more complex layout the presence of a colonnade or arcade creates transitional space and helps mitigating the effect of the scorching sun. The basic modular unit of the project stems from the analysis of this unit (Figs. 10.9 and 10.10).

The Natural Ventilation System

The air movement has an effective role in local climatic control and is often manipulated by using lattices, screens, and awning. In vernacular architecture the *badgir* becomes the most diffused and distinguished element in enhancing the climate of the traditional Baghdadi house. *Badgir* has been shaped precisely to pull down the ruling north-westerly breezes in Baghdad region. The air is conveyed and pushed towards the lower floors, driving out the calming breezes through the room's

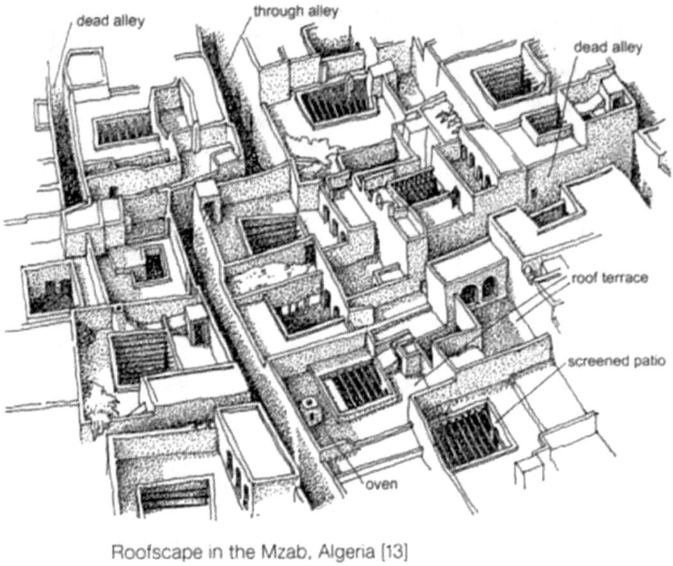

Fig. 10.10 The roof scape in the Mzab, Algeria

opening. Though the air has been cooled in its channel down the shaft, yet it will still be warmer and drier than the air in the basements. Waving out through the room will cool the inhabitant's evaporative and enhances the atmosphere. While the basement rooms would be unpleasantly cold in winter and in the afternoon the open shutters of the *badgir* would serve to bring in some warmed air to ventilate and make moderate the otherwise unusable basements (Figs. 10.11 and 10.12).

The easier solutions to improve natural ventilation in vernacular house building are the wind scoop (*malqaf*). As descripted by Ragette (2003) the wind scoops are ventilation flues carried above the roof to catch the cool and clean air, redirecting the fresh air to the lower-floor rooms, thanks to a reversed chimney effect. The size of a *malqaf* depends on the outside air temperature, reducing the section with increasing temperature. In Iraq the typical *malqaf* pipes are narrow, placed on the northern wall with a small inlet to allow the increase of air cooling (Fig. 10.13).

Lesson Learned from the Past for a Sustainable Concept Design

Mosul from above, like any other city in the Arab and Mediterranean region, shows two distinct fashions of city planning. We can see the compact dense organic irregular shape of the historic city and the regular cartesian orthogonal grid of recent developments. These represent two very different approaches: the first is irregular

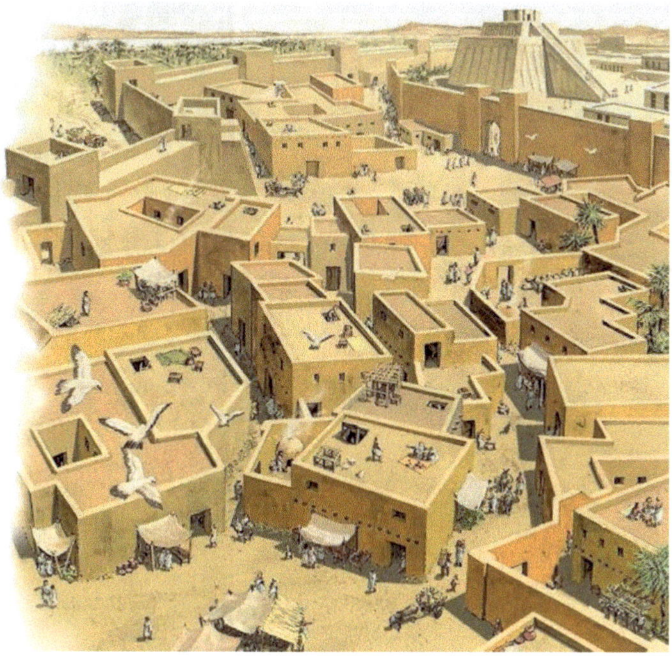

Fig. 10.11 Graphic representation of Iraqi city of Ur

Fig. 10.12 Ajman museum—UAE. The Iraqi building typology is diffused in all Arabic countries

Fig 10.13 (**a**) Example of *badgir* in Bin Ali house, in Muharraq, Bahrain. (**b**) Detail of *badgir.* Internal view

and dynamic, grants customization, and is very attractive to people. The typical example is the medieval city. But it grants little or no control over the growth of the city, because it is basically unplanned. So if the pro is that it's a process, therefore always under constant re-adaptation and elaboration, displaying an extraordinary visual variety, the con is that it lacks planning, definitely undesirable in the XXI century with the numbers at stake in large cities. This is particularly evident in a city like Mosul that is also facing major reconstruction; the latter grants control over the spreading of the city in order to avoid indefinite sprawling generating informal settlements, but generates homogeneity and is often incompatible with the traditional spaces and customs.

The first responds to the planner's need for control, the second to the residents' desire of tradition, visual variety, and familiarity (Fig. 10.14).

Layout and Pattern of Traditional Towns

The traditional urban fabric of Arab towns, and Mediterranean region alike, is made of an intricate network of narrow streets, blind alleys, and passageways as a result of the haphazard union and juxtaposition of houses, mostly courtyard houses. Having a very compact form, this mitigates heat gain or loss and serves as sun

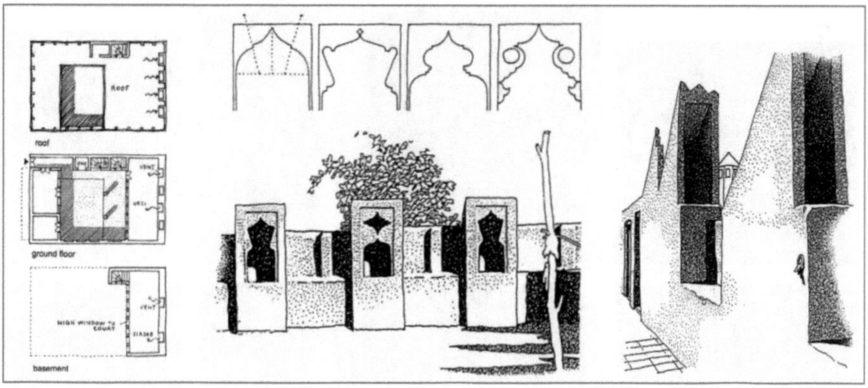

Fig. 10.14 Different solutions of *malqaf* used in the houses in Baghdad to increase the natural ventilation

shields, generating a complex system of private, semi-private, and public spaces. Neighbors have squares, of course, but it's worth remembering a more traditional communal open space (the closest thing to a formal piazza in the western sense): a yard called sahn (or saha when it's a communal private space) typically located inside the local Mosque and serving several quarters. It may be surrounded by arcades (Riwaq) and contain a fountain. Around the local sahn there is a network of alleys, through streets and *culs de sac*, some with a distinct trading character, that gives the place its special flair.

Historically, towns were not growing as the result of a conscious act of planning and control, but following the needs and wants of its inhabitants. Houses were growing simply adding new basic cells, and a process of agglutination, almost organic in nature, enlarged the town. Seen from above these towns "resemble organic concretions: forests or a coral reef or insects' nests. A Moroccan courtyard house is certainly the result of conscious planning, but when we see 500 of them bundled in a texture of alleys and streets we are tempted to believe that they grew together naturally, like the population of individual organisms, all similar but different, like members of the same species" (Davies 2011).

As well described by Ragette "the similarity to a living organism is obvious: like a system of blood vessels, street branch out from a central square (*maidan*) to become lanes and narrow alleys and finally come to a dead end. As each house has a central space, which unites the family by being a neutral zone of contact, groups of houses have some common space and are organized by clan relationship, or along ethnic or religious lines, creating a quarter (*hara*)" (Ragette 2003).

This aggregation creates an intricate system where spaces are semi-public, public, or private. Particularly, in the semi-public zone, outsiders are recognized, so it is also a system of control. Through streets are public and are meant to connect among quarters. The Arabic town has no formal central square as in the European city center. The largest open space is the yard (*sahn*) of the Friday Mosque which serves

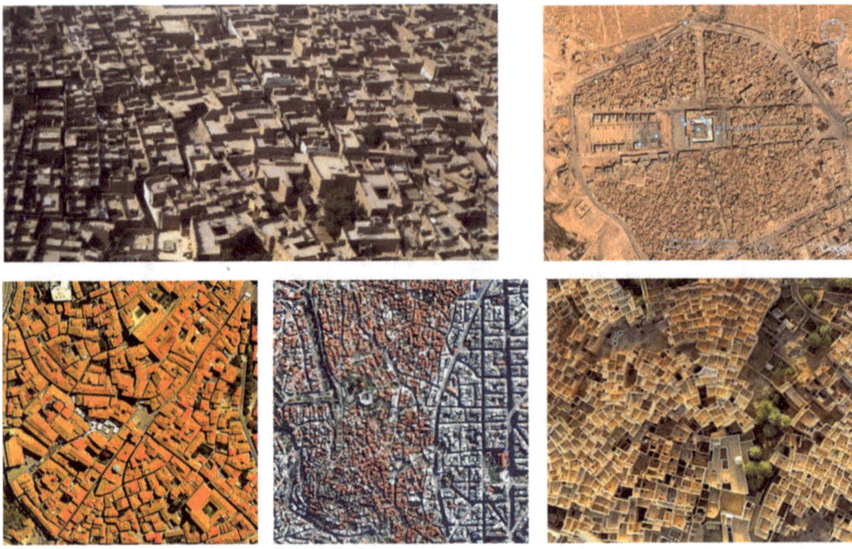

Fig. 10.15 View of Mosul city and Najaf city in Iraq (*on the top*), compared with three different examples of aggregation by agglutination in the area of the Mediterranean Sea: Gadamesh—Libia, Cortona—Italy, and Salonicco—Greece

several quarters. "Trading takes place in the narrow streets, often protected by movable canvas shades or permanent roofing. Different trades occupy specialized streets and sounds of wood or metalworking, or smells of spice or tanneries guide the visitor" (Al-Azzawi 1996) (Figs. 10.15 and 10.16).

Flexibility and Customization Over Time

The project keyword for developing a prototype for Mosul reconstruction is sustainable realism. A prototype is very convenient, as it can be repeated, but it has some problems: it may lack variety and the possibility of customization. To grant residents' customization and visual variety while keeping the standardization advantages, the stratagem is simple and efficient: a non-finished layout.

Hence the proposal to have two separate "layers": the general "finished" layout responding to the need of governance and planning, and the unfinished one, customizable in a later time by the residents. It is based on the use of a simple modular unit (3.8 × 7.6 m) articulated in two types: the finished one and the structural one. The latter is a simple grid to be finished in the future by the occupants or left as it is: a simple open loggia. These units are easy to build with local materials and know-how,

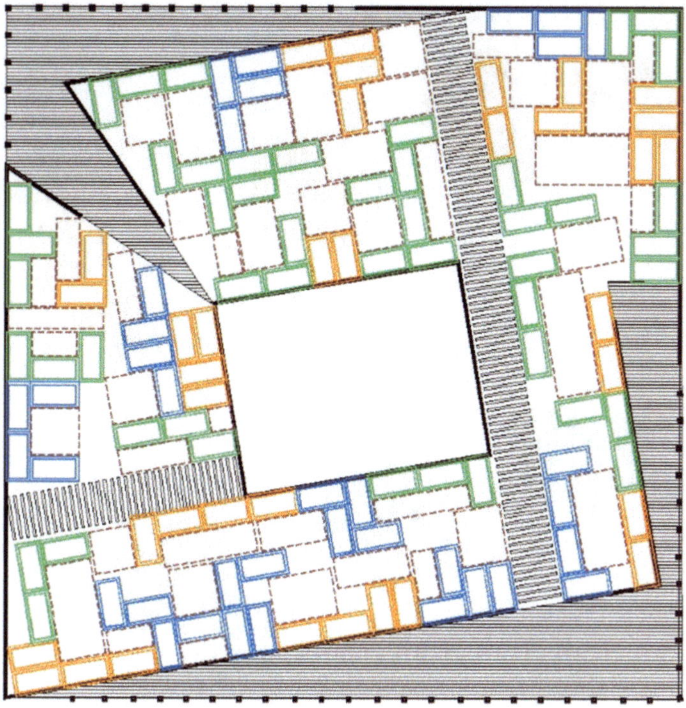

Fig. 10.16 The project plan of the new formal urban expansion zones for housing development

and allow numerous possible aggregations granting wide variety of types, sizes, and esthetics. In this project there are three dimensions: small, medium, and large dwellings. Also, there are additional single floor small apartments meant for people with special needs, the elderly and the impaired (Fig. 10.17).

Bioclimatic Strategies, Climate and Solar Study, and Energy Performance Simulation

The response to sustainable solution's needs is articulated into three domains: social, environmental, and economic. The costs involved are limited since construction is easily accomplished with local materials and manpower.

Like many cities in Arab countries, Mosul shares the same unfriendly environment. It's located in a desert area, although such arid condition is mitigated by the presence of the Tigris river. The summers are sweltering, arid, and clear, and the winters are mild and partly cloudy.

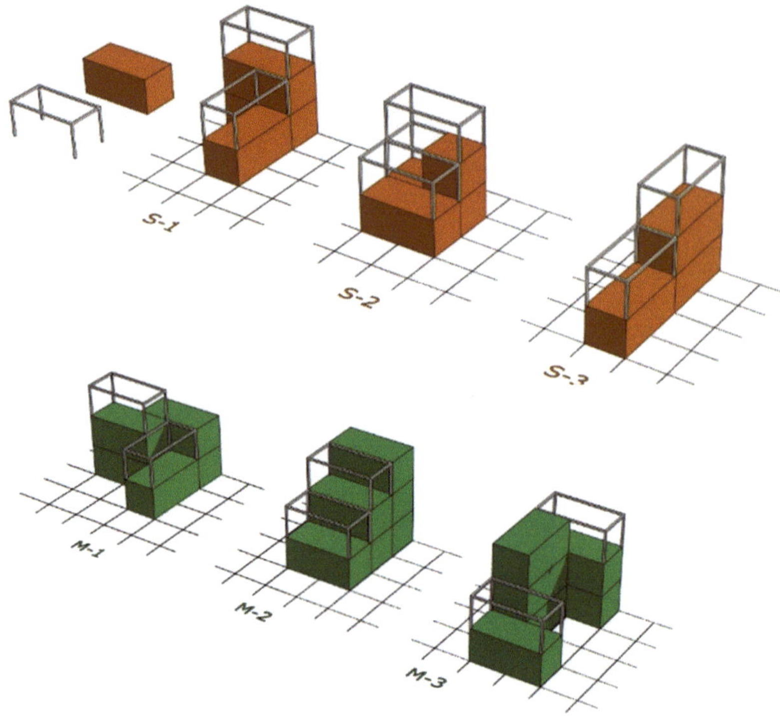

Fig. 10.17 The different solution of units combination and possible aggregation

Summers are long and torrid, with an average daily high temperature above 37 °C (July is the hottest month) and winters are mild, with an average daily high temperature in the range of 18 °C (January is the coldest month). The highest temperature is usually in July (43 °C) and the lowest in January (4 °C). The average amount of precipitation for the year is below 500 mm, typical of arid climate. In a nutshell, the area is characterized by high temperatures, scarcity of precipitations, and a high amount of insulation (3500 ca hours per year). Shamal winds from the north raise dust and sandstorms.

A solar simulation study for a single house run in an area of modern expansion is a very effective tool for checking to what extent summer insulation can cause distress in the interior and consequently the types of interventions to consider for adequate living conditions (Fig. 10.18).

The solar study intends to analyze the different scenarios in different lighting conditions and evaluate if the design choices taken look satisfactory. With the contribution of Revit software, it is possible to have accurate evaluations of the shadows cast at different times of specific standard test days: winter and summer solstices, respectively, 21st of December and 21st of June.

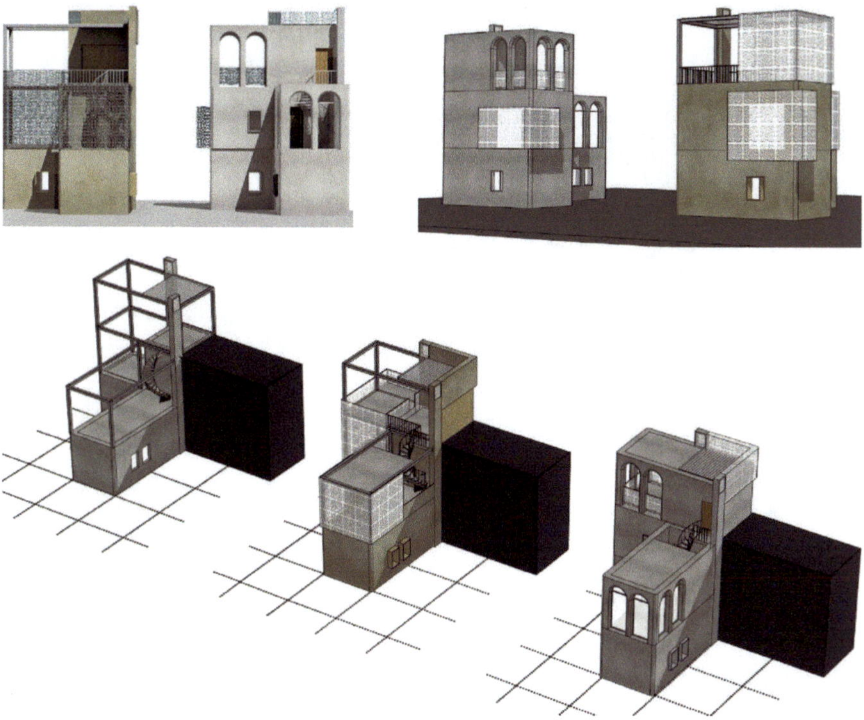

Fig. 10.18 View of the different aggregation of house units

As expected, in winter, the sun is very low, shadows are very long and the sun enters the building at all times of the day, which is desirable. On the other hand, in summer the sun of late afternoon is going to enter the building thoroughly, in the absence of proper solar shields. Similarly, this happens also in the early morning, but seems to be a lesser problem. In the afternoon, on the contrary, the heat accumulated during the day makes the exposition to the sun more problematic. It is required to protect the building with appropriate sun shading systems on the west side (Fig. 10.19).

The environmental aspect of the project is granted by the use of systems and methods, both technological or derived from the local tradition, all aiming at mitigating the effect of heat and high degree of insulation: wind scoops, photovoltaic and solar panels on roofs, water storage and harvesting, and strong thermal mass for walls to prevent loss of thermal energy (the wall unit is 40–45 cm, and consists in a stratigraphy of masonry with 20 cm of wood fiber layer). Also the presence of local sun shading systems and perforated panels help mitigating the scorching sun effect.

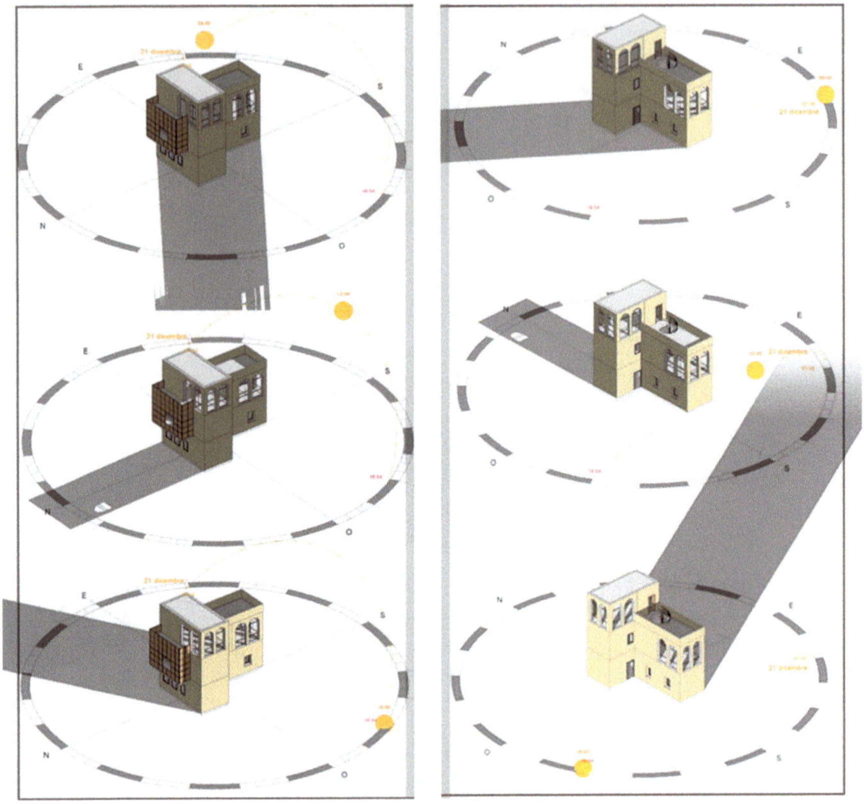

Fig. 10.19 Solar paths analysis, daylighting and shading in the north and south facades during the winter time (December)

The Design Solutions of Apartment Compound

The general layout of the apartments is planned on a square, a grid adapted on the standard Mosulian plots (around 40 m) and can be repeated in arrays (the project shows a double square). The aggregation is very free, can be apparently random, as shown in the project, to create a feeling of uniqueness and overcome the flatness of a grid, or it can be more regular (Fig. 10.20).

Apartment Closed Compound. The Ring

There are three kinds of spaces: the public (the ring), the semi-private (the alleys and paths in the dwellings squares), and the private (the blind alleys leading to the yards next to the dwellings). The squares are surrounded by a ring, whose form is

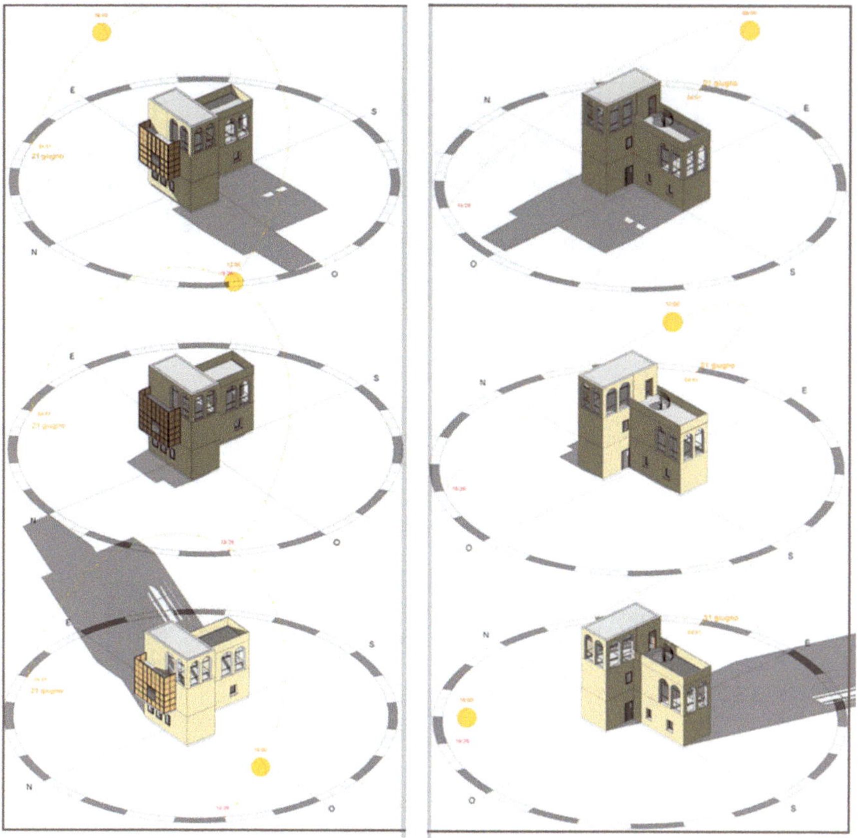

Fig. 10.20 Solar paths analysis, daylighting and shading in the north and south facades during the summer time (June)

easily adaptable to the inclination (if any) of the streets. This is a transitional public space. It doesn't have a specific connotation yet. The ring function will be defined by residents: local market, or even shops, safe playground for children, gardens, orchard, promenade. Its semi-open character will allow protection and privacy to the dwellings inside yet keeping a feeling of openness.

Apartment Open Compound. The Deconstruction of the Ring

The second phase of the project calls for a more open layout, aligned with the characteristics of fluidity typical of the traditional Mosulian urban texture, where areas are overlapping and boundaries are blurred. A sort of "liquid architecture," as said by Zygmunt (2007). The ring is deconstructed and the resulting shards dislocated in

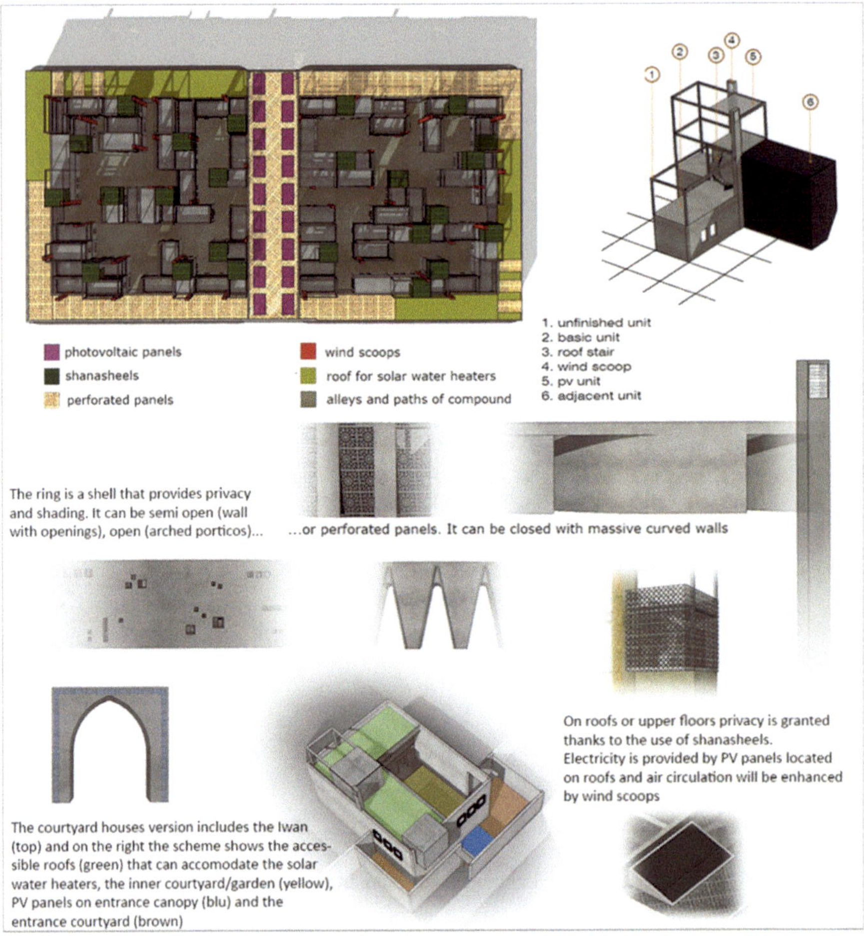

Fig. 10.21 The selection of design solution for apartment compound

different areas, still keeping their character of urban communal containers where social functions other than living are located, such as small local offices or shops, kindergarten, in short social functions at the scale of the neighborhood (Figs. 10.21 and 10.22).

The Courtyard Houses

The module can be used for the apartment's lots where high density is required and/or for courtyard house compounds. The cluster organization allows more openings and *mashrabiyas* on the upper floors. There are three different sizes for the plots and

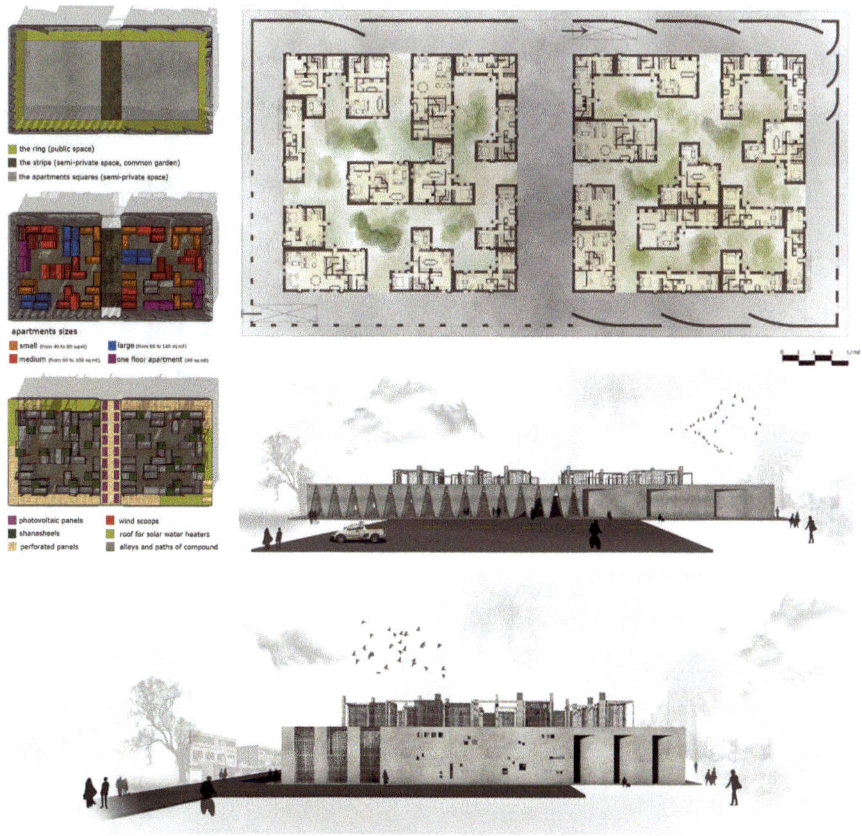

Fig. 10.22 Architectural and technological solutions of apartment compound and residential district. Plan and external view

there are entrance areas enclosed by walls of different heights to grant maximum privacy. These walls at the entrances (having no insulation or structural requirements) are made with the rubble coming from the destroyed areas of the city to ensure memory and continuity with the recent history. Following the local tradition of privacy, the houses show a marked introvert character, are closed to the outside, and the windows on the first floor are located exclusively facing the interior courtyard (Figs. 10.23, 10.24, 10.25, and 10.26).

Towards an Open Layout: The Deconstruction of the Ring

The last version of the project idea proposes the deconstruction of the ring, generating a network of green public space as green corridors to improve the microclimate indoor comfort.

Fig. 10.23 Render of the general view of apartment compound and residential district project in Mosul

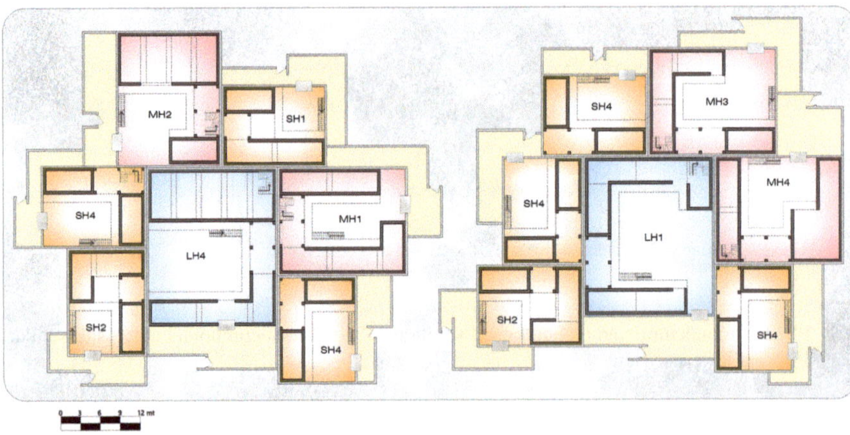

Fig. 10.24 Different combination of courtyard apartments and general plan

Fig. 10.25 Street view of courtyard apartments

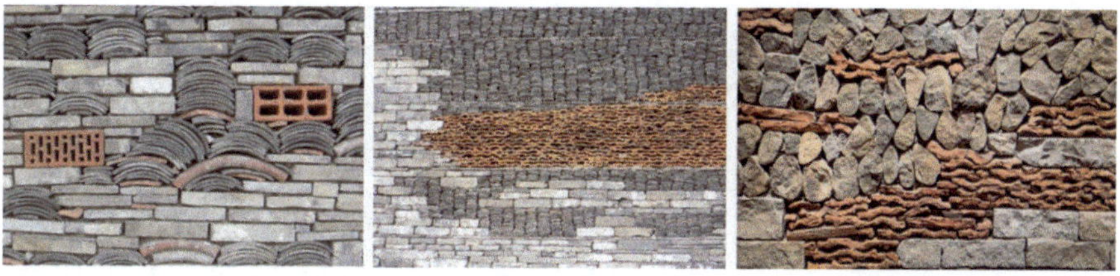

Fig. 10.26 Details of the external wall made with the rubble coming from the destroyed areas of the city to ensure memory and continuity with the recent history. Reference: Wang Shu—Ningbo History Museum

Fig. 10.27 Rendering and general view of the courtyard houses solution

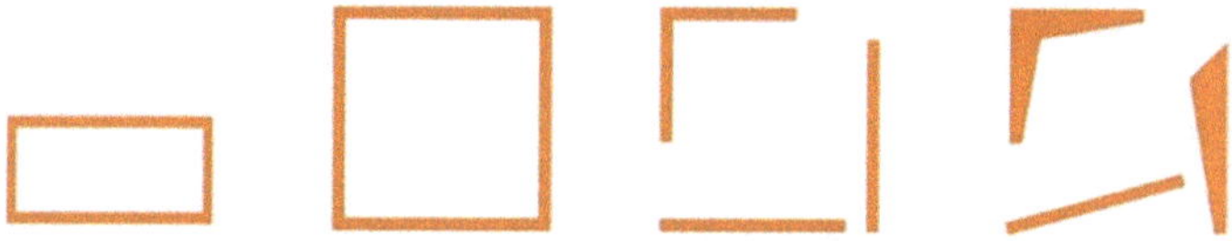

Fig. 10.28 The open layout of *ring* fragmentation. Plan and general view

The objective is to enlarge the ring, fragmenting the main walls, creating different rules and priority among apartments and public spaces, in order to reconnect the surrounding neighborhoods with the new courtyards and gardens, as well as responding to the need to create a space dedicated to the market (Trombadore 2016) (Figs. 10.27, 10.28, 10.29, and 10.30).

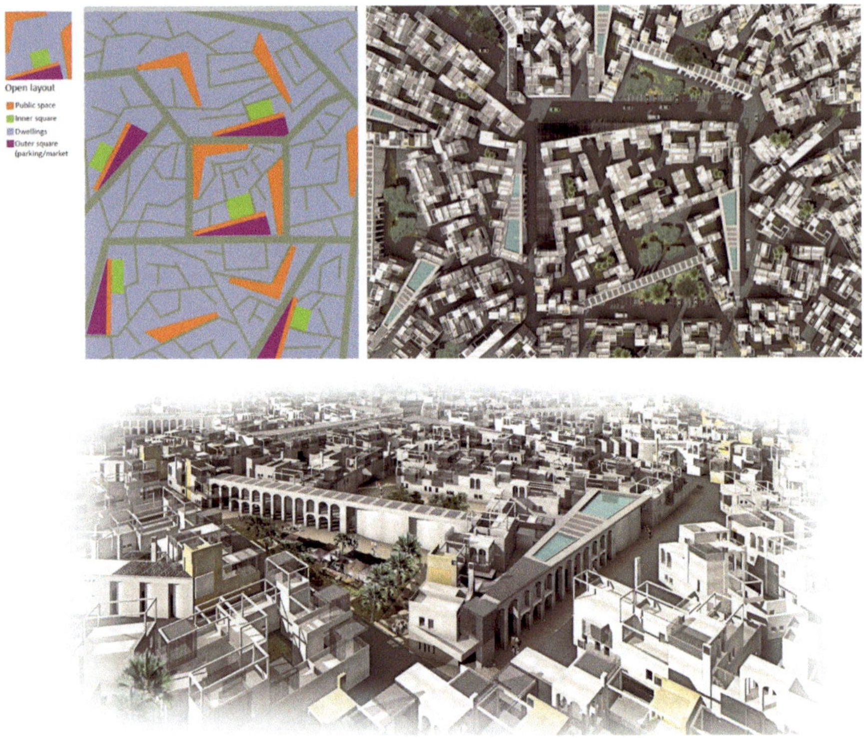

Fig. 10.29 Render of the last version of the apartments with *ring* fragmentation

Fig. 10.30 Street view of the *ring*. Details of the shaded effect walls

References

Al-Azzawi S (1996) Seasonal impact of climate on the pattern of urban family life. Renew Energy 8:283–288

Davies C (2011) Thinking about architecture. Laurence King, London

Ragette F (2003) Traditional domestic architecture of the Arab Region. Edition Axel Menges, Stuttgart

The Rifat Chadirji Prize, An open international prize focusing on ideas responding to local challenges in Iraq Official Announcement of the Competition. www.rifatchadirji.com

Trombadore A (2016) *Mediterranean Smart Cities. Innovazione tecnologica ed ecoefficienza nella gestione dei processi di trasformazione urbana* Altralinea - Firenze

Warren J, Fethi I (1982) Traditional houses in Baghdad. Coach Pub. House, Horsham

Zygmunt B (2007) Modus vivendi. Inferno e utopia del mondo liquido. Laterza, Roma-Bari

Chapter 11
Hausa Traditional Architecture

Amina Batagarawa and Rukayyatu Bashiru Tukur

Introduction

Generally, in Contemporary Nigerian Architecture, manufactured materials like concrete, sandcrete, aluminium roofing and frames, glazing and particle boards are popular. Yet, there remains a significant influence of Traditional Hausa Architecture (THA) in contemporary buildings, especially in residential buildings.

The materials used in THA are rarely company manufactured, processed or fabricated. THA utilizes readily available materials in the immediate vicinity such as soil/earth, timber, reeds, grass and stone (Dmochowski, 1990).

Use of stones/rock for dwellings in THA began with the habitation of naturally occurring caves. Stones are the most utilized materials in building foundations, in layers of un-coursed rubble, usually used for its ability to reduce the amount of the moisture movement from the foundation upwards, walls and fences between the outermost houses of the compound (Denyer, 1979).

Earth for building is called *birji* in a large part of Hausa land. Earth construction that is plastered and properly covered with overhanging roofs has proven to be structurally firm, environmentally sound and could exist for years as long as the periodic maintenance is adhered to in simple building construction. Houses, perimeter walls and roofs are built of mud.

Tubali is the Hausa name referring to mortar. Whereas the most generally used plaster for covering walls is plaster made from *makuba* or the acacia tree *gabaruwa*. Another type of plaster is called *laso*.

The timbers used in construction are commonly called *azara* beams. They serve as wooden reinforcement to strengthen the structures of the wall and pillars, and to make frames constructions, beams, brackets and corbels as elements for carrying flat and domed roofs.

A. Batagarawa (✉) · R. B. Tukur
Department of Architecture, Faculty of Environmental Design, Ahmadu Bello University, Zaria, Nigeria

A. Sayigh (ed.), *Sustainable Vernacular Architecture*, Innovative Renewable Energy, https://doi.org/10.1007/978-3-030-06185-2_11

Grass thatch is one of the oldest of building materials known. It is used to reinforce earth construction and for roofing in the form of thatch or straw possessing tensile strength while the earth has compressive strength. Ropes made from reed of great strength are produced from twisted bark from the roots of trees like *dokwora* (*Acacia senegal*) and *kista* and *geza* (*Combretum micranthum*) are used for making walls, roofs of buildings and mats (*lagara*) which surround whole establishments of up to 15 ft high. In the villages, the roofs may be of grass and boundary walls of matting or corn stalks.

Knowledge of iron-working was well developed early in Hausa culture. The most impressive Hausa iron products are the famous city gates, made of long strips of hammered metal joined together on sturdy frames and set on pivots instead of hinges. Otherwise, iron was used for complementary items mostly for nails with decoratively worked iron heads usually applied to the rails of the outer doors of houses.

THA construction in roofs, walls, columns, slab, beams, foundations, doors, windows, plastering and process of renovations is covered in the following sections.

Roofs

THA in suburban areas has simple thatch roofs over circular planes, while in urban areas has mud roofs on polygonal walls. The climate, human physiology and geography led to the development of curvilinear, conical and mud-roofed structures in the THA. The wide use of flat or vaulted mud roofs in Hausa land evolved due to urban fire prevention, minimizing the use of thatched roofs within the urban areas. On the other hand, mud roofs helped maintain temperature equilibrium in the extreme temperature differences between days and nights, as well as buffer between interior and exterior environments.

The introduction and acceptance of new and more conventional building materials such as cement and corrugated iron sheet further minimize the use of traditional roofs. In addition, since it is very difficult to roof circular structures with rectangular iron sheets, the basic circular plan gradually transformed into a rectangular one.

The simplest way of roofing large rectangular interiors was to reduce the span by placing brackets on one or both sides of the room. The brackets are flat-topped projections made from one or more layers of rods set in rows. Jutting out horizontally from the walls, the brackets supported *tauyi*, wooden beams running parallel to the walls.

Another method of roofing wider interiors was to lay diagonally fitted triangles of the azara beam in the four corners of the room. On top of the triangles, parallel to the two opposite walls, two further layers of azaras were placed, tauyi fashion, reducing the span of the top crosswise azara joists to the desired length. This method was used in both rectangular and square interiors.

The different levels of the individual layers and especially the rhythmical display of azaras, set in four directions, transformed the flat roof into a three-dimensional composition; in consequence, the builders increased the number of superimposed layers above purely structural needs. They created in this way a decorative pattern culminating in the centre of the ceiling, the apex of the whole, which were several thicknesses of azara above the first, bottom layer. This structure provided a support for the slightly domed exterior surface of the roof, through which the rainwater could run down.

Even larger interiors required internal pillars (*alamudi, ginshiki*) to support the beams of the roof. The capitals of these pillars were topped with brackets that were not cantilevered like those set into the thickness of the wall, but, being made of longer rods, projected over both sides of the pillar, thus balancing the weight of the two opposite beams.

There are many ways of waterproofing flat roofs. The most frequently used material for joists was rods of *azara*, which provided both an attractively textured ceiling (*rufi*) and a base for the heat-insulating and rain-proofing slab of the roof immediately over the *azaras* was an overlay of *zana* mats, plaited from zana grass; or a course of neatly arranged cornstalks or straw; or, when there was the danger of termites, of twigs of native plants which are very strong, as well as being resistant to termites. This overlay prevented the plaster, which constituted the next level of the slab, from leaking through the gaps between the underlying *azaras*.

It was important to provide the roof surface with a slope to carry the rainwater towards the parapet. Large flat roofs were occasionally divided by a grid of low internal parapets into rectangular panels: a central parapet formed the ridge, and each panel sloped gently towards the outside.

The structure of domed roofs called tuluwa combines a number of elements found in flat roofs. The simplest tuluwa covered a square or almost square interior and was supported by dauringuga. The crown of the walls at the four corners of the room was covered with triangles of diagonally laid azaras (tauyi). On top of these were laid beams, also called tauyi, although they ran not diagonally but parallel to the four. The apex of the crossed bakawas plastered and next covered with four triangular layers of azara. The layers were necessarily triangular, and were set parallel to the walls of the room. Thus, the upper support for the final layer of the dome's joists was created.

Tuluwa were rainproofed by constructing the dome with the outer edges of the tauyi inserted into the top layers of the bakuna. The joists of the rufi were set radially from the apex, in the same way as in two-bakuna type. The waterproofing of the dome was not much different from that of flat roofs. Instead of flat and stiff zana mats, much more flexible asabari mats were used, made chiefly of strong grass (tsaure). The domes provide better water protection as the rain water runs off the dome surface quicker.

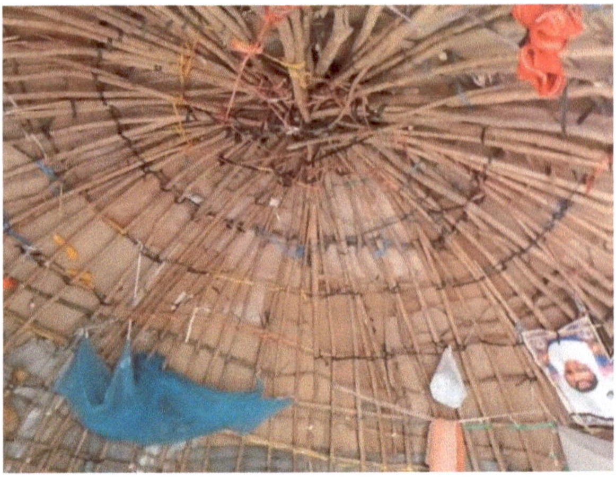

Nowadays in northern Nigeria, metal roofing sheets are used such as zinc and aluminium. The main purpose of the roof as stated is to enclose the house and protect it from sun, rain, wind and dust. The main basic functions of a roof according to are:

Weather resistance which is provided by the covering of the roof. For the covering to be properly effective, the framework of the roof provides a proper pitch or in other words the roof should have a proper slope. All of these factors contribute to a proper weather resistance.

Strength and stability is provided by a proper roof structure. It depends on the materials used and also upon the methods and techniques used to put up the structure together. The roof is able to resist all types of environmental loads of any magnitude due to rain, snow and wind.

Thermal insulation for people living in the hot-humid climatic conditions of the north hugely considers this factor because the regulation of heat is very important for the comfort of the inhabitants of the house especially in rainy seasons. The roofs also provide sound insulation up to a certain extent.

For the interior, different types of finishes are used and are usually selected based on cost and their physical properties. The three most common ceilings are the wooden ceiling, poly vinyl chloride (PVC) and plaster of Paris (POP).

Walls

The construction of a *tubali* wall involves the moulding of pear-shaped sun-dried mud bricks made from birji. Moulding the egg-shaped units of tubali wall bricks involves mixing the laterite soil with water and when properly dried, the bricks are then laid in regular courses with egg points facing upward. Plastering gives the courses a monolithic appearance, and mortar is placed by the builder mainly by throwing the lump of mortar served to him from the ground.

When the cost of the building is not the main consideration, and when the walls are to be high, particularly in two-storeyed buildings, horizontal bracings made from *azara* grids are set within the walls. Each bracing consisted of two layers of timber: a bottom one of short timbers set transversely across the wall and an upper one of longer rods laid longitudinally. These grids are usually fixed at about 1 m and again at 2 m above the ground.

The walls decrease in thickness towards the top through the reduction, one by one, of the number of tubali used in their cross section. When the height of a wall is so great that the tapering reduced its thickness too much, the uppermost part of the wall is often protruded towards the interior on brackets set into its thickness or the walls are strengthened on the outside with a slightly projecting buttress.

These days' walls are constructed with various types of materials ranging from blocks, bricks, concrete and stones. The most common walls erected in northern Nigeria are the sandcrete hollow blocks.

Sandcrete walls provide sufficient sound insulation. These walls are plastered and finished with a paint of choices in the residential houses ranging from different materials and colours. Whereas in the traditional homes, the walls are finished with modernized designs of motifs whereby various symbols are illustrated on the walls.

THA facade decorations include the horns of mud (*Zanko*) shown along the parapet of their buildings, which gives the Hausa structure a lighter and picturesque appearance.

Many palaces are bright and colourful, including intricate engraving or elaborate symbols designed into the façade. Engraving is the practice of incising a design onto a hard, usually flat surface, cutting grooves into it. The wall engravings are designed by traditional builders, professional artisans and highly experienced hand engravers who are able to draw out minimal outlines directly on the wall surface just prior to engraving.

Doorways

Buildings are formed with small door space with the intention to eliminate the day-time hot, dry and dusty air, bright daylight and entry of cold night air prevalent in Hausa land.

Doorways in THA houses are of two kinds: outer doorways enclosed by wood or iron and hung on pivots or hinges, and inner doorways are doorless and occasionally screened with grass-plaited curtains. Outer doorways are generally rectangular in shape, with a horizontal lintel made of a wooden beam supported at each end. The length of the flat lintel beams is restricted by their limited resistance to bending.

All these were fixed together with nails (*kusa*), which had wide, frequently decoratively shaped heads. Occasionally, the outer surface of a door was covered with horse's hide, or narrow strips of iron.

A typical doorless entrance consists of two parts. The lower part was rectangular, and this was covered by a semicircular arch. When the floor of the room is not raised above ground level, the lower part of the doorway was barred with a threshold (*dan-garama*) which prevented the entry of rainwater. The *gemu* protects the doorway from the rain by projecting some distance from the wall above the opening as well as acting as a decorative element in the façade by emphasizing the entrance to the building.

Metallic doors are used in THA at the main entrance to the *zaure*, a lobby, which can be secured at night. After the *zaure*, an open arched doorway is usually provided which is decorated with motifs. Then, another recess/courtyard is provided before which the second *zaure* is provided with another metallic door which when accessed will lead you to the main compartment of the house.

Metallic doors are also used in the exterior when a room is provided in the exterior of the house thereby adding security to the room.

Modern doors are mostly paneled doors or solid wooden doors. These solid cores are generally particle board or laminated timber, which can be faced with sheets of plywood or Medium Density Fibre to provide a flat surface that can be painted, veneered or faced in laminate. The edges may be edged with hardwood or laminates.

While also providing modern features like fire resistance, flush doors, which tend to be cheaper than paneled, are more modern in appearance.

Windows

The windows in THA are simple openings almost always set in the uppermost part of walls on the lee side of the building, which were less affected by driving rain. The location and small size of the windows help in curtailing the incursion of dust and flies. Most often, the shape of windows is rectangular, although in more elaborate buildings they are often topped with an arch, *kandame*, or with a triangle. They can be hinged to open outward and angled in order to direct breezes into the building.

More modern windows usually consist of four elements which include the main window, insect protection, burglar proof and some form of interior or exterior shading device. Interior shading devices usually consist of curtains or blinds, adding an aesthetic element.

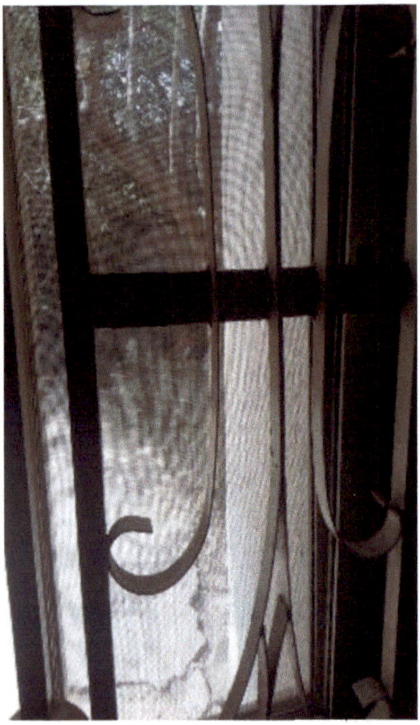

Floors

Floors in THA are usually laid by women. A surface of laterite about 3 cm thick is laid on top of a layer of beaten earth. This laterite, *dabe*, is sprinkled with gravel, then watered and beaten with a tool called *madabi*, a thick, slightly curved branch, flattened a little on the outer side. When it is dry, the floor is further hardened with *makuba.*

Today, tiles play an important role in interior design of Hausa homes, setting the look and feel of the house. Tiles are used to create the desired ambience in various colours and size combinations. Sometimes, combination of tiles and rugs gives texture to the room, while in other cases the floor is made of concrete slab then a carpet/mat is laid all over the room on top of the slab.

Conclusion

The Hausa Housing system is well developed over many years with a very well-established heritage. All the materials are locally produced. Several examples were included to illustrate the construction. The roofing is thatched to allow air movement and space to insulate heat and cold sever climate from disrupting the indoor climate. The walls are made of bricks from birji. Moulding the egg-shaped units of tubali wall bricks involves mixing the laterite soil with water and when properly dried, the bricks are then laid in regular courses with egg points facing upward.

Doorways in THA houses are of two kinds: outer doorways enclosed by wood or iron and hung on pivots or hinges, and inner doorways are frames only and occasionally screened with grass-plaited curtains. Outer doorways are generally rectangular in shape, with a horizontal lintel made of a wooden beam supported at each end. The length of the flat lintel beams is restricted by their limited resistance to bending. Most often, the shape of windows is rectangular, although in more elaborate buildings they are often topped with an arch, *kandame*. The floor is compact dirt so it can be used sometimes in large gathering occasions.

Acknowledgment Photo Credits: Muhammad Musa Danraka, Amina Batagarawa.

Reference

Denyer S (1979) African Traditional Architecture: An Historical and Geographical Perspective. Holmes & Meier Pub.

Dmochowski, ZR (1990). An Introduction to Nigerian Architecture: Northern Nigeria. Michigan: Ethnographica.

Chapter 12
The Intangible Resources of Vernacular Architecture for the Development of a Green and Circular Economy

Marco Sala, Antonella Trombadore, and Laura Fantacci

The Fragility of Vernacular Architecture as a Resource for a Sustainable and Responsible Tourism

Through the years, a lot of inner small towns, especially in Italy and south Mediterranean countries, had suffered heavy phenomena of exodus in order to supply the lack of development opportunities and possibilities. This had caused a deep demographic impoverishment, with serious negative consequences on the public building heritage and economic-cultural activities of historical settlements. This depletion, together with the inadequacy of skills and services management, contributes to reduce further the touristic "attractiveness" of these areas, so that instead have great potentialities thanks to their "uniqueness", belonging to the extremely beautiful surrounding landscape and the artistic and cultural value of their small villages.

Despite the serious problems described above, recently it has been assisted to a general tendency to the growing of touristic request in the inner Mediterranean areas, as well as a crescent research of high quality services (especially for the quality of life and wellness during the travel). These aspects have led to a major awareness of what the traveller wants to find in the place that has chosen, making the vacation a real "experience" based on particularity. Nowadays people (thanks to the opportunity provided by the internet and media) can be aware also of the importance of ecosystems preservation, natural and cultural resources and of the possibility of having a real 360° involvement.

The key action for the renovation of the small realities in the inner villages of Mediterranean areas, as a consequence the revitalisation of small village buildings countries, has to be found in the solutions given by the vernacular architecture, which is the real way to find sustainable and responsible answers to the abandonment and death of these places (Fig. 12.1).

M. Sala (✉) · A. Trombadore · L. Fantacci
"ABITA", Università degli Studi di Firenze, Florence, Italy

A. Sayigh (ed.), *Sustainable Vernacular Architecture*, Innovative Renewable Energy, https://doi.org/10.1007/978-3-030-06185-2_12

New Scenarios of Environmental and Social Innovation

Together with the implementation of local activities and the encouragement of a new way for the revitalisation of small realities, there is the necessity of individuate and develop new scenarios of innovation to support communities in inner settlements and to promote the competitiveness of small and medium companies; the vocation on territorial innovation raises through material and immaterial interconnectivity, linking together socio-economic development implications and local attractiveness. There is also the necessity of aim to the accessibility and safety, using environmental and ecological aspects as leverage, to reach a complete integration between places and infrastructures, in a developing model in which the man (communities) can be less "costumer" and more "user". This model has to be the more interactive and integrated as possible, in order to individuate new viewpoints of operability in a well-rounded perspective.

In this sense the importance of local architecture is well seen; in a world of hyper-technological solutions for every aspect of life, there's the need of coming back to simplicity. As well known at all levels, the architecture of the past in every single part of the earth teaches us optimal solution for the local problem solving. In every specific context at every requirement corresponds a real approach that leads to a particular situation? Also, in the name itself "vernacular" (from the Latin "Vernaculus" which means domestic, familiar), term coined for the first time in 1964 by Bernard Rudofsky in his "Architecture without architects" (Rudofsky 1964), is clearly evident in the message of this type of method: all the problems regarding the primary necessity of most communities (first of all the need of repairing from weathering, feeding, sleeping, living together) are strictly related to the place where it grows. It can be seen as an answer to all the social and environmental

Fig. 12.1 Typical small village in inner Mediterranean areas

requests. In this sense we need to be aware of a simple but not trivial assumption: the vernacular architecture, just because of its adaptive character, deeply linked to the construction (in its most imminent meaning), possesses a strictly fragile character but not in the weak sense (as the term might suggest) but rather resilient, adaptive and sensitive to change. It corresponds to a vision of the world (proper to every human group), declined according to particular needs. These characteristics suggest an idea of architecture (understood as technique, constructive wisdom), absolutely strong and capable of giving a strong message of identity. In contrast to the standardising tendency of the modern vision of the world (the great metropolises are an example), we try to give new life to an idea of individuality based on the enhancement of uniqueness; uniqueness of the small realities of which the Mediterranean area is full, which, however, are likely to remain crushed forever. In this sense, this idea can be extended to all fields of knowledge, in order to create a model (continuously evolving precisely because of its empirical character) economically and socially sustainable to revitalise the most depressed areas of the hinterland. This model can trigger virtuous processes that could be repeated in subsequent experiences, so as to create a real network between the various realities, focusing attention on the theme of tourist enjoyment (Fig. 12.2).

New Hypothesis of Renovation Processes for the Inland Areas of Mediterranean Countries

Among the positive aspects that involve thinking about the territory as an environmental system capable of resilience, one of the main is the reference to an organic vision that allows integration, enhance and direct towards common objectives, solutions and interventions that, alone, risk to create discontinuity. In a dialogue with the different stakeholders, we share the idea of a "Eco sustainable territorial development model", aiming to combine architectural features solutions to new productive, tourist and commercial functions.

It will be important to identify the possible drivers of revitalisation and imagine the profile of communities capable of triggering the proliferation of a network of social relations (stimulating new economic connection with the local productive fabric) and at the end to redevelop its existing building heritage; these are to be considered the first steps to start a regeneration of the anthropic fabric and a sustainable use of the territory.

The VIVIMED is transnational cooperation project financed by Interreg Programme It-Fr. The activities focus on the enhancement of common, tangible and intangible assets, both in this regard to their natural potential (sun, wind, sea, water, land, habitats, plants, animals, ecosystems and landscapes) and in terms of their cultural heritage, which over the centuries has strengthened their uniqueness and the territorial identity. This is the basis for the dissemination of a competitive model, for the development of a network of eco-green infrastructure and sustainable use, a

Fig. 12.2 Two different approaches of tourism development. Handcraft marketing (*on the top*) and albergo diffuso (*bottom*)

model that can be exported to other contexts present in the Italian and Mediterranean hinterland, where naturalistic potentialities and resources linked to the authenticity of the places are dominant, but not adequately valued in the logic of a green-circular economy (and/or conscious tourism) (Fig. 12.3).

Fig. 12.3 The *logo* and web page of Vivimed Project

The Experience of VIVIMED Project

The research on the method can be focused on a strongly transversal sector: the tourism/fruition sector.

A mapping of the tourism SMEs of the chosen territories, with the profile of the activities and services, and the typology of the receptive structures (farmhouses, B & Bs, widespread hotels in the historical villages), will be elaborated. To elaborate the scenarios of innovation, the sectors potentially complementary to the tourism sector, often not considered, as agricultural activities and related historical building heritage (crushers, mills, chestnut dryers, etc.), floriculture activities and forestry, fish farming in the vicinity of rivers, industries or their historical branches related to the territory (educational path "Via della carta" [the paper road] developed with the paper industry in Lucca), craft enterprises (carpentry, glassworks, etc.) or activities completely extinct but of elevated archaeological-cultural interest (system of the ice houses, obsolete after the invention of the refrigerator), will also be included.

Innovative scenarios will be developed for the integrated and multi-theme tourist offer, according to the specificity of the hinterland territories, also in collaboration with regional trade associations and agro-food and social cooperatives and their consortia, through:

- The enhancement of the territorial resources for the development of an integrated and multi-theme tourism offer, flexible in the themes and methods of development by type of user (e.g. single, families, old people);
- The configurability of the type of involvement with the desired territory: experiential, social, naturalistic/sporting, cultural, historical, artistic tourism;
- The promotion of innovative forms of hospitality such as the widespread hotel, network of bike hotels, hospitality in parks (e.g. zero-impact accommodation facilities) and pathways of ecological brands;
- The enhancement of educational aspects as promoters of the environmental theme;
- The rethinking of the seasonality of tourist packages to ensure affluence throughout the year (e.g. promotion of events related to agricultural activities, e.g. collection of grapes, olives, chestnuts by combining landscape/cultural routes);
- The activation of partnerships with marinas and terminals of sea ports, cruise activities and tourist villages along the coast through rural tourism packages/excursions;
- The promotion of a model of infrastructure recovery (road network, paths, integrated village/territory/landscape systems, historic buildings) and construction of green structures to support tourism" (Fig. 12.4).

Through a process of cross fertilisation, it could be easier to reach a convergence of territorial development policies that increases the advantages of the alignment of the respective socio-economic systems. In this way, a virtuous process which will lead to the revitalisation of the territory could be triggered, but on the other hand favours an improvement in the competitiveness levels of local SMEs operating in the green-circular economy chain, focusing above all on technological innovation, multicultural inclusion, the pursuit of quality of life and the application of the concept of "Design for all" (people with reduced mobility, children, old people) as the common thread of the new vision of territorial and business development.

To implement an efficient and operational governance mode where all the stakeholders of the territories in question are actively represented and involved, the project envisages a process of public-private participation articulated within Living Labs that play the role of territorial catalysts.

Fig. 12.4 Different solutions of *Albergo diffuso*, as sustainable accommodation and responsible tourism

Fig. 12.5 Bagno Vignoni, amazing little village in Tuscany with *thermal bath* and *Albergo diffuso*

The Living Labs will involve the four main actors: public administration, local businesses, scientific researchers and the inhabitants of the territory (Fig. 12.5).

The Mediterranean hinterland has as common thread the low competitiveness of the commercial sector and the lack of inter-connections (the strong connotation of territorial "islands" predominates with respect to the potential of being an archipelago). The territories of the hinterland need to:

- increase the awareness of the territorial potential and skills of SMEs in the sector;
- diversify into the types of services and products to offer;
- specialise towards innovative, attractive and tailor-made activities and infrastructures for end users, both in terms of services and eco-friendly reception facilities;
- enhance in terms of the green-circular economy of the resources present in the territory, through the enhancement of the authenticity of the historical cultural heritage, the protection, management and sustainable accessibility to environmental resources.

Vernacular Architecture as Baseline of Bioclimatic Technology

Vernacular architecture is the definition of buildings designed on the base of local needs, with local materials, generally without formal design (*architecture without architects*) (Rudofsky 1964) and reflecting local traditions. At least originally, vernacular architecture did not use formally schooled architects, but relied on the design skills and tradition of local builders.

The tradition of this kind of architecture developed different stile in different countries and culture, is so important for the identity of communities that many administrations impose as rule for new buildings and many professional architects work in this style even today, to preserve the urban landscape and the homogeneous vision of the town. This approach is similar in countries very different, all over the world, but in Italy is so radicated because the strong value for our history and its architectural expression.

We have to admit that building materials, construction tools and know-how did not change a lot in the past, while architects and craftsmen were used to share the same knowledge how to rise buildings wand in the meantime the society identified himself in this architectural models produced and only slightly modified during centuries. Only with the beginning of the nineteenth century started a rapid modification of this scenario, with the availability of new building materials, the establishment of scientific competencies, the enlargement and recently the globalisation of the market and the rapid spread of different cultures through the world provided by new communication tools.

This new scenario deeply modified design approaches and building activities, and architects and engineers have at their disposal a large database of available technologies, building materials and energy choices that allow them to overcome almost any problems, and feel that only economic issues and aesthetic inspiration are the criteria to achieve success. But 20 years ago, a new general concern developed and soon became a great ethical approach, up to be the international paradigm that will mark our century: the need of sustainable future as a strategy to be applied in every activity aimed to modify and transform our world. The increase of building activities and the rapid transformation of our cities and villages need new rules and limits to proceed further, benchmarks and certification to evaluate the results and guidelines to address design activities (Fig. 12.6).

The holistic approach in architecture means the art of integrating different competencies, technologies and materials, to meet project requirements with a sustainable strategy. In this new approach the value and the lesson from vernacular architecture are evaluated as an important root to understand how to reach a climatic condition and how to use local materials (0 km materials) to reduce wasteful energy use. They are the set of strategies of design and construction, through which it seeks to achieve the realisation of a sustainable architecture, an architecture that using its formal configuration, technologies, components, materials and equipment tries to establish an optimal relationship with the surrounding environment so as to reduce energy consumption and provide the best comfort to the occupants, using as far as possible natural systems and reducing the use of mechanical systems of the building. This design approach can be applied in the transformation of the natural and built environment at different operational scales: from the size of regional and urban planning to town and district design, architectural concept, building detailed design, mechanical plant, up to the choice of building component and to sustainable building

Fig. 12.6 View of *Sassi of Matera* (on the left) compared with Shiban city in Yemen (on the right)

materials and even applied to rules setting and post realisation management (Figs. 12.7 and 12.8).

Definition of "bioclimatic technologies" can be misleading because it does not constitute an independent category of technologies applied to construction, but is

Fig. 12.7 *Shady street* in Sicily—Italy

the result of a more comprehensive, holistic approach that involves all issues related to sustainability, and in this meaning includes the reduction of impacts on the environment, the amount of energy contained in the components (embodied energy), the evaluation of their active role in the life of the building, until the final disposal (life cycle analysis).

However, there is a tendency to define some applications as "bioclimatic technologies" where the aims to pursue bioclimatic architecture are more relevant, and these technologies are mainly related to the building envelope, such as ventilated roof or façade, winter garden, solar shading systems, light chimneys, ventilation towers, and so on, but the concept of "bioclimatic technology" must always be compared and validated by the analysis of its appropriateness with respect to the local context in which it is used, i.e. the climatic context and the specific environmental design intervention and not as a model or a fashion design. For this reason, a bioclimatic technology that is properly applied in Berlin is certainly not very appropriate when applied in Dubai and vice versa. The choice of appropriate technology may rise from the needs of the specific building under the current climatic condition.

Fig. 12.8 View of the typical *trulli* in Alberobello—Italy

A possible articulation of bioclimatic technologies in line with the conditions of the environment could be made in relation to the prevalent objectives to be achieved, with regard to the lesson learned from vernacular architecture. The traditional urban morphology typical of Italian rural villages, with narrow pedestrian passages and street between buildings, tends to reduce energy losses, to break the wind, to provide shadow in summer while promoting stronger proximity relation amongst inhabitants (Fig. 12.9).

New and Traditional Technologies for Urban and Environmental Quality

Ecosystems and Mediterranean environmental assets represent a heritage on which to develop knowledge about living by stimulating the cooperation capacity on themes of a sustainable development for regional identity evolution. It is about the

Fig. 12.9 The typical *dammuso* in the island of Pantelleria and a semi-hypogean in the countryside of Marostica

distinctive architectural, climatic and cultural elements to which research centres, academia, companies and public administrations must aim to boost competitiveness on an international scale.

The small towns of the hinterland are still today chests full of stories, memories, and symbols, which, in recent years, have undergone a process of transformation, often uncontrolled, which has altered the profile and urban structure (in favour of big cities).

Today the construction sector is still one of the most important assets of the global economy. This has strong impact on the local businesses and the use of indigenous constructive materials. Activity of local SMEs (most of the companies in this sector are SMEs, with 97% of them having less than 20 employed).

For a long time now, the building renovation activity has surpassed the new construction in terms of investments amount, and it is foreseen that this trend will be improved, due to both the increased demand of interventions of ordinary and/or extraordinary maintenance on existing buildings (increasingly older and in need of care) and the mandatory request to adapt to energy consumption standards, particularly high for buildings built up to the 1990s.

The aim of reducing energy consumption passes necessarily through the renovation of existing buildings measures, and also because the current national and regional regulations in the field of new construction have adopted very strict limits for energy saving. The true goal is reachable only by acting on the revitalisation of existing buildings (performed not only on a single building level, but also on large urban areas) and on the smart management of the building envelope (Figs. 12.10 and 12.11).

Federico Butera, in his assay *Dalla Caverna alla casa ecologica: storia del comfort e dell'energia* (2014), clearly indicated what are "smart" properties of an ecological building in the Mediterranean, but which can also be repeated for small inland areas: "[…] an ecological and dynamic system, characterized by an extreme complexity, equipped with sensors, actuators, reaction […] The ecological house is a well-designed house, as a biological organism, with control systems, which is not necessarily indifferent to what happens outside. The ecological house of the future is, abusing of the word "smart", intelligent, able to react moment to moment. Attention then, do not dive into the past except to get ideas and principles" (Figs. 12.12 and 12.13).

The architectural traditions of the past, rich with potentialities and very interesting cultural contamination, represent an important heritage for the communities, and they are the bases to take as inspiration for developing new building components with high energy performance. It's necessary to "refer to the ancients" to recover and reprocess the traditional constructive principles and to increase the level

Fig. 12.10 General view of Santa Fiora, small village in Tuscany

Fig. 12.11 The amazing view of Riomaggiore, little town of *Cinque terrre*, in Liguria, very popular and touristic destination

Fig. 12.12 The real risk of inner small towns, especially after disasters, is to lose the community and become ghost cities

of environmental competitiveness, for the design of new constructions and redevelopment interventions.

Faced with the growth in demand for the constructions market concerning products and technologies able to guarantee effective energy performances through their integration into the building system, there's the need to optimise manufacture and construction processes and promote the development of "products" that are really able to lower the buildings energy consumption. Nowadays, this is of paramount

Fig. 12.13 Castello del Reschio. Luxury resort realised in an old country estate (village) in Umbria. One of the best examples of regeneration of vernacular architecture and rural heritage with a sustainable touristic approach

importance for companies that want to enter the market with innovative and truly effective proposals to address environmental issues.

What we need to do is try to create a winning synergy between the main stakeholders involved in the building process (from companies to professionals, from individual to public administrations), for the promotion and development of innovative processes and products from the point of view of the environmental and energy saving through the establishment of integrated platforms that, working in collaboration with the industrial sector, can achieve the definition of innovative components and architectural systems that can be globally applied. Compared to the northern European countries, Italy is lagging behind in the adoption and promotion of strong energy policies in the national area.

Some regions, such as Lombardy, Piedmont, Trentino Alto Adige, Tuscany, Emilia Romagna, Marche, Puglia and Sicily, have for a long time paid special attention to the issue of sustainability and energy production through renewable sources, as emerges clearly from the promotion of development plans, regional guidelines and awareness of public opinion on the choice of products and lifestyles oriented towards environmental compatibility (Fig. 12.14).

The solutions of "Mediterranean living" linked to traditional housing models gives rise to the aforementioned "vernacular architecture" where the use of simple technologies linked to the constructive traditions (especially in Italy; examples of this are quite frequent), passed down from generation to generation without solid scientific basis, is reinterpreted and repurposed to ensure high quality levels. These characteristics, common to many Mediterranean countries, both on the North industrialised coast and on the South and East coasts where agricultural activities are still prevalent, have not been mapped into the design, construction and even use of materials in the contemporary architecture (Fig. 12.15).

At the European level research centres, such as BRE, have tried to find answers in terms of environmental performances through the use of energy simulation software.

At a national level, the need for more and more performance products in the energy sector pushes the companies producing materials for the building envelope

Fig. 12.14 Examples of *Mediterranean living*: good practice of retrofitting actions of vernacular architectures. The valorisation of materials, daylight and integration with natural environmental

Fig. 12.15 Santa Fiora village in Tuscany. View of rehabilitation actions of courtyards, towers, walls and house buildings, according to the valorisation of environmental performance

to invest in research for understanding how to optimise products and solutions. The complexity of urban transformation processes therefore presents us today with an ethical challenge: building in the built environment through soft and light interventions that respect the local traditions and the identity of places, heave preserving the historical legacy for benefits of the nowadays communities. This also represents a precious resource and a stimulus for designers, who have nothing to invent, but who can and must tap into history, through what we can call a principle of "conscious imitation" (the past that builds the future).

Moreover, this enables the transitions to a sustainable city that has a strategic importance especially as regards the need to revitalise the territory. In this context a main problem to be considered in order to cope with population growth and the consequent increase in the energy consumption is to identify appropriate solutions able to satisfy all the new social needs and based on the use as much as possible of construction materials that reflect local traditions.

Within the framework of the European Nearly Zero Energy Building (NZEB) directive, the architects must be able to guarantee a suitable trade-off between architectural quality of design solutions and technological innovation in the project, paying particular attention to the implications on the structure of the envelope and on the changes that new materials and new technologies induce the language of architecture. The architect's privileged point of view therefore becomes the "sustainable project" for environmental well-being, which aims at a bioclimatic approach focused on the control of green parameters, energy saving and the use of appropriate technologies and materials (Figs. 12.16 and 12.17).

Solutions to Improve Microclimate Comfort

Pergolas in Mediterranean countries were the most natural way to reduce solar radiation in outdoor living places, exterior spaces of houses and pedestrian passages, but only after the EXPO in Seville in 1992 became a modern building component designed with scientific approach.

The pergola in Seville was designed to protect millions of visitors in their visiting pilgrimage through to the national Pavilions. The contingency problem to solve

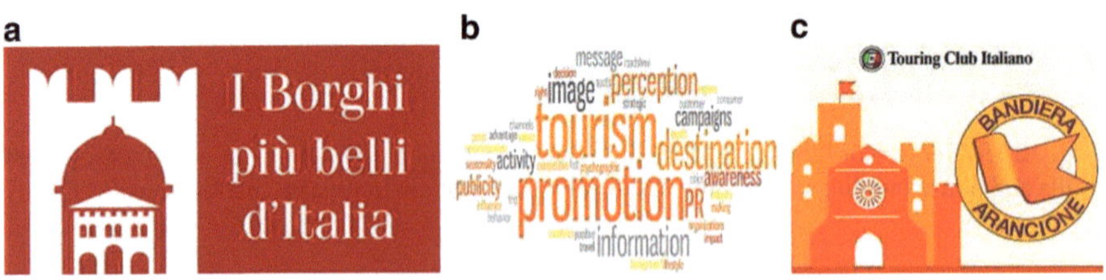

Fig. 12.16 (**a–c**) Different logo to evaluate and define the architectural and environmental quality of historical village

Fig. 12.17 Gradara—Pesaro Urbino. The winner of 2018 evaluation rating as *the beautiful historical village in Italy*

was the contradictory working conditions of the alleys, which until the very last moment was used as working road for the hundreds of independent construction operation of the Pavilions. The solution was in a prefabricated industrial series of elements, realised in a different place, the concrete bases of the columns serving also as sitting places, the column made of steel frame reticular element to be fitted in the bases and the steel reticular spatial and finally the long basin with green was growing since five months before the opening of the EXPO.

The week before the opening the bases was placed in right position, the column and the trusses connected together and finally the basins with the grown hanging green were inserted, creating immediate shadowed or Pedestrian Avenue. Moreover, hydraulic pipes system with micronizes inserted in the roof creates artificial fog that though evaporative process absorbs heat and produces a fresh green shadowed environment for the people. Modern technology and material for the most ancient shadow device created by human civilisation in hot and arid regions (Figs. 12.18 and 12.19).

Fienile The traditional rural building in Tuscany, named “fienile” (hay storage, “fieno” means dry grass), is a stone structure with brick grid, once used to provide dry and ventilated space for agricultural products (usually hay for animals). The vernacular heritage protection laws do not allow to demolish or modify this kind of ancient buildings, but the new owners are allowed to close from inside with large glass panels and use as common room or big living space. Sometime this building constitutes an independent volume, generally very close to the main house, other time are integrated as part of a larger house. A typical vernacular architecture in Tuscany, newly renovated, with the brick grid of the “fienile” was used as sun blind of a large living room. The whole house is an example how additional volumes provide space for new functions. The stone exposed today trend is a bad interpretation

Fig. 12.18 Three different solutions of *pergola*: courtyard integration or urban environmental solution reducing direct solar radiation

Fig. 12.19 Typical *fienile*, the traditional rural building in Tuscany, with stone structure and brick grid to allow ventilation

Fig. 12.20 Different solutions of bioclimatic and energy retrofitting of typical *fienile* buildings

of ancient constructive technology, since at the when was built, nearly all country houses was plastered (the only way to protect from wall and inside room from rain infiltration). Later on when the houses loosed the plaster (due to the poor materials used by the countryman, the "new rural style" interpretation appreciate the stone aspect and carefully restored using good sealing against rain infiltration. The roof chimney identifies the position of the only fire place of the house, used for heating and cooking (Fig. 12.20).

Double Roof: Ventilated Roof

In tropical countries, to improve the summer comfort, the first strategy is to create screens and roof to avoid the direct exposition to solar radiation, better if this exposed roof is also highly reflective, as white colour (like in all Mediterranean villages). The roof tends to be overheated by solar radiation and the strategies to avoid thermal radiation from it to the interior of the space are to have the roof as high as possible, to realise with thick air containing soft material, placed in different layers, such as dried grass, palm leaves, and other thick natural material that reduces the heat transfer and allows a certain air movement in the roof itself without creating a thermal mass.

With this system a continuous insulation can be achieved for the exterior of the building, protecting the interior sheet as well as the slab edges. In the ventilated chamber, due to the heating of the air layer of the intermediate space compared to the environment air, the so-called "chimney effect" is produced which generates a continuous ventilation in the chamber. Appropriately dimensioned the air entry and exit, a constant evacuation of water vapour coming from both the interior and the exterior of the building is achieved, keeping the insulation dry and obtaining a better performance of the insulation and big savings in energy consumption (Figs. 12.21, 12.22, and 12.23).

Ventilated Wall: Ventilated Facade

We can consider the technological solution of ventilated façade as the contemporary architectural application of *malqaf*, to implement the envelope performance both for new buildings and renovations of existing buildings, giving considerable advantages of wall durability over time and of energy saving, especially where tall, exceptionally exposed, isolated buildings are concerned.

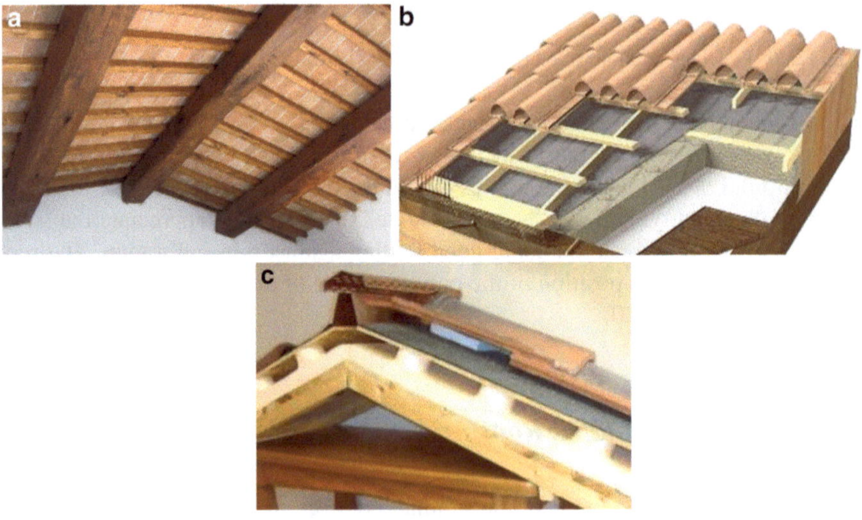

Fig. 12.21 (**a–c**) The traditional wood ventilated roof: internal view, the different layers and the new insulated version

Fig. 12.22 Modern apartment house near Siena, with traditional brick cladding and the replica of brick grid as sun blind for the staircase. PV panels integrated into the roof, and not superposed to it. Large terraces protect windows of lower floor from direct sun radiation. (Project in Castelnuovo Berardenga, by Marco Sala Associated)

A ventilated wall consists of multiple layers and, thanks to the "chimney effect", these cause an ongoing process of natural ventilation through the facade, removing excess humidity and playing a vital role in keeping it cool in summer and warm in winter by controlling heat loss, guaranteeing a high level of living comfort. In terms of energy performance, thanks to the partial reflection of solar radiation by the covering and the ventilated air gap and to the application of insulating material, the ventilated walls can reduce the passive contribution of solar energy and the amount of heat that buildings absorb in hot weather and summer season, allowing a considerable reduction of the costs of air conditioning. Besides, in winter, ventilated walls manage to retain heat, resulting in savings in terms of heating.

In addition, ventilated walls tend to increase the reflection of external noise as the particular construction, consisting of layers of facing, air gap and insulating material, ensures a certain level of acoustic absorption (Figs. 12.24, 12.25, 12.26, 12.27, and 12.28).

Fig. 12.23 Banca di Lodi, Italy. Brick solution of ventilated façade and shading devices, design by Renzo Piano

Conclusion

Vernacular architecture is a major attraction to tourists and foreign visitors to the country. It combines the appeal and the comfort living during its building period. Architects normally use their innovative technique to design an iconic building to fulfil a specific function such as a castle, museum or palace for royalty to live in. However vernacular architecture is defined as buildings for the masses, in most cases simple and functional.

Several examples representing villages in Europe have those features. These villages were explained and documented well in this chapter.

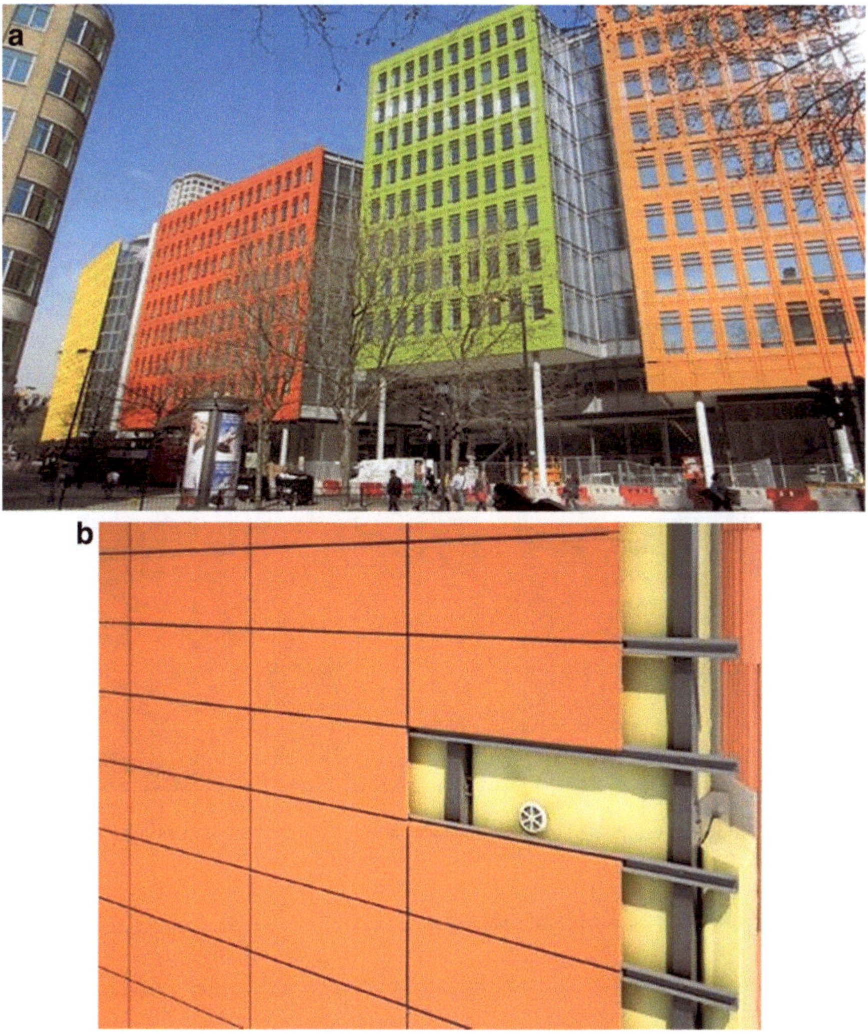

Fig. 12.24 (**a**, **b**) Buildings in Berlin. Different aspects of ventilated façade design by Renzo Piano. Detail of brick solution

Fig. 12.25 Double envelope and ventilated facades realised with different materials. Metallic and brick shading devices

Fig. 12.26 New concept design for shading and translation of vernacular architecture to control direct solar radiation

Fig. 12.27 New challenge of dynamic envelope and kinetic architecture. The Arab Institute in Paris had a southern façade made up of hundreds of light-sensitive diaphragms that automatically regulate the amount of light entering the building

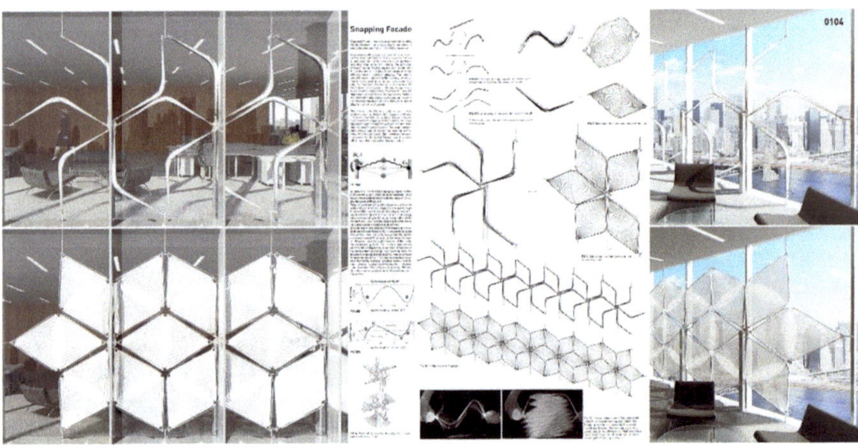

Fig. 12.28 First prize of competition for snapping façade 2018. Snapping Facade explores a sustainable building envelope design strategy that utilises elastic instability to create dynamic motion at the building envelope. Designed by Jin Young Song and Jongmin Shim

References

Butera F (2014) Dalla Caverna alla casa ecologica: storia del comfort e dell'energia. Edizioni Ambiente

Rudofsky B (1964) Bernard Rudofsky architecture without architects: a short introduction to non-pedigreed architecture is a book based on the MoMA exhibition of the same name by Bernard Rudofsky originally published in 1964

Sala M, Ceccherini Nelli L, D'Audino E, Trombadore A (2007) Schermature solari. Alinea, Firenze

Trombadore A (2016) Mediterranean smart cities. Innovazione tecnologica ed ecoefficienza nella gestione dei processi di trasformazione urbana. Altralinea, Firenze

Chapter 13
Enhancing Environmental Performance of Vernacular Architecture. A Case Study

Luca Finocchiaro

Introduction

Vernacular architectures spread throughout the world represent the result of an evolutionary process in which buildings' form and construction have been continuously refined with the aim of adapting to local climate and providing optimal living conditions for the human habitat. These buildings were built recurring to local materials and resources available on site. This ensured a tight connection with site and context, determining also identity and diversity of the built environment throughout the world. As Andrea Deplazes observes in his book, titled "Constructing Architecture", "where different cultures have had access to the same resources of usable materials, they have been developing surprisingly similar forms of building more or less independently of each other" (Deplazes 2013). Similarities among vernacular architectures spread throughout the world have been analysed in studies developed by the brothers Victor and Aladar Olgyay in the mid-twentieth century (Olgyay 1963). In the book "Design with Climate", published in 1963, Victor Olgyay traced a series of maps giving evidence of how similar architectural solutions for climate adaption had been developed whenever climatic conditions were comparable (Fig. 13.1). The Olgyay brothers focused on analysing specific building features developed for climate adaption. Their studies, however, were not only based on a mere visual observation of historical constructions but included also several numerical analyses about building morphology and climatic data.

L. Finocchiaro (✉)
NTNU - Department of Architectural Design, History and Technology, Trondheim, Norway
e-mail: luca.finocchiaro@ntnu.no

A. Sayigh (ed.), *Sustainable Vernacular Architecture*, Innovative Renewable Energy, https://doi.org/10.1007/978-3-030-06185-2_13

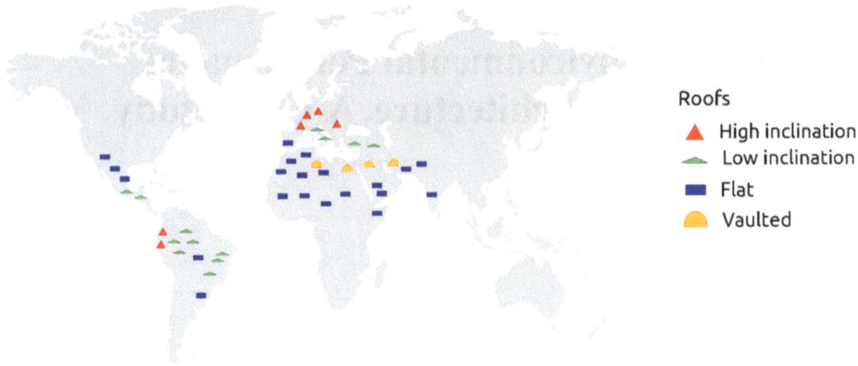

Fig. 13.1 Vernacular architecture. A map developed by Victor Olgyay in Design with Climate stresses the connection between construction systems in vernacular architectures and climatic conditions throughout the world (image elaborated starting from data reported in "Design with Climate")

Climate Adaption. From Tradition to Science

Principles and strategies for climate adaption have been extensively used and adjusted throughout the century with the mere purpose of ensuring optimal living conditions in a very variable set of climatic contexts. Technical expertise acquired in this regard was transmitted and refined from generation to generation. It is, however, only in the second half of the twentieth century that the ability to design and construct buildings as climatic shelters for indoor comfort assumes the characteristics of a science. Bioclimatic design, as a regionalist approach to architectural design, is based on a reasoned use of qualitative and quantitative data (Olgyay 1963). As such, it goes beyond a mere visual perception of architectural spaces, to include a whole series of quantitative concerns related to the buildings' environmental performance and physiological requirements for human comfort. As buildings based on passive strategies are less reliant on active energy systems, bioclimatic design was soon embraced by a large community of designers who saw in this discipline the opportunity to forefront challenges arose from the energy crises of the second half of the twentieth century.

Beside the fundamental contribution of the Olgyay brothers collected in the aforementioned "Design with Climate", this transition was possible thanks to the fundamental contribution of scientists such as Baruch Givoni (1969) or Steven Szokolay (2008), among others, giving an analytical basis to the understanding of climate as a source for making architecture.

Victor Olgyay developed also a numerical model for quantifying the thermal stress on a building and used this as the basis for defining a methodology for optimizing buildings' morphological characteristics. This methodology, called the sol-air approach (Olgyay 1963), took into account both conduction heat exchanges throughout the whole building envelope—depending on air temperature differences between outdoor and indoor—and also thermal stress due to solar radiation. According to the Sol-Air approach, an optimal form should be able to maximize solar heat gains throughout the cold season while minimizing thermal stress coming from the solar radiation in the hot one. Victor Olgyay referred, in describing this methodology, to vernacular architectures as a relevant example of optimization of morphological characteristics of buildings. As he observes in the book, "compact aggressive forms were characteristic of extreme climatic conditions while articulated ones were common in temperate climates", being the first able to minimize exposure to adverse climatic conditions and the second taking maximum advantage of available natural resources (Olgyay 1963). Throughout his studies, numerical analyses of different kind strengthened moreover the analogy between nature and the built environment in the effort of adapting to climate and context.

Empirical analyses conducted on perception of environmental parameters made it possible for Victor Olgyay to design the first bioclimatic chart where human comfort requirements were related to the contribution of sun and wind. In this bioclimatic chart, the human comfort zone was traced within specific boundaries included between 20 and 25 °C and a relative humidity of 20–80% circa. Olgyay recognized that people could comfortably conduct light work activities within this area. Different strategies could be applied in order to re-establish comfortable conditions whenever environmental parameters are slightly outside the comfort zone boundaries, sun or simple breezes above the skin activating evaporative cooling effects and changing the perception of environmental conditions (Fig. 13.2).

Olgyay's studies about climate and architecture put the premises for the diffusion of a new architectural regionalism based on a balanced use of quantitative and qualitative parameters related to climate and context. A few years after Victor Olgyay's bioclimatic chart, Baruch Givoni (1969) transposed the comparison between climatic data and comfort requirements within the psychometric chart defined by Willis Carrier in 1904. Givoni titled this chart the *Building bioclimatic chart*, distinguishing it from Olgyay's bioclimatic chart, because of its potential for the development of integrated design processes of climate adaptive buildings. The comparison between climatic data and human comfort requirements, now plotted on the psychrometric chart (Fig. 13.3), could not only be used for quantifying thermal challenges related to a specific climatic context but also as a tool for identifying passive strategies that could have been used in the bioclimatic design of energy-efficient buildings. Baruch Givoni traced on this chart a set of perimeters around the comfort zone, each corresponding to a different passive strategy: passive solar heating, thermal mass, natural ventilation, humidification or evaporative cooling. The perimeter of each zone defined the conditions under which a specific passive strategy could be considered as an effective solution. The boundaries of such areas, later named by Steven Szokolay as control potential zones (CPZ) (Szokolay 2008), depended both on the technology adopted and on the environmental conditions under which a specific strategy was applied.

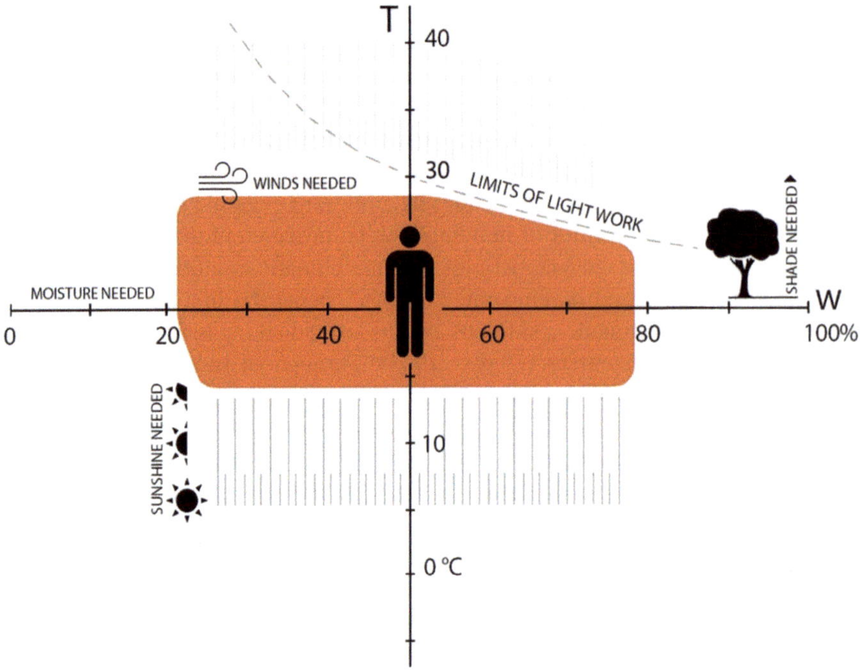

Fig. 13.2 An abstract representation of the bioclimatic chart, developed by Victor Olgyay and published in 1963 in Design with Climate, relating human comfort requirements with the contribution of meteorological factors such as sun and wind (elaboration from Design with Climate, Victor Olgyay 1963)

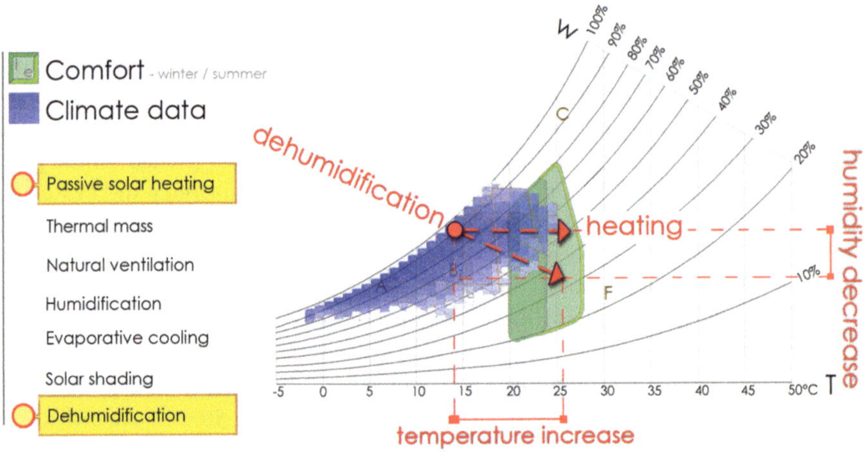

Fig. 13.3 A climate/comfort comparison on the psychrometric chart, considered as the basis for bioclimatic design and a fundamental tool for identification of passive strategies and calculating their potential for indoor comfort

Enhancing Environmental Performance of Historical Buildings

The twentieth century was a period of significant advancements for natural sciences and the development of building technologies. A growing community of scientists, observing empirical phenomena in real case studies, identified in those years numerical equations behind environmental performance of buildings. With the advent of personal computers, those equations started to be used as the numerical basis for the development of simulation software. Moreover, technological development in manufacturing and production of building materials made it possible to conceive advanced components for energy efficiency. Thanks to these materials, potential of traditional systems for climate adaption, such as thermal mass or natural ventilation, could be further enhanced. Their potential for optimizing buildings' environmental performance could be tested, during the early stage of the design process, thanks to the use of digital simulation tools (Finocchiaro and Lobaccaro 2017).

Beside significant advancements in architectural science, technical solutions for energy efficiency are today striving to have an effective impact on the existing building stock. European buildings are therefore still today responsible for approximately 40% of the energy consumption and 36% of CO_2 emissions released in the atmosphere (EU energy efficiency Directive 2012). Energy efficiency programmes developed by the European Union—such as the Energy Performance of Buildings Directive 2010/31/EU—addressed a significant effort into defining clear targets for energy efficiency while harmonizing criteria among the different state members. These programs, however, focused largest part of their attention into new constructions, "often without taking into account the impact of the urban environment where they are located" (EFFESUS 2016). However, circa 23% of the European building stock was built before 1945 (LEAF project 2016), is located in large cities and is often characterized by significant historical value. The EFFESUS project was developed in order to reduce environmental impact of the existing building stock, giving particular attention to those buildings, built before 1945, that were characterized by a significant historical and cultural value. The project took care of understanding the implication of urban structure on the value and performance of selected case studies and developed a methodology for selecting and prioritizing energy efficiency measured on historical buildings.

According to a recent research study, in the last two decades—signed by a deep global economy crisis—attention on energy efficiency of historical buildings has increased significantly. A research conducted by Martinez-Molina, analysing the number of publications between 1978 and 2014 in comparison with GDP values, showed that "economic stagnation tends to result in increases in saving policies". In times of economic crises, the focus is shifted "from new construction to refurbishment and restoration and the implementation of energy efficiency measures to existing buildings" (Martínez-Molina 2016).

Researchers involved in the project seem to agree that historical buildings, or vernacular architectures in general, are relevant examples of bioclimatic design, embedding site-specific solutions for climate adaption. This facilitates their transition into energy-efficient structures, being able to fulfil human comfort requirements with a limited energy demand. "Masonry buildings throughout Europe have shown—moreover—great environmental adaptability. Interesting challenges are therefore the reuse and rehabilitation, in respect to safety, linked to seismic risk and energy efficiency" (Dipasquale 2016). Surveys conducted through post-occupancy evaluation showed greater thermal comfort satisfaction among users in naturally ventilated historical buildings rather than in air-conditioned modern dwellings (Ealiwa et al. 2001).

Potential of natural ventilation strategies or passive solar heating systems in historical buildings can be further enhanced thanks to the use of advanced material and components. Employed hybrid systems can moreover rely on the energy provided by integrated renewable energy systems. Because of this, in the last years, relevant examples of climate adaptive design throughout Europe have demonstrated that effective measures can be taken in order to reduce significantly energy consumption of historical buildings.

Energy Efficiency of Historical Dwellings in Mediterranean Climate

According to the Köppen–Geiger climatic chart, regions surrounding the Mediterranean Sea are characterized by hot-dry or cool-dry summers with less than 30 mm of precipitation (classified as *Csa* and *Csb*, respectively). Precipitations are generally concentrated in winter and spring and account for a total of 500 mm circa. Average temperatures in these regions are generally with winter values generally included between 0 and 18 °C. Average temperatures in the warmest month can exceed the 22 °C, while peaks over 40°.

A building bioclimatic chart, elaborated according to Givoni–Milne for this climate area, suggests the use of heavy construction systems characterized by high thermal capacity and relatively low thermal conductivity (Fig. 13.4). The use of such an envelope would make it possible to stabilize temperature fluctuations on a daily basis, delaying chilly night temperatures to the following hours of the day, while still avoiding large thermal losses during winter. Besides thermal mass, natural ventilation strategies and passive solar heating systems are the two most relevant strategies in this climatic context. Those strategies should be sufficient to guarantee temperatures within the comfort zone for most of the year. A study, conducted by Mike Coillot et al. (2017), exploring the potential of double windows in historical buildings, showed how a supply air window can preheat the incoming air and reduce heat losses by 134% in the Mediterranean climatic context.

Several studies, conducted by Philokyprou (2013), Milne (1979), and Shaviv (2001), among others, gave evidence of the importance of the cooling potential of natural ventilation strategies in the Mediterranean areas. According to Milne and

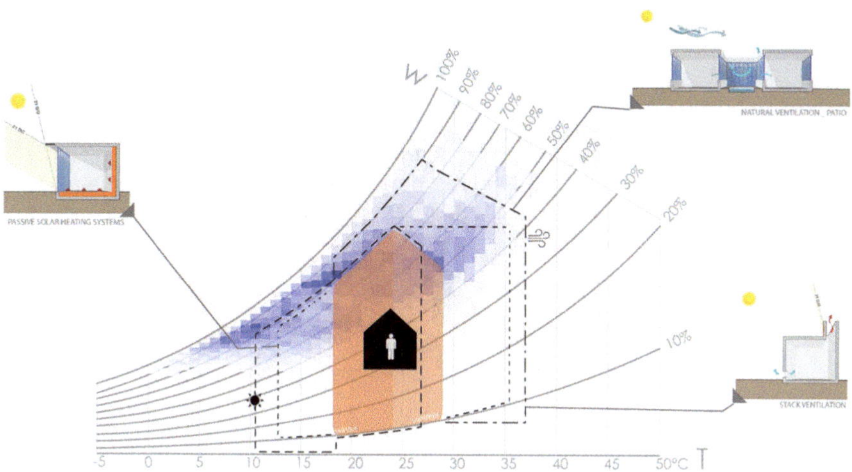

Fig. 13.4 A building bioclimatic chart in the Mediterranean climatic context underlines the potential for passive solar heating systems, and natural ventilation strategies in combination with thermal mass (data elaborated using climatic data in the context of Catania, Sicily)

Givoni, direct natural ventilation strategies can be used during the day whenever environmental conditions are within the comfort zone, both in terms of temperature and relative humidity. Direct ventilation can be also used for cooling purposes, whenever temperatures are slightly above the comfort zone, if an air speed of 1.5–2 m/s can stimulate evaporative cooling effect altering the perception of indoor conditions. According to the same study, still, direct ventilation during daytime is beneficial whenever amplitude of temperature fluctuations is included within 10°. In spring and autumn, when temperatures are slightly below the comfort zone, a single-side ventilation strategy represents the most effective solution for re-establishing temperatures within the comfort zone range (Philokyprou et al. 2013).

Indirect natural ventilation systems, such as night purge ventilation, can effectively re-establish comfort conditions whenever nocturnal outdoor air temperature is around 20° and diurnal temperature fluctuation is more than 10 °C (Milne and Givoni 1979). The implementation of such strategy could effectively improve environmental conditions during the following day (Santamouris 2006). A study conducted by Amilios Michael showed how the location of small openings at the upper level of the external walls—similar to the arseres used in traditional dwellings in Cyprus—would facilitate the extraction of hot air and thus contribute to a better indoor thermal comfort during the day after in the overheated season (Michael et al. 2017). According to this study, night purge ventilation can reduce indoor peak temperature up to 3°—whenever temperature fluctuations are greater than 6°—significantly improving indoor thermal conditions during the following day. This strategy takes full advantage of the relatively low outdoor air temperatures during the night as an effective way to cool down the building envelope. As such, its efficiency is mostly depending on two fundamental factors: the outdoor environmental conditions

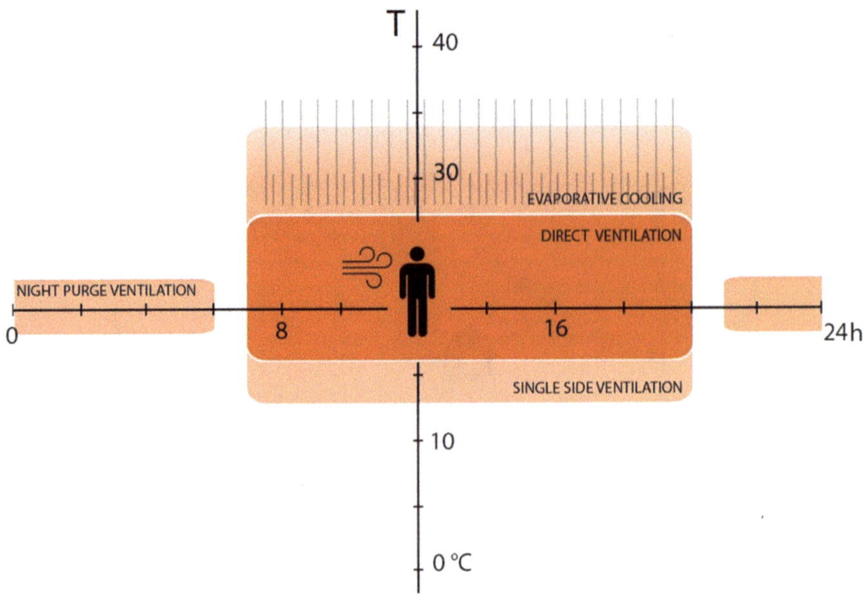

Fig. 13.5 A conceptual diagram about the potential of natural ventilation strategies in the Mediterranean climatic context

and the thermal mass of the building envelope. Air should be let in whenever temperatures are below the comfort zone. Under these conditions, natural ventilation is not desirable during the day but its potential increases in the night. Potential of alternative natural ventilation strategies in relation to different hours and temperatures has been resumed in Fig. 13.5.

Vernacular Architecture in Sicily

Throughout the last three thousand years, the island of Sicily was occupied by different civilizations, starting from the Greek one of the eighth century B.C., and following with Romans, Saracens from North Africa and Normans. These dominations resulted in an extremely rich mix of cultures, readable also in the variety of architecture styles and construction systems throughout the island. During the nineteenth century, when the Reign of the two Sicily—including the regions south of Rome—was under the crown of Aragon, the Spanish culture had a major influence on the island. At that time, economical power was concentrated in the hands of few rich families, owning large pieces of land in the form of *latifundia*. This was also the case of the territories surrounding the mountain Etna, where cultivations of grapes, citrus, olives and almonds were domain of a few people.

Fig. 13.6 The interior of a winery in Sicily with the torchio holding the mill stone on the right side

Wine production required precise climatic conditions and infrastructural systems. As such, it determined both architecture of buildings and landscape. Wineries generally included a compound of buildings destined to a series of different activities. The production building was generally characterized by a sequence of cells of different sizes, where grapes were first stored, then crushed with large millstones and finally stored again, in form of wine, inside wooden barrels. The millstone was supported by a tall piece of wood, called *torchio*, carved as a screw (Fig. 13.6). This sequence determined the section of these buildings, where the pendant among the different cells ensured that wine could naturally flow from one side to the other of the building. The height of the torchio, generally included between 3 and 5 m, required that spaces where tubs were placed had a considerable height, often above 6 m. The room where barrels were stored was often located below the ground level. This ensured a slightly cold and humid environment, appropriate for the conservation of the wine. Besides the production building, wineries included also one or two houses: one for the family owning the land and one for the family of the "massaro", as it was called the man taking care of the land.

Residential buildings were characterized by very simple plans, where rooms of regular form were connected in row without any service space or corridor. Double doors of generous height—called *bussole*—divided the different spaces. Independent access to each room was eventually guaranteed through the courtyard. Outdoor spaces represented a natural extension of inner spaces, where many fundamental activities, such as eating or leisure, were hosted. Porches, terraces, roof gardens or pergolas were recurrent architectural components whose aim was that of minimizing the impact of the abundant solar radiation on the building envelope. The reasoned use of these components, in combination with a heavy construction system finely tuned to climate and context, made historical buildings in Sicily excellent examples of bioclimatic architecture, embedding different features for optimizing the building's environmental performance throughout the whole year.

Rubble masonry walls, with a thickness included between 50 and 90 cm, determined environmental conditions in these houses, stabilizing temperature fluctuations throughout the day while taking advantage of cool temperatures of the night. These thick masonry walls were realized using rough stones of lava from the mountain Etna of different sizes. Large stones were generally arranged in courses of rather equal height on both sides of the wall, while spaces in between were filled with lime mortar and smaller stones or crushed bricks. Large stones of regular shapes were used as cornerstones or for framing doors and windows. This ensured a more effective coupling of perpendicular walls and a better resistance in case of earthquakes. Masonry walls were most of the times plastered with a compound called cocciopesto. This compound was obtained mixing crushed bricks, lava sand and natural lime with water. Sand used in the mix determined the colour of the façade, often in contrast with the black lava stone or the more expensive white ones of Syracuse, generally used for decoration purposes. Façades were generally very simple, with little or poor ornaments with few exceptions in the owner's house.

Horizontal partitions were realized in wood, a material particularly abundant throughout the eighteenth and nineteenth century, when large forests were converted into farming fields. Because of the high solar contribution from the zenith, light construction roofs were the most reasonable solution for avoiding overheating issues. Wooden roof structures supported curved tiles of 15–25 mm thickness. Light vaults, made of canes and plaster, were used as false ceiling enclosing the most important spaces of the house with a structure resembling load-bearing vaults of more noble structures.

Because of the rather favourable climatic conditions, most rural buildings were not equipped with any heating system, if it was not for a stove in the main living area. Rain water, generally rare during the summer period, was stored in large cisterns for irrigation purposes. Because of the rather stable and colder temperature, when compared with the exterior environment, cisterns were occasionally used to store food (Caltabiano 2006).

Building and Climate

The east coast of Sicily, in the part included between the cities of Catania and Giarre, has been shaped by the clash of lava coming from the volcano Etna and the water of the Mediterranean Sea. As such, it is characterized by black rocks of different nature and fertile lands rich of minerals. Throughout the eighteenth and nineteenth century, this was called *the coast of lemons*, covering with its production large part of the European demand for citrus fruits.

Aci Platani is a small town in the municipality of Acireale, a few kilometres north of Catania and a couple of kilometres away from the east coast of Sicily. The town, counting 2000 inhabitants circa, lays at an altitude of 130 m. A steep green wall called Timpa—today a protected natural reserve—lifts the whole territory. As Acireale and other seven towns, Aci Platani takes its name from the river Aci providing fresh water to the whole territory. The town flourished during the nineteenth century, in connection with farming activities. Most inhabitants were employed in the large fields surrounding the town, destined mostly to the production of lemons, but also oil and grapes.

Temperatures in summer are often above 30°, with peaks in July and August above 40. In winters, temperatures are always above the zero with average values included between 5 and 18°. Daily temperature excursion is included between 7 and 15° with chilly nights and warm temperatures at midday, supporting the use of thermal mass and natural ventilation as the most effective strategies for cooling purposes.

As large part of eastern Sicily, the town was built after the earthquake of 1693, one of the most devastating events of the modern history of Italy. The territories around the village were inhabited since the roman time, as it demonstrates the thermal baths placed just outside the village and several findings around the town. The whole city centre, structured along a north–south-oriented axis, is today protected by law because of its historical value. The structure of the village determines the orientation of most of the buildings, having a main façade oriented towards east or west, along the main road, and a courtyard on the backside. The winery of the family Puglisi Cosentino represented an exception when it comes to structure and orientation. This building was in fact characterized by a structure resembling the ancient roman townhouses, where access to the different cells was given through a narrow street leading into a large hall or atrium, representing the core of the house. The private access road, giving access to the backyard and to the large fields, where grapes were produced, generated the whole plan of this building. This east-west-oriented road divides the residential buildings, on the north side, from the production buildings, on the south one (Fig. 13.7). As such, the production building has its main façade towards the north, while the living area of the house faces south. The Massaro, taking care of the land, occupied the tiny rooms on the southeast corner of the building, just next to the cantina where wine barrels were placed. A large millstone, today removed, was used to press grapes in four large tubs directly connected to the barrels through a channel in plastered terracotta running along the walls. This production sequence defined the building section ending with the room where chestnut barrels were stored, characterized by a height of 11 m (Fig. 13.8).

Backyards play an essential role for the environmental performance of the house, being able to create specific microclimates that can markedly differ from the regular outdoor climatic conditions. Small doors and windows, accounting for a fifth part of the floor area, give access to limited amount of solar radiation while still letting in air from courtyards characterized by different environmental conditions. The access of air and sun is further regulated through the use of inner shutters and outer louvres. The alternations of buildings and courtyards represent a perfect adaptation to the heat of the Mediterranean region, providing circulation of fresh air among the

Fig. 13.7 Plan and facade of the winery in Aci Platani. Red lines on the plan show alignment of doors and windows in order to facilitate view, daylight distribution and air across the plan

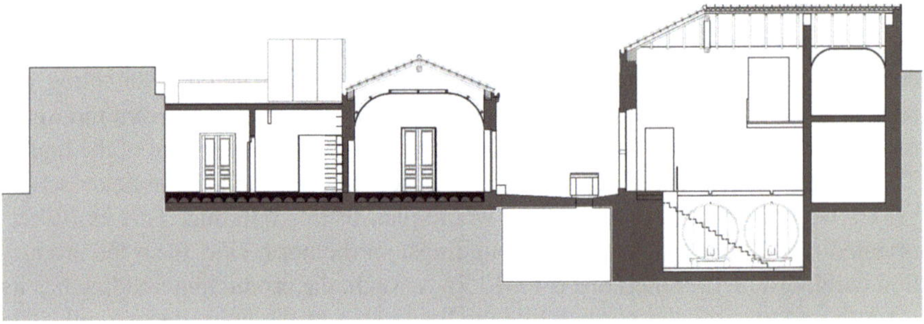

Fig. 13.8 A section of the winery shows the house on the left side and the production building on the right side. A cistern is included between the two buildings

corridors and rooms. Besides giving access to air, doors and windows are almost always aligned with the purpose of letting in views while distributing daylight throughout the cells.

Each building in the winery was originally covered by a pitched roof supported by a wooden structure in chestnut. Light construction of the horizontal partitions ensured that as little as possible solar energy from the zenith was stored in the building fabric. In the latter part of the nineteenth century, however, two large roofs have been changed into terraces, using steel profiles and lava stone. This affected the environmental performance of the building resulting into warmer spaces suffering from overheating problems for a short part of the summer period. Almost every drop of rainwater reaching the roof surface is today still stored in two large cisterns, one placed in the access road and one in the tiled courtyard in the north side of the house. Precipitations, concentrated between October and March, account for an average of 500 mm per year, providing water enough for irrigating the last piece of land still owned by the house, accounting for circa 400 sq m.

Project Scope and Vision

With the creation of the free global market, the production of citrus fruits began to suffer the concurrence of South American producers. This soon turned into a progressive devaluation of the products and an economic crisis of the sector leading farmers to abandon their land and seek for other kinds of occupations in larger cities. As in most villages spread throughout *the coast of lemons*, also inhabitants of Aci Platani started to abandon the fields and seek for other jobs, often without necessarily moving away. The village turned therefore more and more into a dormitory suburb of Catania, slowly losing its identity and social cohesion among the inhabitants. Because of this, the old winery—converted in the beginning of the twentieth century into a storage for lemons—was soon abandoned. Only the northern part of the house close to the church was inhabited until the seventies. In the meanwhile, in order to give access to larger trailers, part of the main façade was demolished, substituting the large arch with a wide gate.

After over 50 years of abandonment, the old winery lied, at the beginning of the twenty-first century, in poor conditions and without connection to any urban infrastructural system. The large terrace above the production building started to leak, deteriorating both the steel profiles and the mortar of walls and roof. Humidity from the ground had provoked the detachment of the plasters in the lowest part of the façades. Part of the facade was demolished weakening the walls towards the main road. Most of the windows and doors had been vandalized and, in a few cases, sill stones and tiles inside the house were stolen.

A refurbishment project was therefore initiated in 2011 in order to stop the process of decay and ensure the building a new life. The defined project embraced regenerative design principles in accordance to which architects could not any longer limit themselves into minimizing environmental impact of the built environment but rather understand how each project could positively contribute to the development of site and context under different contexts (Brown et al. 2018). Such a vision clearly requires a radical shift in the way green buildings are designed and produced, to integrate concerns related to the interrelation between humans and natural ecosystems. This certainly presumes an inedited set of analyses that go far beyond energy efficiency, to include concerns related to materials, water cycle and human perception and behaviour. Materials used, characterized by low embodied emissions, should ensure a long life cycle of the building. The use of local building techniques, whose preservation and enhancement is fundamental for preserving the cultural identity of a place, should also be incentivized in order to enrich local community. The hydrologic system of the building should be based on closed loops and should not affect availability for neighbourhood buildings and the natural ecosystem.

Refurbishment activities have been planned in three phases, corresponding, to a large extent, to the three different volumes constituting the winery. A few interventions were anyway made in order to ensure structural stability and to stop decay processes throughout the whole building fabric. Services and technical systems realized in the first phase of the project have moreover been dimensioned in order to be able to serve a large part of the building structure when the next stage of the refurbishment will be finalized.

The developed project did not aim to restore the building to its original conditions but rather to recognize those characteristics of the building whose potential could have been further enhanced thanks, also, to the use of modern technologies. On an environmental performance perspective, this meant recognizing the positive role of thermal mass and natural ventilation strategies and implementing their potential through the use of more advanced components for energy efficiency. The heavy stone walls, of a thickness included between 50 and 70 cm, have therefore been preserved while intervening on the floor and roof of the building. A naturally ventilated gap of around 25 cm of height has been obtained excavating for half a metre circa below the existing floor. The floor, equipped with a water-based radiator system, was in this way detached from the cold and humid earth. The wooden floor was glued over a thermally activated layer of concrete, 8 cm thick. This was coupled with an air-to-water heat pump able to provide air at an adequate temperature for heating or cooling the building (Fig. 13.9).

Solar radiation in Sicily is generally abundant. Global radiation values can be as big as twice the sunniest location of Germany. Southern Italy is, for this reason, one of the regions in Europe where integration of solar technologies has the largest potential. Water heated through a south-oriented solar thermal system is stored in an accumulation tank and provided to the heat pump, when needed. During summer time, cold water from the cistern is sent to the radiator floor for cooling purposes. A temperature of 17° circa is controlled through the heat pump running on electricity provided by an integrated photovoltaic system. Arsenes at ceiling height facilitate stack ventilation for extracting exhausted air. These valves, regulating the amount of air extracted, have been connected to a south-oriented chimney, integrated to the solar thermal system. The chimney aims to extract air enough to limit condensation phenomena that might be caused by the use of the floor as cooling system.

When refurbishment activities were started, services in the house were no longer functioning and the house had been detached from the city infrastructural systems. Existing toilets had been built as extensions of the building structure towards the courtyard. The bathroom, made of brick walls of 15 cm thickness circa, had a direct access from the kitchen. The volume, containing the bath tab, was built out of light bricks of cement and could be only accessed through the courtyard. In defining the new plan, two different criteria were therefore identified. The existing toilets, characterized by poor construction, were dismantled and replaced by two new service

Fig. 13.9 A section of the building showing the integrated systems for enhancing the environmental performance of the existing structure. Wooden surface of the floor has been obtained cutting wood obtained by disassembling chestnut barrels

cores. One of these was placed in the inner part of the building, and one outside it. Both of them had direct access from the inner side. In adding the new services, position of doors was slightly adjusted in order to create a series of alignments through the building plan (Fig. 13.7). This aimed to ensure a better distribution of daylight, a stronger visual connection with the outdoor and the possibility of cooling the building through cross ventilation.

Materials deriving from the disassembly of the service cores and the movement of doors within the building plan—such as stones and bricks—were sorted according

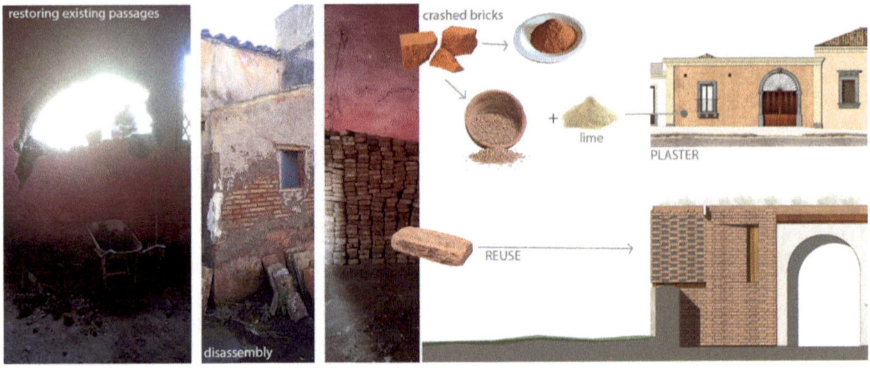

Fig. 13.10 Materials provided by disassembly of existing structures and by opening existing passages have been sorted for nature and condition. This material was reused for building the new service core and sorted for producing plaster to be used in the next project stage

Fig. 13.11 Five barrels of 5000 L and 2.4 m diameter have been disassembled in order to provide materials enough for producing the wooden floor. Elements have been cut in a way to minimize waste

to nature, conditions and size (Fig. 13.10). This operation provided materials enough to build the new service core towards the courtyard. A wall made of recycled old bricks was coupled with porous bricks, characterized by lower thermal transmittance, on the inner side and a layer of wood fibre insulation in between them. This layers aims to limit heat losses in the winter while still ensuring a buffering of temperature fluctuations on a daily basis. Windows and doors have been substituted with new handmade ones, produced with short travelled chestnut coming from the mountain Etna and now equipped with a double glass filled with argon gas. Materials for covering the floor surfaces have been obtained by disassembling five wooden barrels purchased in a winery, six kilometres away from the construction site (Fig. 13.11).

References

Brown M, Haselsteiner E, Apró D, Kopeva D, Luca E, Pulkkinen K, Vula Rizvanolli B (eds) (2018) Sustainability, restorative to regenerative. COST Action CA16114 RESTORE, working group one report: restorative sustainability

Caltabiano I (2006) The traditional architecture in Sicily: the rural area around the volcano Etna. In: Proceedings of conference PLEA2006, Switzerland, 6–8 September 2006

Coillot M et al (2017) Heating, ventilating and cooling impacts of double windows on historic buildings in Mediterranean area. Energy Procedia 133:28–41

Deplazes A (2013) Constructing architecture: materials processes and structures. A handbook. Birkhauser Verlag AG, Basel

Dipasquale L (2016) Ancient stone masonry constructions. In: Non-conventional and vernacular construction materials. Woodhead, Amsterdam, pp 301–332

Directive 2012/27/EU of the European Parliament and of the Council of 25 October 2012 on energy efficiency, amending Directives 2009/125/EC and 2010/30/EU and repealing Directives 2004/8/EC and 2006/32/EC Text with EEA relevance

Ealiwa M et al (2001) An investigation into thermal comfort in the summer season of Ghadames, Libya. Build Environ 36:231–237. https://doi.org/10.1016/S0360-1323(99)00071-2

Energy Efficiency for EU Historic Urban Districts Sustainability (EFFESUS), 2012 to 2016, cofunded by the European Union's Seventh Framework Programme for research, technological development and demonstration under grant agreement number 314678. http://www.effesus.eu

Finocchiaro L, Lobaccaro G (2017) Bioclimatic design of green buildings. In: Handbook of energy systems in green buildings. Springer, Berlin

Givoni B (1969) Man, climate and architecture. Elsevier, Amsterdam

LEAF project – Final report 3 (2016) Improving the energy performance of apartment blocks: Low Energy Apartment Futures (LEAF). http://www.lowenergyapartments.eu/wp-content/uploads/2016/03/LEAF_Final_Report_D1.1%20Mar16.pdf

Martínez-Molina A (2016) Energy efficiency and thermal comfort in historic buildings: a review. Renew Sust Energ Rev 61:70–85

Michael A et al (2017) Natural ventilation for cooling in mediterranean climate: a case study in vernacular architecture of Cyprus Aimilios. Energy and Buildings 144:333–345. Contents lists

Milne M, Givoni B (1979) Architectural design based on climate. In: Watson D (ed) Energy conservation through building design. McGraw- Hill, New York, pp 96–113

Olgyay V (1963) Design with climate: bioclimatic approach to architectural regionalism. Princeton Press, Princeton

Philokyprou M, Michael A, Thravalou S et al (2013) Assessment of the bioclimaticelements of vernacular architecture. The historic centre of Nicosia, Cyprus. In: Argenziano P (ed) Le Vies dei Mercanti XI Forume Internationalle di Studi, Aversa, Capri, Italy, 2013, pp 666–675

Santamouris M (2006) Ventilation for comfort and cooling: the state of the art. In: Wouters P, Santamouris M (eds) Building ventilation: the state of the art, 1st edn. Earthscan, London, pp 217–245

Shaviv E, Yezioro A, Capeluto I (2001) Thermal mass and night ventilation as passive cooling design strategy. Renew Energy 24(3-4):445–452

Szokolay SV (2008) Introduction to architectural science: the basis of sustainable design. Architectural Press, Amsterdam

Chapter 14
Climatic Adaptations of Colonial School Buildings in Malaysia

Aliyah Nur Zafirah Sanusi, Aida Kesuma Azmin, Fadzidah Abdullah, and Mohd Hisyamuddin Kassim

Introduction

The architecture of Malaysia was developed throughout the years in various styles. Heritage of Malaysia Trust (Heritage of Malaysia Trust 1990) stated that Malaysia architectural style has been classified according to the building styles of different eras. The building styles are as follows (Table 14.1).

The diversity of building styles were the outcome from Europeans, Chinese, Malay and Indian influences, in which the various styles were adopted into building designs and modified to correspond well with the contextual environment. This study focuses on the Colonial building style, built between the seventeenth and twentieth centuries. Colonial building style is found in various building types such as residential, office buildings and school buildings. However, this study focuses on colonial school buildings, with the aim to evaluate its viability in adapting to Malaysia climate and achieving thermal comfort among its occupants.

The first objective is to assess thermal comfort of the occupants in the classrooms of colonial school buildings. Secondly, the study is intended to evaluate the strength of passive design strategies adopted in colonial school buildings. The purpose for this is to acquire passive design knowledge from buildings built in the earlier eras, which are still being used and occupied until today. Three colonial schools in Kuala Lumpur have been selected for this study and they are Victoria Institution, Methodist Girls Secondary School and Maxwell Secondary School.

A. N. Z. Sanusi (✉) · A. K. Azmin · F. Abdullah
International Islamic University Malaysia, Kuala Lumpur, Malaysia
e-mail: aliyah@iium.edu.my

M. H. Kassim
Malaysia Multimedia University, Cyberjaya, Malaysia

A. Sayigh (ed.), *Sustainable Vernacular Architecture*, Innovative Renewable Energy, https://doi.org/10.1007/978-3-030-06185-2_14

Table 14.1 Building styles in Malaysia of different periods (Heritage of Malaysia Trust 1990)

Building styles	Period
Indian Kingdoms	Seventh to fourteenth centuries
Traditional Malay Vernacular	Pre-fifteenth century until present
Straits Eclectics	Fifteenth to mid-twentieth centuries
Chinese Baroque	Nineteenth to early twentieth centuries
Colonial	Seventeenth to twentieth centuries
Modern	1950s until present

Background History of Malaysia School Buildings

Government school buildings in Malaysia consist of naturally ventilated classrooms, and some school libraries and teachers' meeting room are exceptionally air conditioned. Early school design dates back to before the eighteenth century, where schools were called "Sekolah Pondok", which is an informal religious class (Syed Fadzil 1995). "Sekolah Pondok" was built in Traditional Malay Vernacular Style and it is commonly small with lightweight walls and large openings. Most of these schools were demolished and some were replaced with new ones. Between the eighteenth century and Malaysia Independence Day, year 1957, new schools were built, which segregated the students into four school types; Malay, Chinese, Tamil and English Schools. This was during the Colonial period and each of these schools has their own aim. During the Colonial period, some "Sekolah Pondok" were retained as Islamic Religious Schools. The Religious Schools had better infrastructure and were more organised than "Sekolah Pondok". However, the Religious Schools were built in Traditional Malay Vernacular style (Mohamad 2008).

Malay schools were set up by the British to teach Malay children into becoming better farmers than their parents. Malay schools were built in Malay Vernacular style with brick walls (Mohamad 2008). Meanwhile, Chinese schools were established in the early twentieth century. It was built differently than the Religious and Malay Schools (Mohamad 2008).

Tamil schools were initially built by the Indian workers who were brought into Malaya to work in rubber estates and plantations. The schools then developed alongside with the rubber estates, coffee, sugar-cane and coconut plantations for the children of the Indian workers (Mohamad 2008). Tamil schools were built with timber structures but lack in facilities due to minimal aid and assistance from the British. Over the years, some parents opted to send their children to English schools where they had better facilities. English schools were established in Malaya, led by the British government and Christian missionaries, with the purpose of spreading Christianity among the locals (Mohamad 2008). The colonial school architecture reflects influence of British architecture with thick walls and high ceilings.

After Malaysia Independence Day in 1957, National Schools were built to contain all students regardless of their background (Fig. 14 .1). National Schools were designed by the Public Work(s) Department. Typically, a National School consists

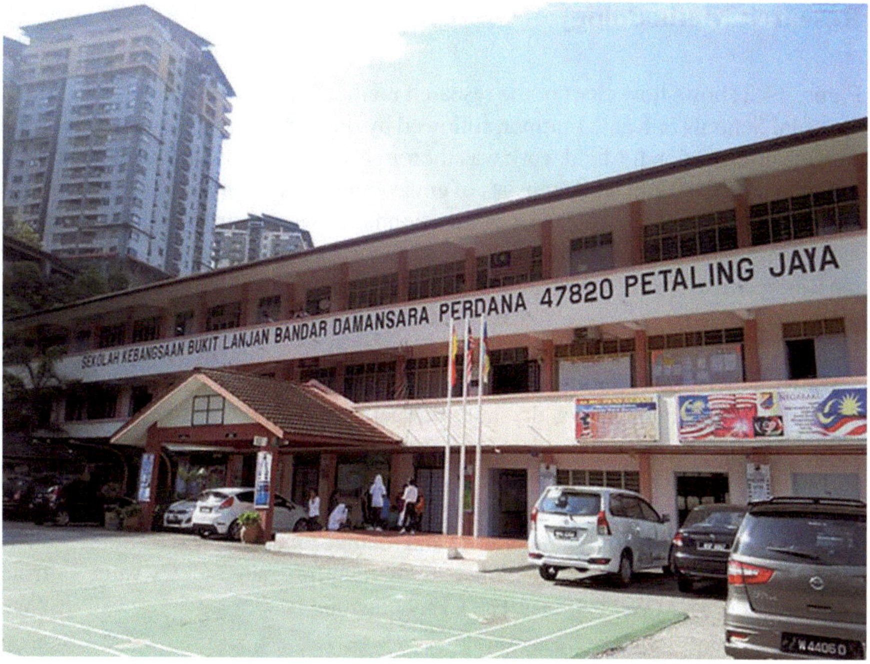

Fig. 14.1 An example of National School

of rows of classrooms, linked with a long but 6-ft.-wide corridor. Its roofs are pitched with slope at 20° (Syed Fadzil 1995). A study carried out in 1971 has concluded that National School classrooms were uncomfortable particularly in the afternoon (Sheath and Vickery 1971). On top of the empirical measures, the teachers and students also complained about the classrooms' thermal discomfort. It was detected that the thermal discomfort arised from un-shaded large opening, which had caused direct solar radiation penetrating into the classrooms.

In 1990, a similar study was led by Keumala where four National Schools and one Colonial School were evaluated. Keumala (Keumala 1990) used Predicted Mean Vote (PMV) index to determine the comfort level of the classrooms. It was found that the national schools were uncomfortable but the colonial school classroom achieved thermal comfort at most time during school hours in the morning as well as the afternoon (Keumala 1990). However, Keumala concluded that the colonial school achieved thermal comfort in its classroom due to its location, being surrounded by greenery of Lake Garden, a butterfly park and a bird park (Keumala 1990). Since the Colonial School was proven to be more comfortable than the National School, this study investigates the potential of other Colonial Schools in adapting to Malaysia climate and achieving thermal comfort.

Research Methodology

Figure 14.2 shows flow chart of the research methodology. It starts with selection of Colonial Schools in Kuala Lumpur, followed by on-site observations of the selected schools. Preparation for field work was thoroughly made before the commencement of field work. The field work consists of environmental measurement of outdoor and indoor environment. Comfort survey questionnaires were distributed to the students occupying the classroom while measurements were taken. The raw data from the measurements were collected, and comfort analysis was carried out with two thermal comfort indexes, Corrected Equatorial Comfort Index (CECI) and Tropical Summer Index (TSI).

Environmental measurements were taken during field work and they are outdoor dry bulb temperature, indoor dry bulb temperature, indoor relative humidity and indoor air velocity. These raw data were used in the calculation of CECI and TSI.

In 1951, Webb (Webb 1960) carried out a study on comfort scales for warm and humid regions. The comfort scale involves dry bulb temperature, wet bulb temperature (derived from dry bulb temperature and relative humidity) and air velocity. The neutral value of CECI is 26 °C, whereas the comfort range CECI value is from 24.5 °C to 26.7 °C (Webb 1960). The CECI equation is

$$\mathrm{CECI} = \tfrac{1}{2}(t + t_w) - \tfrac{1}{4}v^{\frac{1}{2}}$$

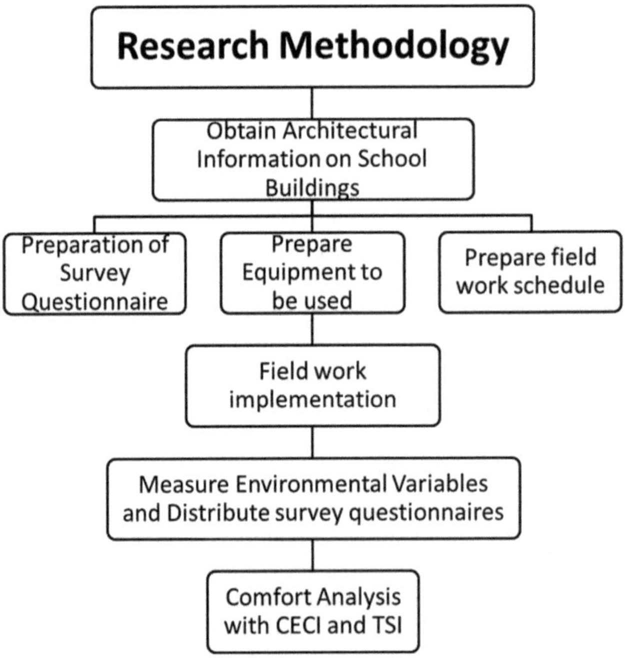

Fig. 14.2 Research methodology chart

where t = dry bulb temperature, °C, t_w = wet bulb temperature, °C and v = air velocity, m/s.

Years later, in 1986, Sharma and Ali (1986) came out with an index called Tropical Summer Index (TSI). TSI is created to predict thermal comfort level for warm-humid and hot-dry climate. The optimum value of TSI is 27.5 °C, and the comfort range value of TSI is from 25.0 °C to 30.0 °C. The following is the calculation for TSI:

$$\mathrm{TSI} = \tfrac{1}{3} t_w + \tfrac{3}{4} t_g - 2v^{1/2}$$

where t_w = wet bulb temperature, °C, t_g = globe temperature/dry bulb temperature, °C and v = air velocity, m/s.

Reflection of Vernacular Design in Colonial Schools in Malaysia

Vernacular Approaches to Providing Thermal Comfort

Vernacular architecture is a shelter designed according to local needs, using local products and responsive to local climate. Meanwhile, traditional Malay vernacular design concepts are purely responding to the warm and humid Malaysia climate. Originally, local villagers designed the vernacular houses with absolute understanding on local climate and their own needs for living. Figure 14.3 illustrates the primary concepts of a Malay Vernacular House and their responses to the local climate.

In hot-humid tropical region, vernacular architecture is well known as the best example of building designs that adapts successfully to the environment. Therefore, any contemporary building that intents to apply bioclimatic design should take precedent of traditional building. Hence, in Malaysia, colonial and modern buildings could use the vernacular Malay houses' passive design strategies to achieve the best solution for human comfort level.

Most of the configuration and design elements of the vernacular Malay houses are dedicated for the provision of promoting airflow in the interior space. Houses are usually arranged in random manner and located far apart from each other to minimise the obstruction of air stream and maintain the velocity of the wind, so that the whole neighbourhood is properly ventilated. The provision of air inlet and air outlet in the roof construction of vernacular Malay houses enable the convective cooling to work effectively. The vernacular Malay houses usually have high ceiling to allow warm air to rise, and then escape out near the ridge. The escaped warm air will be replaced by cooler air coming through the eaves so that the whole house is well ventilated. The houses also have floor-to-ceiling openings to allow air ventilation, especially at the occupants' body level. The openings are located opposite to

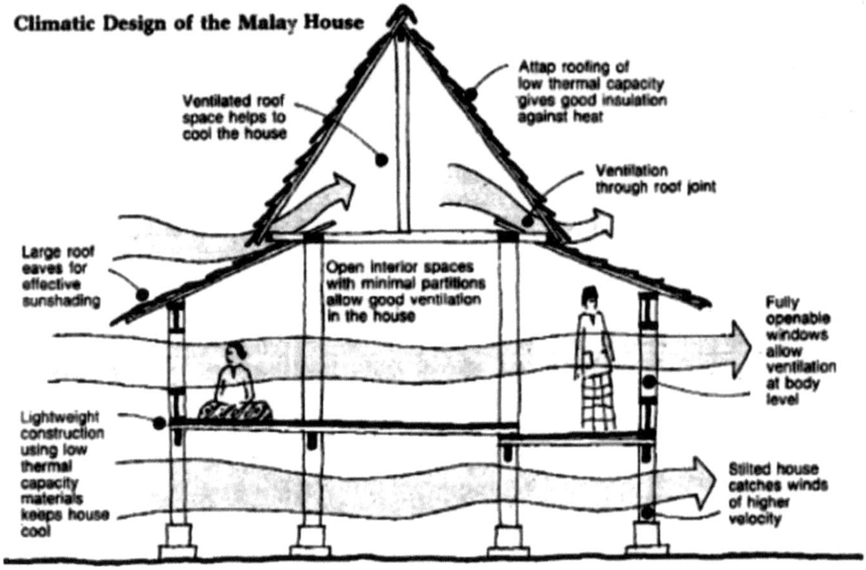

Fig. 14.3 A cross section of a typical Malay Vernacular House (Lim 1987)

each other and there is minimal obstruction in the space to allow cross ventilation. The natural low-density building materials for the construction of the houses also permit the penetration of air to further cool the interior spaces.

Lightweight materials such as timber walls are used for low thermal mass or capacity. This provides comfortable environment during resting and sleeping hours. The roofs are pitched to reduce solar heat gain while creating air space in the roof. The pitch roof also provides fall for heavy rainfalls, which is a common tropical weather. Vernacular houses are raised on stilts to capture higher air velocity on the higher level and to protect occupants from flood and wild animals. The houses are best orientated along the East West direction to minimise solar heat gain. They have large overhangs that shade the walls and windows from solar radiation and rain. Due to large overhangs, lighting level in the house is low lit, giving the occupants cooling effect psychologically.

In a housing estate, building blocks are usually spread out to allow uniform and adequate wind flow around each building block. Around the house compound, tall trees such as coconut trees are planted to provide natural shade and allow air flow under the trees canopies (Lim 1987).

As mentioned earlier, Colonial Architectural style in Malaysia began from the seventeenth until the twentieth century of British colonisation (Heritage of Malaysia Trust 1990). The colonist had attempted to adopt their own architectural individuality and styles in adapting to the local climate and culture of the multi-racial Malaysian society. Hence, these colonial buildings are combination of styles from other cultures, such as Indian and Chinese migrants, and the local Malay traditions.

Fig. 14.4 Victoria Institution

Among others, school building design is one of the many architectural typologies involved in portraying its architectural identity of the colonial period. The study took samples from three distinguished neo-classical architecture Colonial School buildings within the vicinity of Kuala Lumpur. These schools are Victoria Institution, Methodist Girl Secondary School and Maxwell School. Since these colonial schools were built in response to the Malaysian climate, they were assumed to be thermally comfortable. The study attempts to measure the truth to the former inference.

Victoria Institution, Kuala Lumpur

Victoria Institution (V.I.) was first established in 1893 and opened in July 1894 (Fig. 14.4). Victoria Institution was initially located in High Street (now Jalan Tun H. S. Lee). Due to reoccurring flood from nearby Klang River, the school building was relocated to Jalan Hang Tuah (Fig. 14.5), which was formerly known as Shaw Road then. The relocated V.I. is the present V.I. building, which dates from 1929.

Located in the urban site of Kuala Lumpur, it is surrounded by busy main roads with a stadium to the North West of the school and grass field on the South East (Fig. 14.5). Its surrounding buildings are at least five storeys high. There are big trees along the South West, South and North East side of the classroom block, which give shade to the classroom facing towards the South West, South East and North East. In addition to that, there is a grass field on the South East of the classroom block, separated by tall trees that are surrounding the classroom block and there is the National Stadium to the North West. The school is located on a hill, at 59.7 m

Fig. 14.5 Site location of Victoria Institution

above sea level. This is the highest location among the three selected colonial schools. Its classroom block is linear shaped as wide as one classroom with corridors along both sides of the classrooms wall (Fig. 14.6).

Methodist Girls Secondary School, Kuala Lumpur

Methodist Girls Secondary School (Fig. 14.7) was founded in 1896 by a group of Christian missionaries who came over to Malaya with the intention to expand Christianity through the education of young people.

It was one of the earliest all-girls schools in Malaysia. During the Second World War, the school was used as Japanese Military Headquarters. It was reopened and operated back as school in 1945 after the war. This school is also located within the urban context of Kuala Lumpur (Fig. 14.8). However, it differs slightly from Victoria Institution since it is surrounded by small roads. In the bigger context, the school is a neighbour to the National Mosque to the South, Lake Garden, as well as bird and butterfly parks to the west. The school is located at 49.1 m above sea level on a hill, at a higher level than the compounds surrounding it. The school compound comprises of several building blocks. Designed with environmental and climate consideration, the buildings blocks are narrow with the width of one classroom and corridors along both sides of classroom walls. On top of that, trees are planted around the perimeter of the school compound.

Fig. 14.6 Classroom block in Victoria Institution

Maxwell Secondary School, Kuala Lumpur

Maxwell Secondary School was initially established in the year 1917 and it was named after a road in honour of Sir William George Maxwell. The school consists of two blocks, which are the main block for office and classrooms and a library block. The design and features of the two blocks are typical of the colonial era structures bearing some similarities with the Malayan Railway station and Sultan Abdul Samad buildings, both with Moorish designs (Fig. 14.9). The school is located within the urban context of Kuala Lumpur at 37.8 m above the sea level. The school compound is quite compact with surrounding buildings located very near to the school building. There are also thick layers of trees to the North West of the school site, an upward inclining road to the South West, a flyover to the South East, and a tall building to the North East of the school site (Fig. 14.10). These natural and manmade urban textures surrounding the site could easily obstruct the wind flow

Fig. 14.7 Methodist Girls Secondary School

Fig. 14.8 Site location of Methodist Girls Secondary School

Fig. 14.9 Maxwell Secondary School 1917 to present

Fig. 14.10 Site plan of Maxwell Secondary School

Fig. 14.11 Maxwell Secondary School floor plan of the classroom block

into the school compound. However, these urban elements could also provide shade from direct solar radiation.

The school consists of a two-storey structure which was later extended to a block of solid masonry construction with arched openings along the corridors, typical of many of the school buildings that were built that time. There is a veranda on both sides of each storey (main block), high ceilings and broad stairways which depict typical colonial architecture (Fig. 14.11).

Summary

Table 14.2 summarises the reflection of vernacular design strategies in Colonial school buildings in Kuala Lumpur. There are five vernacular design strategies found in the three colonial schools; pitched roof, long overhangs, openings opposite to each other, corridors acting like verandas and floor-to-ceiling openings. However, the openings in Maxwell Secondary School were not fully utilised which affected the actual comfort level inside the classroom.

The colonial schools have other passive design strategies that are different from traditional vernacular design, which the British believed that they could contribute in creating indoor comfort in the classrooms. Contradicting to the traditional

Table 14.2 Adoption of vernacular design (Fig. 14.9) in colonial school buildings in Kuala Lumpur

Vernacular design	Victoria Institution	Methodist Girls Secondary School	Maxwell Secondary School
Pitched roofs to reduce solar heat gain and create air space. Long roof overhangs	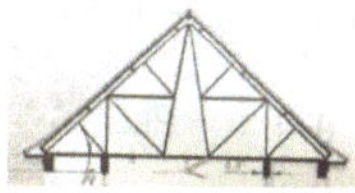 Roof angle = 43° Overhang length from classroom wall = 3.75 m	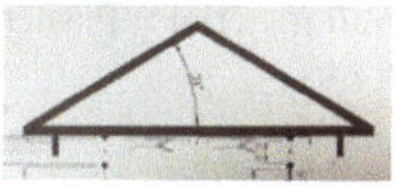 Roof angle = 31° Overhang length from classroom wall = 3.20 m	 Roof angle = 28° Overhang length from classroom wall = 3.90 m
Openings opposite to each other to allow cross ventilation	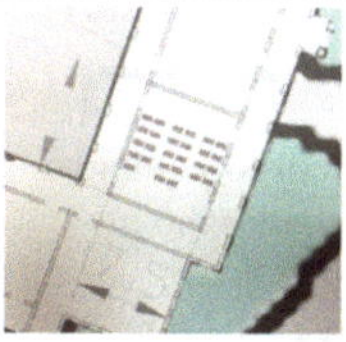 Five doors and 6 grilled openings above door level	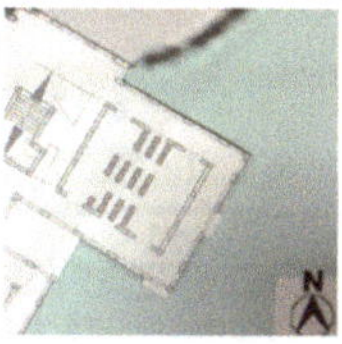 Six doors and 6 grilled openings above door level	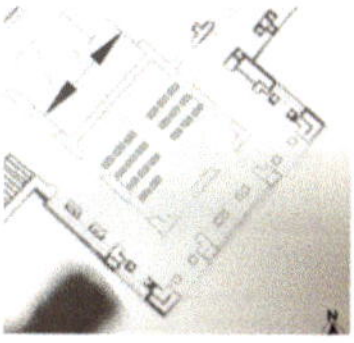 Four doors (only 2 were used) and 2 windows
Floor-to-ceiling openings to allow air ventilation at all height level	 Opening-to-wall ratio = 1:5 Total area of opening = 5.85 m^2	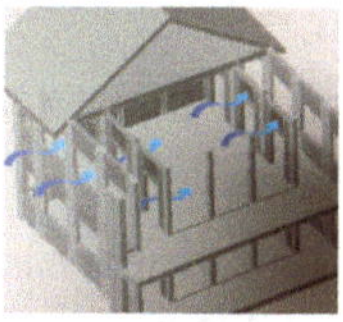 Opening-to-wall ratio = 1:4 Total area of opening = 4.76 m^2	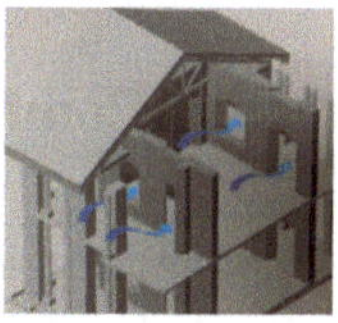 Opening-to-wall ratio = 1:13 Total area of opening (used) = 5.14 m^2
Corridors as transition zone between outdoor and indoor, similar to veranda	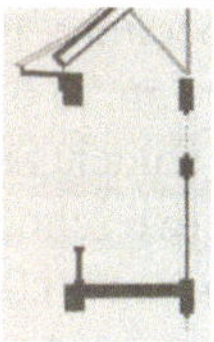 Corridors are located along all exposed classroom walls Width of corridor = 2.17 m	 Corridors are located along all exposed classroom walls Width of corridor = 1.75 m	 Corridors are located along all exposed classroom walls Width of corridor = 2.4 m

vernacular design, the British believe that thick walls could be a better option as it can delay solar heat transfer into the classroom and high ceiling creates larger volume which could provide better air exchange rate. Table 14.3 summarises the passive design strategies introduced by the British to colonial school buildings.

Table 14.3 New passive design strategies found in colonial school buildings

New passive design strategies	Victoria Institution	Methodist Girls Secondary School	Maxwell Secondary School
Thick thermal mass to delay heat transfer into the classrooms	Wall thickness = 0.30 m Wall material = Plastered bricks	Wall thickness = 0.25 m Wall material = Plastered bricks	Wall thickness = 0.50 m Wall material = Plastered bricks
High ceiling and high volume for better air exchange rate	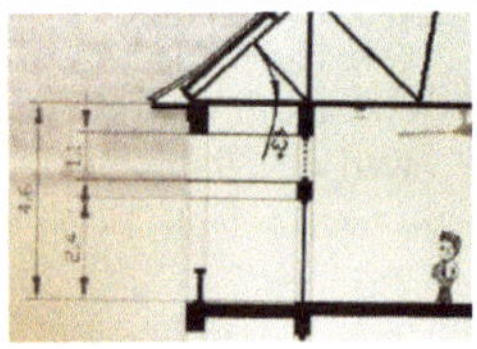 Floor-to-ceiling height = 4.6 m Total floor area = 62.6 m^2 Total volume = 288.1 m^3	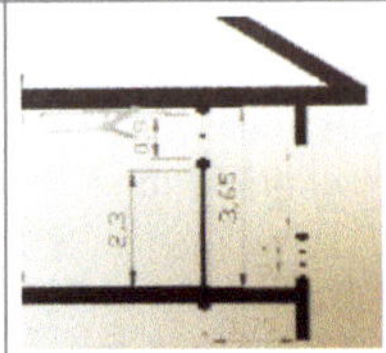 Floor-to-ceiling height = 3.6 m Total floor area = 62.6 m^2 Total volume = 228.6 m^3	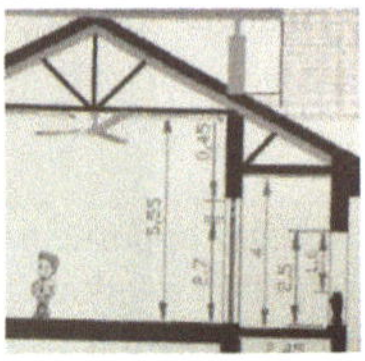 Floor-to-ceiling height = 5.55 m Total floor area = 50.8 m^2 Total volume = 281.75 m^3

Thermal Comfort Analysis of Colonial School Buildings

As stated above, thermal comfort analysis was carried out using Corrected Equatorial Comfort Index (CECI) and Tropical Summer Index (TSI). Thermal comfort range differs slightly between the two Indexes; CECI comfort range is between 24.5 °C and 26.7 °C and CECI neutral is 25 °C. Meanwhile, TSI comfort range is between 25.0 °C and 30.0 °C and TSI optimum comfort value is 27.5 °C. Both indexes have similar empirical variables, which are dry bulb temperature, relative humidity and air velocity. All measurements were taken from 7:30 a.m. to 6:00 p.m. The schooling hours in which the classrooms are occupied with students are from 7:30 a.m. to 1:30 p.m. for VI and Maxwell. Meanwhile, MGS has two schooling sessions, morning from 7:30 a.m. to 1:30 p.m. and afternoon 2:00 p.m. to 6:00 p.m. In addition to empirical measurements, thermal comfort surveys were conducted occasionally and simultaneously.

Each school were measured in 2 days. Figures 14.12 and 14.13 are graphs of CECI for Victoria Institution, Methodist Girls Secondary School and Maxwell Secondary School.

CECI predicted all three classrooms' experiences of thermal comfort during most of the schooling hours. The comfort evaluation survey results show that the students occupying the classrooms were all agreeable to CECI prediction (Figs. 14.12 and 14.13). There is one disagreeable in Fig. 14.13, in Victoria Institution at 11:30 a.m. Only 60% students felt comfortable due to direct solar radiation from morning sun that enters into the classroom. This could be solved by adding appropriate shading devices according to sun angles. In Fig. 14.12, there are hours when CECI predicted

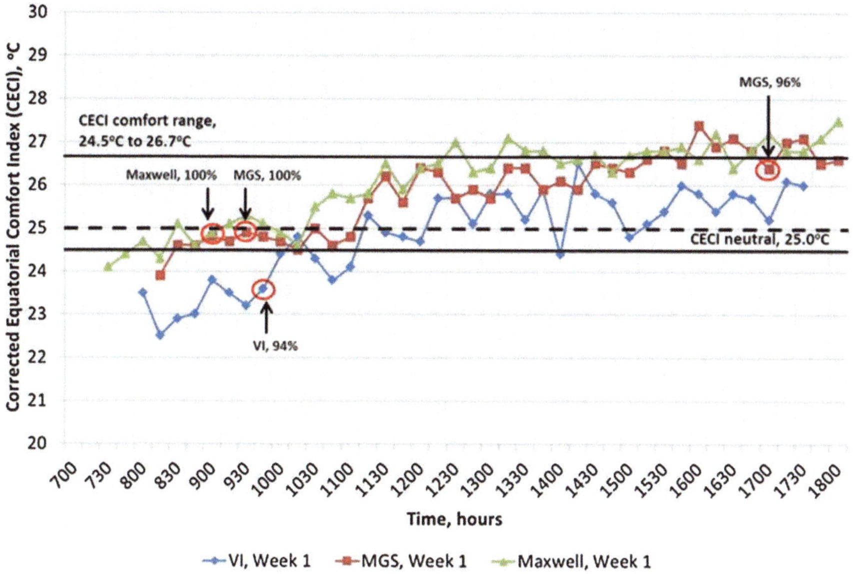

Fig. 14.12 CECI and thermal comfort survey in Week 1 for Victoria Institution (VI), Methodist Girls Secondary School (MGS) and Maxwell Secondary School

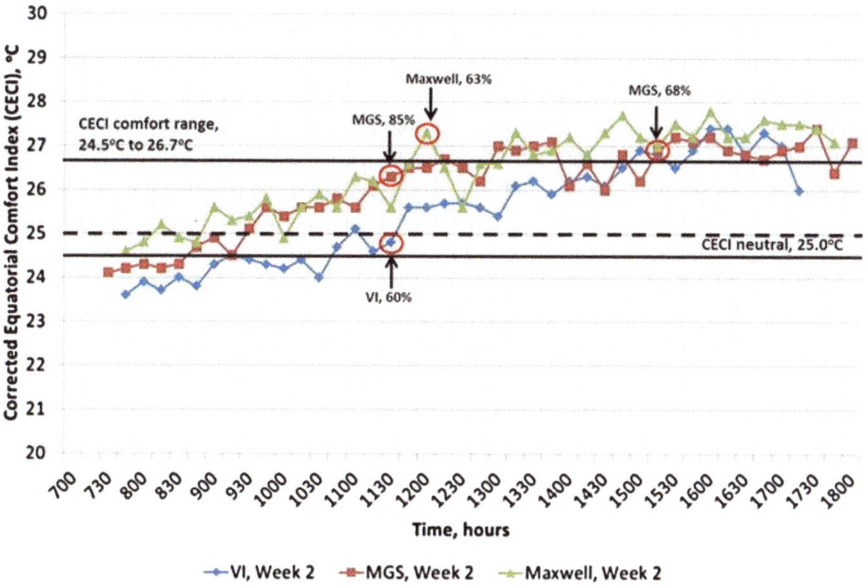

Fig. 14.13 CECI and thermal comfort survey in Week 2 for Victoria Institution (VI), Methodist Girls Secondary School (MGS) and Maxwell Secondary School

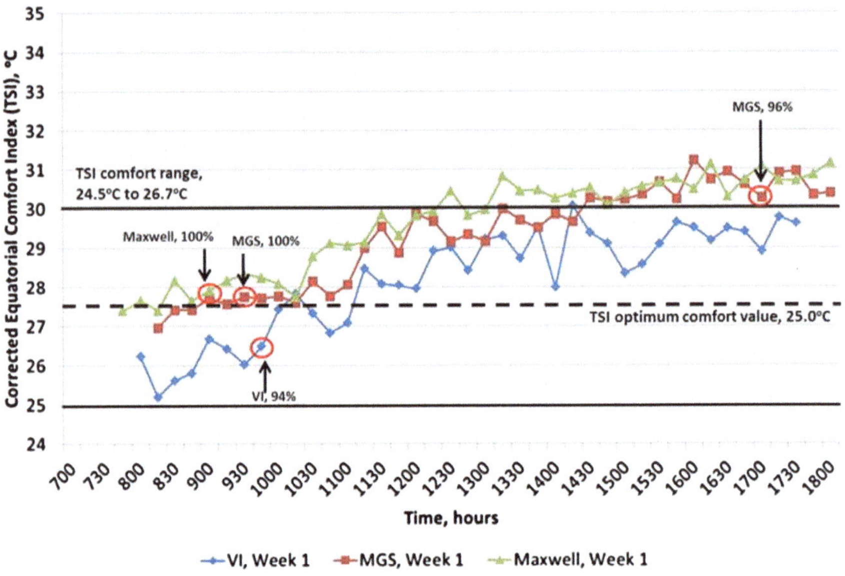

Fig. 14.14 TSI and thermal comfort survey in Week 1 for Victoria Institution (VI), Methodist Girls Secondary School (MGS) and Maxwell Secondary School

the classroom to be uncomfortably cold. However, the occupancy survey shows that 94% of the occupants find the classroom comfortable.

Total schooling hours of morning and afternoon sessions are 11 h. In average of Week 1 and 2, Victoria Institution classroom achieves thermal comfort for 6 h within the schooling hours. Meanwhile, Methodist Girls Secondary School classroom achieves thermal comfort in average for 7 h within the schooling hours and Maxwell Secondary School classroom achieves thermal comfort in average for 6 h within schooling hours. Therefore, according to CECI, which was agreeable with occupancy thermal comfort survey, colonial school classrooms are comfortable for more than half of the schooling hours.

Figures 14.14 and 14.15 are graphs of TSI for Victoria Institution, Methodist Girls Secondary School and Maxwell Secondary School.

TSI predicts that all classrooms were comfortable in the morning session, which the classroom occupants agree most of the time (Figs. 14.14 and 14.15). In Victoria Institution classroom, some students experienced discomfort at 11:30 a.m., due to direct solar that penetrates into the classroom. In average, TSI predicts that Victoria Institution classroom experienced thermal comfort for 9 h, Methodist Girls Secondary School classroom experienced thermal comfort for 7 h and Maxwell Secondary School classroom experienced thermal comfort for 5 h. According to TSI, Victoria Institution and Methodist Girls Secondary School classrooms are thermally comfortable for more than half of the total schooling hours. Maxwell Secondary School classroom is thermally comfortable only in the morning session

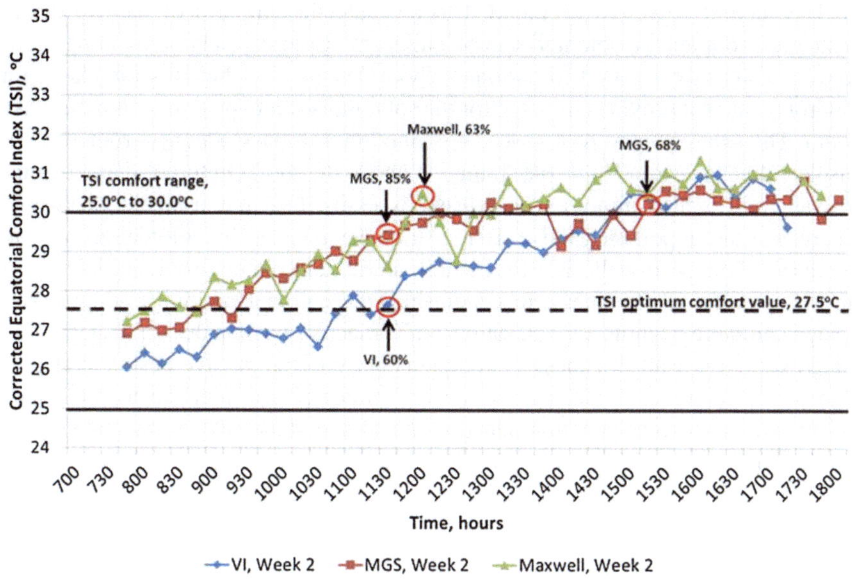

Fig. 14.15 TSI and thermal comfort survey in Week 2 for Victoria Institution (VI), Methodist Girls Secondary School (MGS) and Maxwell Secondary School

of the schooling hours. Maxwell School classroom has a pitched roof, coherent with vernacular passive design strategy as the Victoria Institution and Methodist Girls Secondary School. However, Maxwell School has the least roof angle and roof air space that acts as insulation from solar radiation above the roof. Furthermore, Maxwell classroom has the least opening used during field study, 1:13 window-to-wall ratio, as compared to Victoria institution 1:5 and Methodist Girls Secondary School 1:4. Roof design and ventilation have made an impact to thermal comfort of the classrooms.

Conclusion

The first objective is to assess thermal comfort of the occupants in the classrooms of colonial school buildings. Thermal comfort was measured empirically and qualitatively through post-occupancy surveys. Corrected Equatorial Comfort Index (CECI) and Tropical Summer Index (TSI) were used to predict thermal comfort of each school classroom. According to CECI, all three schools achieved thermal comfort for more than half of the schooling hours. Meanwhile, TSI predicted all three schools were thermal comfortable during the morning schooling hours and only Victoria Institution and Methodist Girls Secondary School are thermally comfortable in more than half of the afternoon schooling hours.

The second objective is intended to evaluate the strength of passive design strategies adopted in colonial school buildings. Colonial school buildings have thicker walls as compared to national school design. This benefits the classroom environment as it provides time lag for the solar radiation to penetrate into the classroom. The same goes to roof space. It is important to have adequate roof space to insulate the classroom from solar radiation above the roof. On top of that, opening-to-wall ratio is also important to release excessive heat out from the classroom. Therefore, the important passive design strategies adopted by colonial schools are adequate window-to-wall ratio for natural ventilation, pitched roof, sufficient roof air space and wall thickness. In conclusion, the colonial buildings adopted roof design and large opening-to-wall ratio from vernacular passive design strategies and added sufficient wall thickness as complementary to the passive design strategies. The newly formed colonial passive design strategies are intended to protect the classrooms from solar radiation penetration and simultaneously allow good natural ventilation.

References

Heritage of Malaysia Trust (1990) A handbook Malaysian architectural heritage survey. Badan Warisan Malaysia, Kuala Lumpur

Keumala NIM (1990) Post occupancy evaluation on classroom thermal comfort level of schools in the federal territory of Kuala Lumpur during hot seasons using predicted mean vote index. MPhil thesis, Faculty of Built Environment, Universiti Teknologi Malaysia

Lim JY (1987) The Malay house, rediscovering Malaysia's indigenous shelter system. Institut Masyarakat, Malaysia

Mohamad M (2008) Chapter 1 Education in Malaysia: a historical review in education in Malaysia, a journey to excellence, Ministry of Education, Kuala Lumpur

Sharma MR, Ali S (1986) Tropical summer index – a study of thermal comfort of Indian subjects. Build Environ 21(1):11–24

Sheath RH, Vickery DJ (1971) A study in the utilization, design and cost of secondary schools in Malaysia, UNESCO, Asian Regional Institute of School Building Research

Syed Fadzil SF (1995) Thermal and lingthing consideration for the prototype schools of Malaysia. PhD thesis, Faculty of Built Environment, The Welsh School of Architecture, University of Wales College of Cardiff

Webb CG (1960) Thermal discomfort in an equatorial climate – a nomogram for the equatorial comfort index. J Inst Heating Ventilating Eng 27:297–304

Chapter 15
Sustainability Principles and Features Learned from Vernacular Architecture: Guidelines for Future Developments Globally and in Egypt

Mohsen Aboulnaga and Mona Mostafa

Introduction

Vernacular architecture is representing native, domestic (local) and indigenous buildings in a country. It reflects the environmental, cultural and historical context in which it existed. Vernacular architecture uses available materials to fulfil the local needs. It is almost by definition "A sustainable architecture" since it utilized locally available resources without exhausting them. Also, vernacular architecture is defined as the architecture that utilises methods of construction and traditions to address local desires and culture. In addition, vernacular architecture tends to evolve over time to reflect the environmental, cultural and historical context in which it exists. Nevertheless, it is influenced by a wide range of different aspects of human behaviour and environment, leading to different building forms for almost every different context as shown in Fig. 15.1. A vernacular structure is characterized by two main issues: (a) inexpensive materials and (b) practical design. Box 15.1 outlines the meaning of Vernacular architecture

Review of Leading Historical and Global Vernacular Architecture

Vernacular architecture has spread all over the globe; from Latin America to Europe and Asia and Africa. Figs. 15.2, 15.3, 15.4, 15.5, and 15.6 depicts several examples from different parts of the world, mainly Peru, UK, Denmark, Italy, China and Japan.

M. Aboulnaga (✉)
Department of Architecture, Faculty of Engineering, Cairo University, Giza, Egypt

M. Mostafa
Modern Sciences and Arts University, Giza, Egypt

A. Sayigh (ed.), *Sustainable Vernacular Architecture*, Innovative Renewable Energy, https://doi.org/10.1007/978-3-030-06185-2_15

Fig. 15.1 Vernacular buildings in Ghent, North of Belgium. *Photo credit*: Author

Box 15.1: Vernacular Architecture
"Vernacular Architecture is the one that reflects local traditions and culture, yet built by local citizens using local materials".

Historically, vernacular architecture was originated when mankind was forced to make use of the natural resources around him/her and to provide a comfortable shelter, which responds to their needs and the microclimate. Such architecture is a mere reaction to individual's or society's needs. Vernacular architecture was even constructed, as a shelter to meet inhabitants' needs, even before architects started their attempt to do so.

Vernacular Architecture and Green Building
By looking at vernacular architecture, it adheres to basic green architectural principles of energy efficiency and utilizing materials and resources in close proximity to the site, yet meets climate-responsive building (Box 15.2). Such adherence was centuries before the establishment of green building principles in the USA around the year 2000 and the emerged rating system in 2002. In principle, these structures capitalize on the native knowledge of how buildings can be effectively designed as well as how to take advantage of local materials availability and natural resources. Even in an age where materials are available well beyond our region, it is essential

Chan Chan, a Mud-Brick adobe city in Trujillo

Fig. 15.2 Vernacular architecture in Peru. *Image Source*: http://amazingdata.com

A farm house in London

Fig. 15.3 Vernacular architecture in the UK. *Image source*: http://cdn.wn.com

to take into account the embodied energy lost in the transportation of these goods to the construction site.[1] The vernacular building construction techniques and specifications are more based on knowledge achieved by trial and error rather than conventional practices. This provides a good solution to the climatic constraints. Vernacular

---

[1] https://www.archdaily.com/155224/vernacular-architecture-and-the-21st-century (Accessed on 28. 02.2018).

Fig. 15.4 Vernacular architecture in Europe. *Image source*: www.greatbuildings.com

Fig. 15.5 Vernacular architecture in Puglia, Italy. *Image source*: https://whc.unesco.org/en/list/787

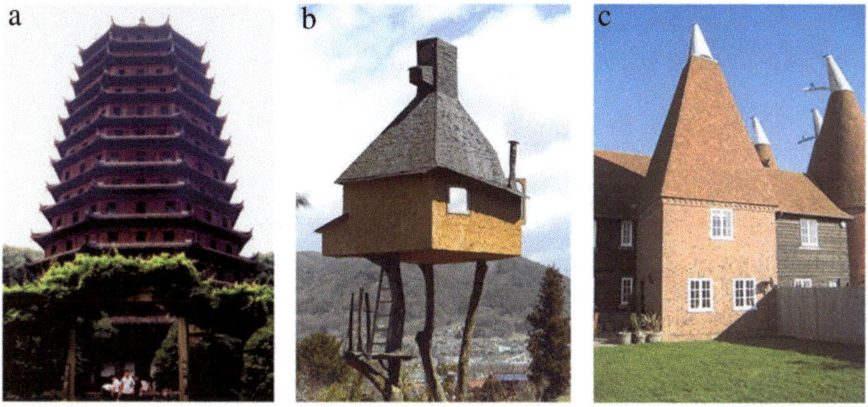

Fig. 15.6 Vernacular buildings in China, Japan and the UK. (**a**) The Liuhe Pagoda, Hangzhou, China. (**b**) Tea house, Chino-Nagano, Japan. (**c**) A house, Kent, UK. *Image source*: http://ashfield-hansendesign.blogspot.com

Box 15.2: Vernacular Architecture Adheres to Green Principles
"Vernacular architecture adheres to basic green architectural principles of energy efficiency, utilizes local materials and resources in close proximity to the site and yet meets climate-responsive building requirements".

Box 15.3: Vernacular Architecture Context
"Vernacular Architecture is sometimes known as local architecture".

"Vernacular built heritage can be perceived as 'the essence of sustainability'".

architecture has more than one approach in solving the same climatic constraint (Soleymanpour et al. 2015).

Vernacular architecture—sometimes is referred to as traditional or local architecture—is the fundamental expression of the culture of the community. In more specific terms, it reflects the relationship with territory, and meanwhile it depicts the expression of the world's cultural diversity, which can be seen in its forms, colours and structure as well as heritage features (Box 15.3). Moreover, the vernacular built heritage can be perceived as "the essence of sustainability", being constructed with local materials and the minimum waste of resources. In most countries, vernacular built heritage is generally neither protected nor considered worthy to be conserved (De Filippi 2006).

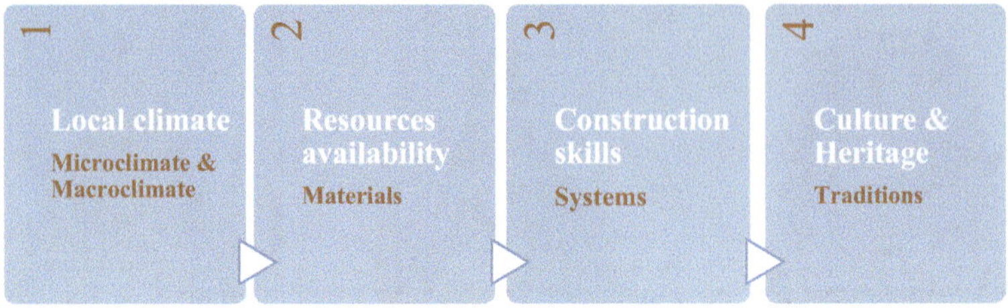

Fig. 15.7 Factors affecting vernacular architecture

Globally, vernacular architecture is extremely vulnerable and facing serious problems of obsolescence, internal equilibrium and integration. This is mainly due to the fact of homogenization of culture and global socio-economic transformations. Furthermore, these buildings are vulnerable to climate change risks (De Filippi 2006). Thus, protecting vernacular architecture as a symbol of communities' cultural heritage is essential and must take priority nowadays.

Factors Influencing Vernacular Architecture

Factors affecting vernacular architecture vary from region to another but nonetheless they have many factors in common as shown in Fig. 15.7. These factors are interlaced somehow.

From the literature, we can note that factor number 1, 2 and 3 are the most prime factors influencing most of the vernacular architecture in Africa. However, in the Middle East, factor number 1 and 4 are the most predominant factors.

Regional Vernacular Architecture

Africa and Middle East have witnessed rich examples of vernacular architecture. The vernacular architecture in Africa has been characterized by forms and materials used in developing such unique architecture. These materials depend on natural resources, mainly wood and tree residuals. This section presents examples from the African continent and Middle East. For Africa, it is divided into five zones namely, North Africa, East Africa, Central Africa, West Africa and South Africa (Fig. 15.8, Tables 15.1 and 15.2).

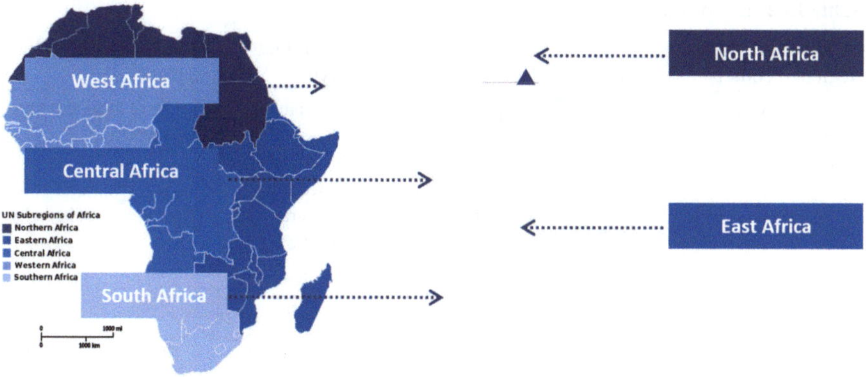

Fig. 15.8 Africa's five regions according to the United Nations Geo-scheme for Africa. *Source*: https://openi.nlm.nih.gov/detailedresult.php?img=PMC3966389_13-1128-F1&req=4#

Table 15.1 List of African nations that exhibit vernacular architecture

North Africa (7)	East Africa (20)		Central Africa (9)	West Africa (17)		South Africa (5)
Algeria	Burundi	Mozambique	Angola	Benin	Mauritania	Botswana
Egypt*	Comoros	Reunion	Cameroon	Burkina Faso	Niger	Lesotho
Libya	Djibouti	Rwanda	Central African RRepublic	Cape Verde	Nigeria	Namibia
Morocco	Eritrea	Seychelles	Chad	Ivory Coast	Saint Helena	South Africa
Sudan	Ethiopia	Somalia	D. Republic of Congo	Gambia	Senegal	Swaziland
Tunisia	Kenya	South Sudan	Republic of the Congo	Ghana	Sierra Leone	
Western Sahara	Madagascar	Tanzania	E. Guinea	Guinea	Togo	
	Malawi	Uganda	Gabon	Guinea-Bissau		
	Mauritius	Zambia	São Tomé & Príncipe	Liberia		
	Mayotte	Zimbabwe		Mali		

**Egypt will be discussed in a separate section.*

Table 15.2 List of vernacular architecture in African countries

African region	African country	Building image	Building type	Materials, form and elements
Zone 1: North Africa	Algeria	 Oasis Rouge Hotel exterior *http://freevst.x10.mx/sahara/algeria/ethnography/algeria%2008.jpg*	Hospitality hotel (currently used as a cultural centre)	– Use of local red ochre adobe material[a].
	Libya	 Old Town of Ghadames clusters *http://www.amusingplanet.com/2013/03/the-old-town-of-ghadames.html*  Exterior and interior of Ghadames, Tripolitania *http://endangeredsites.leadr.msu.edu/ghadames/*	Residential	– Clustered mud-brick-and-palm houses resembling a honeycomb, – Typical vertical architecture (ground floor for storage, first floor for family and top roof for women), – Connected rooftops creating walkways to allow women to move freely concealed from men's view, – Overhangs cover the alleys between houses creating shaded pathways[b].
	Morocco	 Kasbah Ruins-Ait Benhaddou *http://openwalls.com/image?id=1333*  The old city wall *https://decideyouradventure.com/ait-benhaddou-southern-morocco/*	Residential	– Group of earthen buildings surrounded by defensive walls and reinforced by corner towers, – Perfectly adapted to the climatic conditions and are in harmony with natural and social environment, – Presence of conserved decorative motifs[c]

(continued)

Table 15.2 (continued)

African region	African country	Building image	Building type	Materials, form and elements
		 Chefchaouen City, Morocco *http://www.socimage.net/user/vicdady/275274010/1634257877203353586_275274010*	Residential	– City's distinctive shades of blue set against white and muted creams, – Visually appealing architecture elements such as doors and gates[d]
	Sudan	 Tukel—Typical Sudanese home *http://naturalhomes.org/african-vernacular*	Residential	– Traditional Sudanese home (Tukel), – Built from strands of small branches bounded together with twine[e]
	Tunisia	 Ksar Ouled Soltane, Tunis *https://i.pinimg.com/564x/7c/6a/45/7c6a45124b127f8954340d734d35ae65%2D%2Dunusual-houses-star-wars.jpg*	Residential	– Simple multi-level adobe mud fortified building, – Cool shelter during arid weather, – Steep outdoor staircases
		 Ouled Ech Cherif exterior—Nefta *http://looklex.com/tunisia/nefta05.htm*	Residential	– Elaborate brick patterns, – White colour has been used, usually painted over the bricks, – Doors here are far less decorated[f]
		 Great Mosque of Sidi Salem *http://looklex.com/tunisia/nefta04.htm*	Religious (Mosque)	– Simplicity is the key as even the inner parts are as unadorned as the exterior[g]

(continued)

Table 15.2 (continued)

African region	African country	Building image	Building type	Materials, form and elements
	Western Sahara	 Iwan Baan photography *http://42mzqz26jebqf6rd034t5pef-wpengine.netdna-ssl.com/wp-content/uploads/2016/06/Screen-Shot-2016-06-03-at-4.46.21-PM.png*	Residential	– Mud architecture, – Small openings, – One-storey buildings
Zone 2: East Africa	Burundi	 Children's Library in Burundi *https://www.designcauseinc.org/single-post/2016/04/12/Blending-traditional-and-modern-to-create-successful-humanitarian-architecture*	Cultural (Library)	– Compressed earth blocks (CEB) and baked clay roof tiles, – Based on study of Burundi's vernacular architecture, – Hallway porch provides shelter from rain and sun, – Blinder controls access, – Roof is slopped with an overhang to protect the CEB, – Ventilation and daylighting[h]
	Comoros	 Moroni waterfront *http://www.wondermondo.com/Comoros.htm*	Residential	– Historical centre of old African-Arabic trade town, – Characteristic maze of narrow streets, – Houses are made of dark basalt plastered with coral lime[i]
	Djibouti	 Dijibouti's Toukouls *http://www.visitdjibouti.dj/explorerEN*	Residential	– Branch-framed, transportable huts (toukouls), – Covered with woven mats or boiled bark is pulled into fine strands and plaited, – Use of local materials[j]

(continued)

Table 15.2 (continued)

African region	African country	Building image	Building type	Materials, form and elements
	Eritrea	 Eriteria's Hidmo dwellings *http://explore-eritrea.com/tour-6-discover-enchanting-culture-nature/tigrinya-tribe-family-and-hidmo-traditional-house/*	Residential	– Local materials construction called "hidmo", – Adobe material dwelling present mainly in the highlands[k]
		 Eriteria's Agudo dwellings *http://www.eritrea.be/old/eritrea-barentu.htm*	Residential	– Local materials construction called "agudo", – Thatch roofed hut that is common in the lowlands[l]
	Ethiopia	 Sidama people house *http://naturalhomes.org/african-vernacular*	Residential	– Split bamboo plaited round plan with pointy dome, – Designed to shed heavy rainfall where a circular dome would have a flat region prone to leaks[m]
	Kenya	 Kenya's traditional architecture *http://www.archidatum.com/articles/the-taita-sustainable-vernacular-house/*	Residential	– Traditional sustainable architecture termed "Taita", – Circular floor plans with conical thatched roof, – Mud walls and the roof ensure suitable temperature, – Deep roof overhangs so that walls are not exposed to direct sun, – Small operable windows[n]

(continued)

Table 15.2 (continued)

African region	African country	Building image	Building type	Materials, form and elements
	Madagascar	 Merina village, Highlands, Madagascar *http://www.travelmadagascar.org/CITIES/Antananarivo-surroundings.html*	Residential	– Red earth dwellings with pitched roof.
	Malawi	 House in Chasesa village *http://www.malawiarchitecture.com/mwanza*	Residential	– A veranda wraps around the home and placed on a platform, – Roof is supported by poles, – Local materials: burnt bricks and white plaster coated, – In the kitchen, a gap is left open on the top of the walls to release smoke[o]
	Mauritius	 Creole-Style house *By Jean Francois Guimbeau* *http://vintagemauritius.org/vintage-houses/old-colonial-creole-style-houses-mauritius-part-1/*	Residential	– A Creole-Style House, – Materials used include rocks, wood and some metals, – Symmetry is an important feature of this architecture[p]
	Mayotte	 Traditional houses in Mayotte *http://www.earthauroville.com/mayotte_as_development_model_en.php*	Residential	– Locally produced shelters: earth material, corrugated iron roof, or "Kripi" shelters which means earth plastered onto stone-filled framework[q]

(continued)

Table 15.2 (continued)

African region	African country	Building image	Building type	Materials, form and elements
	Mozambique	 School in Inhambane  Roof structure *https://www.archdaily.com/144527/educational-building-in-mozambique-andre-fontes-sixten-rahlff*	Educational	– Consists of two rooms, – Exploit RC bearing structure in the closed room, – Use of sandbags in east and north façade (thermal mass), – Roof extension provides shade from summer sun and collects rain water into cistern, – Shaded south facade has a glass-bottle wall for letting in light and keeping dust out, – Roof trusses structure allows natural ventilation[r]
	Reunion	 Reunion Creole architecture *https://atcnews.org/2015/02/20/creole-architecture-all-part-of-reunions-attractions/*	Residential	– It features stunning Creole architecture with high-quality detailed windows and doors[s]
	Rwanda	  Umubano Primary School, Kigali *By Iwan Baan, https://www.archdaily.com/372709/umubano-primary-school-mass-design-group*	Educational	– Use of local materials (brick and papyrus reeds), – Includes environmental strategies such as natural ventilation, and daylighting by the clerestory windows in the verendeel trusses in the roof[t]

(continued)

Table 15.2 (continued)

African region	African country	Building image	Building type	Materials, form and elements
	Seychelles	 Creole house, Victoria *http://www.seychellesnewsagency.com/articles/1652/The+traditional+Creole+architecture+-+An+aspect+of+the+Seychelles+heritage+in+need+of+preservation*	Residential	– Made of local material (wood), – Allows natural ventilation and daylighting by the presence of large doors and windows, high-pitched roof and large balconies[u]
	Somalia	 Somalia's hut *http://wiki.colby.edu/download/attachments/6294655/Houses+in+Somalia.pdf*	Residential	– Circular, dome-shaped dwelling made of local materials, – Roof made of thatching of bundles of grass[v]
	South Sudan	 South Sudan's hut (Tukuls) *http://sudantribune.com/spip.php?article46395*	Residential	– Grass-thatched or mud huts, popularly known as tukuls, – Roof is made of thatched straw[w]
	Tanzania	 Mud house in Dar Es Salaam (Svensson and Ekvall 2012)	Residential	The traditional Swahili houses are built from piles, grass and mud (Svensson and Ekvall 2012)

(continued)

Table 15.2 (continued)

African region	African country	Building image	Building type	Materials, form and elements
	Uganda	 Traditional hut in Uganda *http://www.monitor.co.ug/Magazines/HomesandProperty/The-hut%2D%2Dthe-first-building-in-Uganda/689858-1522996-r03yg4z/index.html*	Residential	– Small and simple hut, – Use of circular forms and local materials, – Types of hut differ according to ethnicities in Uganda[x]
	Zambia	 Zambia's Insaka structure *https://www.insaka.org/pages/about*	Residential	– Hut termed "Insaka" has two roles; socializing area in the village and as kitchen, – Form is round, with a thatched roof and sides can be left open or closed with low-height mud walls[y]
	Zimbabwe	 Zimbabwe's traditional hut *http://www.designindaba.com/articles/point-view/hut-decorating-competition-could-give-urban-homes-run-their-money*	Residential	– Traditional mud huts, – Decoration includes abstract patterns paint drawings[z]
Zone 3: Central Africa	Angola	 Traditional hut in Angola *http://www.africavernaculararchitecture.com/angola/*	Residential	– Round form hut, – Thatched roof made of straw

(continued)

Table 15.2 (continued)

African region	African country	Building image	Building type	Materials, form and elements
	Cameron	 Cllay Obos—Musgum people *http://naturalhomes.org/african-vernacular#southafrica*	Residential	– Local material, – Vents at the top of the building and small entrances with few, if any, windows, – High domes collect the hot air, moving it away from people sleeping providing efficient cooling[aa]
	Central African Republic	 Notre-Dame of Bangui Cathedral *http://www.dailyherald.com/article/20151129/news/311299977*	Religious (Church)	– Use of local red brick, – Symmetry, – Use of arches in Façade
	Chad	 Tradional Chadian Shelter *http://www2.exxonmobil.com/Chad/Library/Photo_Video/web_photoalbum/english/14g.html*	Residential	– Shelter is made of local materials such as: mud, bricks and straw, – Form is commonly round hut[ab]
	Democratic Republic of the Congo	 Village of Banseuba *http://www.endingextremepoverty.org/2010/04/an-upper-class-neighborhood-in-rural-congo.html*	Residential	– Use of brick with tin roofs as the village of Banseuba is considered a rural upper class neighbourhood[ac]
	Republic of the Congo	 La Basilique St-Anne-Congo *http://www.thedigitalglobetrotter.com/country-185196-republic-of-congo-a-wedding-in-a-great-african-family/*	Religious (Church)	– Dazzling green tiles used in roof and carved copper doors, – Unusual shape combining native artistic inspirations (spear heads of Congo, the mud huts of Chad) and local traditional techniques with European innovations (medieval cathedrals)[ad]

(continued)

Table 15.2 (continued)

African region	African country	Building image	Building type	Materials, form and elements
	Equatorial Guinea	 Wooden house in Mbini *http://www.traveladventures.org/continents/africa/mbini10.html*	Residential	– Local wood material is used in houses construction
	Gabon	 Traditional house in Gabon *http://www.gaboncultura.blogspot.com*	Residential	– Use of natural materials in houses construction (fired earthen bricks and bamboo), – Roof made of palm residues
	São Tomé and Principe	 São Tomé wooden house *http://tibarose.com/social-project.php?id=222*	Residential	– Houses made of wooden planks with vivid colours (pink and blue) and raised above the ground are typical of the local building methods[ae]
Zone 4: West Africa	Benin	 Traditional houses built on stilts in Benin, Africa *https://www.citylab.com/design/2015/08/the-beauty-of-african-vernacular-architecture-captured-in-one-big-database/400778/*	Residential	– Using local material (bamboo) to build huts on silts

(continued)

Table 15.2 (continued)

African region	African country	Building image	Building type	Materials, form and elements
	Burkina Faso	 Francis Kere, Gando Primary School in Burkina Faso *https://www.designcauseinc.org/single-post/2016/04/12/Blending-traditional-and-modern-to-create-successful-humanitarian-architecture*	Educational	– Use of local materials, – Roof truss is made of recycled steel bars, – Roof is made of double layers to maximize natural ventilation in hot-humid climate, – Bricks are unbaked which do not need a large amount of embodied energy, and – Windows were made to maximize daylighting
	Cape Verde	 Cape Verde's Vernacular house (Garcia 2016)	Residential	– Use of local stone in walls construction, – Use of natural straw material in roof construction (Garcia 2016)
	Ivory Coast (Côte d'Ivoire)	 Biankouma traditional houses *http://tabisite.com/gallery_af/ivorycoast/westen.shtml*	Residential	– Use of local materials such as reeds, poles and sun-baked clay, with thatch or metal roof, – People live in rectangular structures, while others in circular huts[af]
	Gambia	 Ecological house in Gambia *http://www.archidatum.com/projects/ecological-house-in-gambia-virai-arquitectos/*	Residential	– Use of local natural materials (blocks of pressed earth clay, sand and lime), – Use of passive solutions such as courtyard which helps in improving microclimate, air ventilation, creating shaded area and daylighting, – Patio collects rainwater from roof and stores it in a large cistern, to be reused in irrigation[ag]

(continued)

Table 15.2 (continued)

African region	African country	Building image	Building type	Materials, form and elements
	Ghana	 The Larabanga Mosque, Ghana *https://www.wmf.org/project/larabanga-mosque*	Religious (Mosque)	– Ghanaian architecture driven by available local materials, – Distinct mud minaret with horizontal timber, pyramidal towers and buttresses, – Triangular perforations over entry portals[ah]
	Guinea	 Tukul hut in Guinean village *Photographer: Boris Kester* *http://www.visualgeography.com/pictures/guinea_2_1.html*	Residential	– Small mud huts with thatched roofs are commonly used in countryside, – Dwellings are cool and easy to maintain[ai]
	Guinea-Bissau	 Huts in the Bijagos Islands *http://www.africavernaculararchitecture.com/gallery/guinea-bissau/*	Residential	– Huts made of local materials, – Thatched traditional shelters.
	Liberia	 Liberia schools programme *https://www.aecom.com/projects/liberia-schools-programme/*	Educational (School)	– Load bearing walls made of block work, generally plastered, – Timber trusses roofs covered by aluminium profile sheeting[aj]
	Mali	 Great Mosque of Djenné, Mali *https://www.citylab.com/design/2015/08/the-beauty-of-african-vernacular-architecture-captured-in-one-big-database/400778/*	Religious (Mosque)	– Use of local traditional construction techniques, – Use of sun-dried mud to build mosques, archways and houses[ak]

(continued)

Table 15.2 (continued)

African region	African country	Building image	Building type	Materials, form and elements
	Mauritania	 Satara zone housing, Rosso *https://archnet.org/sites/183/media_contents/10783*	Residential	– Uses of local materials and labour to reduce house cost, – Walls built from local clay and roof is masonry dome, – Construction withstands environmental conditions, – Respects social aspect[al]
	Niger	 Grass hut in Niger *https://www.gwiwestafrica.org/sites/default/files/image00046.jpg*	Residential	– Traditional mud grass hut with thatched roof
	Nigeria	 Clay house in Zaria, Nigeria *https://www.britannica.com/place/Zaria-Nigeria*	Residential	– Traditional clay houses with low-relief ornament, – Use of vibrantly and vivid yellow and orange coloured symbols reflecting Nigeria's heritage
	Saint Helena	 Wranghams House in St. Helena *http://sainthelenaisland.info/houses.htm*	Residential	– Use of local rubble stone and mud mortar with a smooth plaster finish, – Pitched timber truss roof under corrugated tin[am]

(continued)

Table 15.2 (continued)

African region	African country	Building image	Building type	Materials, form and elements
	Senegal	  Cultural hub Senegalese *https://www.dezeen.com/2017/01/25/toshiko-mori-compressed-earth-bamboo-thatch-cultural-centre-senegal-africa-architecture/*	Cultural (Cultural Centre)	– Main structure uses bamboo framework and CEB wall, – Thatched roof with overhang provides shades and helps in collecting rainwater, which is stored in reservoirs, – Perforated sections in roof allow natural ventilation, – Use of broken recycled tiles used for flooring[an]
	Sierra Leone	 House Sierra Leone *http://www.ibike.org/bikeafrica/sierra-leone/essay/09-Tiwai.htm*	Residential	– Distinctive roof height, – Use of local materials (clay walls with thatched roofs), – Traditional architecture creates a cooler and more liveable environment and is economically affordable[ao]
	Togo	 Koutammakou Valley, Togo *http://www.middle-africa.com/blog/togo/togo-architecture*	Residential	– Koutammakou Valley is the home of the Tamberma people, – Famous for beautiful adobe architecture[ap]
Zone 5: South Africa	Botswana	 Rondavel house in Botswana *https://commons.wikimedia.org/wiki/File:Rondavel_house_in_Botswana.jpg*	Residential	– Made of local material (clay and wood) with a thatched roof[aq]

(continued)

Table 15.2 (continued)

African region	African country	Building image	Building type	Materials, form and elements
	Lesotho	 African house "Rondavel" *https://commons.wikimedia.org/w/index.php?curid=3797723*	Residential	– Usually round or oval shaped made of local materials, – Walls are often constructed from stones. – The roofing structure is made of poles taken from tree limbs, while the covering is of thatch that is sewn to the poles[ar]
	Namibia	 Traditional house (for Himba) *http://traditionscustoms.com/people/himba-people*	Residential	– Mud structure – Use of local materials[as]
	South Africa	 Basotho hut in South Africa *http://naturalhomes.org/african-vernacular*	Residential	– Dwelling termed "Basotho", – Strong grass roof lasting 20–30 years, – Roof keeps the inside cool during summer and traps heat during winter without a drop of water seeping through[at]
	Swaziland	 House in Swaziland *https://www.citylab.com/design/2015/08/the-beauty-of-african-vernacular-architecture-captured-in-one-big-database/400778/*	Residential	– The structure is constructed using wooden poles, then rocks fill the walls which will be then plastered over with mud[au]

[a]https://www.dzbreaking.com/2018/03/12/tourism-algerias-sahara-timimoun-oasis-known-red-ochre-color-buildings/ (Accessed on 5.02.2018)
[b]http://www.amusingplanet.com/2013/03/the-old-town-of-ghadames.html (Accessed on 5.02.2018)
[c]https://decideyouradventure.com/ait-benhaddou-southern-morocco/ (Accessed on 5.02.2018)
[d]https://theculturetrip.com/africa/morocco/articles/an-architectural-tour-of-chefchaouen-moroccos-

Table 15.2 (continued)

majestic-blue-city (Accessed on 8.02.2018)
[e]http://naturalhomes.org/african-vernacular (Accessed on 8.02.2018)
[f]http://looklex.com/tunisia/nefta05.htm (Accessed on 8.02.2018)
[g]http://looklex.com/tunisia/nefta04.htm (Accessed on 8.02.2018)
[h]https://www.designcauseinc.org/single-post/2016/04/12/Blending-traditional-and-modern-to-create-successful-humanitarian-architecture (Accessed on 9.02.2018)
[i]http://www.wondermondo.com/Comoros.htm (Accessed on 9.02.2018)
[j]http://www.visitdjibouti.dj/explorerEN (Accessed on 10.02.2018)
[k]http://explore-eritrea.com/tour-6-discover-enchanting-culture-nature/tigrinya-tribe-family-and-hidmo-traditional-house/ (Accessed on 10.02.2018)
[l]http://www.eritrea.be/old/eritrea-barentu.htm (Accessed on 12.02.2018)
[m]http://naturalhomes.org/african-vernacular (Accessed on 12.02.2018)
[n]http://www.archidatum.com/articles/the-taita-sustainable-vernacular-house/ (Accessed on 12.02.2018)
[o]http://www.malawiarchitecture.com/mwanza (Accessed on 12.02.2018)
[p]http://vintagemauritius.org/vintage-houses/old-colonial-creole-style-houses-mauritius-part-1/ (Accessed on 13.02.2018)
[q]http://www.earthauroville.com/mayotte_as_development_model_en.php (Accessed on 13.02.2018)
[r]https://www.archdaily.com/144527/educational-building-in-mozambique-andre-fontes-sixten-rahlff (Accessed on 13.02.2018)
[s]https://atcnews.org/2015/02/20/creole-architecture-all-part-of-reunions-attractions/ (Accessed on 13.02.2018)
[t]https://www.archdaily.com/372709/umubano-primary-school-mass-design-group (Accessed on 13.02.2018)
[u]http://www.seychellesnewsagency.com/articles/1652/The+traditional+Creole+architecture++An+aspect+of+the+Seychelles+heritage+in+need+of+preservation (Accessed on 13.02.2018)
[v]http://wiki.colby.edu/download/attachments/6294655/Houses+in+Somalia.pdf (Accessed on 13.02.2018)
[w]http://sudantribune.com/spip.php?article46395 (Accessed on 13.02.2018)
[x]http://www.monitor.co.ug/Magazines/HomesandProperty/The-hut%2D%2Dthe-first-building-in-Uganda/689858-1522996-r03yg4z/index.html (Accessed on 13.02.2018)
[y]https://www.insaka.org/pages/about (Accessed on 13.02.2018)
[z]http://www.designindaba.com/articles/point-view/hut-decorating-competition-could-give-urban-homes-run-their-money (Accessed on 13.02.2018)
[aa]http://naturalhomes.org/african-vernacular#southafrica (Accessed on 13.02.2018)
[ab]http://www2.exxonmobil.com/Chad/Library/Photo_Video/web_photoalbum/english/14g.html (Accessed on 13.02.2018)
[ac]http://www.endingextremepoverty.org/2010/04/an-upper-class-neighborhood-in-rural-congo.html (Accessed on 13.02.2018)
[ad]https://mobile.nation.co.ke/lifestyle/Seeking-imagination-beauty-and-identity/-/1950774/2241852/-/format/xhtml/item/1/-/u4sxfn/-/index.html (Accessed on 14.02.2018)
[ae]http://tibarose.com/social-project.php?id=222 (Accessed on 07.04.2018)
[af]http://www.everyculture.com/Bo-Co/C-te-d-Ivoire.html%23ixzz5Bvym97bo (Accessed on 14.02.2018)
[ag]http://www.archidatum.com/projects/ecological-house-in-gambia-virai-arquitectos/ (Accessed on 16.02.2018)
[ah]https://www.wmf.org/project/larabanga-mosque (Accessed on 16.02.2018)
[ai]http://www.visualgeography.com/categories/guinea/houses.html (Accessed on 16.02.2018)
[aj]https://www.aecom.com/projects/liberia-schools-programme/ (Accessed on 16.02.2018)
[ak]https://www.citylab.com/design/2015/08/the-beauty-of-african-vernacular-architecture-captured-in-one-big-database/400778/ (Accessed on 16.02.2018)
[al]https://archnet.org/sites/183/media_contents/10783 (Accessed on 16.02.2018)

Table 15.2 (continued)

[am]http://sainthelenaisland.info/houses.htm (Accessed on 16.02.2018)
[an]https://www.dezeen.com/2017/01/25/toshiko-mori-compressed-earth-bamboo-thatch-cultural-centre-senegal-africa-architecture/ (Accessed on 16.02.2018)
[ao]http://www.ibike.org/bikeafrica/sierra-leone/essay/09-Tiwai.htm (Accessed on 16.02.2018)
[ap]http://www.middle-africa.com/blog/togo/togo-architecture (Accessed on 16.02.2018)
[aq]https://commons.wikimedia.org/wiki/File:Rondavel_house_in_Botswana.jpg (Accessed on 16.02.2018)
[ar]https://commons.wikimedia.org/w/index.php?curid=3797723 (Accessed on 16.02.2018)
[as]http://traditionscustoms.com/people/himba-people (Accessed on 16.02.2018)
[at]http://naturalhomes.org/african-vernacular (Accessed on 16.02.2018)
[au]https://www.citylab.com/design/2015/08/the-beauty-of-african-vernacular-architecture-captured-in-one-big-database/400778/ (Accessed on 16.02.2018)

Africa

It is clear from Table 15.3 that vernacular architecture in Africa has many types, varieties and examples. It uses a wide range of local materials, which indicate that indigenous people use materials from the surroundings, including: (a) natural environment (bamboo, boiled tree bark, grass-thatched, papyrus reeds, reeds, straws, timber, trees residuals, wood and woven mats), (b) metals (aluminium, copper, iron, steel and tin), (c) earth materials (basalt, burnt bricks, clay, earthen bricks, glass bottle, lime, masonry bricks, mud grass and mud as well as plaster, rocks, rubble stone, sand, stone, sun-baked clay and sun-dried mud) and (d) sea materials (coral lime) as highlighted in Box 15.4. These materials were used to develop their own simple dwellings and other types of buildings without the dependence on transport of materials from other areas. Hence, vernacular architecture shows the principles of sustainability.

In term of forms, vernacular architecture in Africa used different shapes and types whether it is residential, educational, cultural or religious buildings. These forms were vertical architecture, single-storey dwellings, round or oval plans with pointed domes, circular floor plans with conical thatched roofs, branch-framed,

Box 15.4: Use of Local Materials

Mud, clay, wood, straw, tree residuals, reeds, earth, bricks, stone, basalt plastered with coral lime, woven mats or boiled bark, burnt bricks and white plaster coated, rocks, some metals, sandbags, brick and papyrus reeds, grass-thatched, corrugated iron roof, earth plastered onto stone-filled framework, glass-bottle walls, thatched straw, piles, tin, green tiles, copper, recycled steel bars, fired earthen bricks, bamboo, poles and sun-baked clay, thatch or metal roof, sand, lime, mud grass, adobe, rubble stone, sun-dried mud, timber and aluminium sheets.

Table 15.3 Synergies in African vernacular architecture

Elements of Comparison							
Building Type	Climatic Zone[a]	Building Image	Form	Colour	Material		Sustainability Features
Residential	Tropical and subtropical desert climate "Bwh"	  a. *http://openwalls.com/image?id=133* b. *http://42mzqz26jebqf6rd034t5pef-wpengine.netdna-ssl.com/wp-content/uploads/2016/06/Screen-Shot-2016-06-03-at-4.46.21-PM.png*	Compact rectangular	Earth	Mud	Wall bearing	– Local material – Minimum exposure to sun heat – Minimal openings
		  a. *http://naturalhomes.org/african-vernacular* b. *http://www.visitdjibouti.dj/explorerE*	Cone		Tree residual natural	Branch-framed hut system	– Local materials
		  a. Garcia (2016) b. *http://sainthelenaisland.info/houses.htm*	Rectangular with pitched roof	Beige	Stone, straw, timber	Wall bearing with pitched truss roof	
		 *https://archnet.org/sites/183/media_contents/10783*	Rectangular with domes	White	Clay, masonry dome	Wall bearing	– Local materials – Construction withstand environmental conditions – Respects social aspect

(continued)

Table 15.3 (continued)

Elements of Comparison							
Building Type	Climatic Zone[a]	Building Image	Form	Colour	Material		Sustainability Features
		 *http://traditionscustoms.com/people*	Dome	Earth	Mud	Dome	– Local materials
	Mid-latitude steppe and desert climate "Bsh"	 *https://www.architectural-review.com/pictures/2000x2000fit/5/3/4/1225534_AR05_EE_MUD_LATTICE_IN_GABON.jpg*	Rectangular	Earth with white plaster	Adobe, thatched roof, wood poles	Wall bearing and roof on poles	– Local materials
		  a.*http://wiki.colby.edu/download/attachments/6294655/Houses+in+Somalia.pdf* b.*http://www.africavernaculararchitecture.com/angola/*	Round	Earth	Grass, mud	Hut system	
		 *http://naturalhomes.org/african-vernacular#southafrica*	Round dome shaped	Dark earth	Mud	Dome system	– Local materials – High domes collect the hot air, moving it away so to provide efficient cooling

	Tropical and subtropical steppe "BSk"	 *http://naturalhomes.org/african-vernacular*	Round with hut roof	Earth, grey, beige	Grass, mud	Strong grass roof lasting 20–30 years	– Local materials – Roof keeps the inside cool during summer and traps heat during winter without a drop of water seeping through
	Tropical Savannah "Aw"	  a. *https://www.diva-ortal.org/smash/get/diva2:616686/FULLTEXT01.pdf* b. *http://www.gaboncultura.blogspot.com*	Rectangular with thatched roof	Earth	Piles, grass and mud	Wall bearing	– Local materials – Pitched roof shed rain
	Tropical Savannah climate "Aw"	  a. *http://www.earthauroville.com/mayotte_as_development_model_en.php Brejo Chimundo* b. *http://www.endingextremepoverty.org/2010/04/an-uppe-class-neighborhood-in-rural-congo.html*	Rectangular with metal roof	Earth	Earth and corrugated metal roof	Wall bearing	– Local materials – Pitched roof shed rain

(continued)

Table 15.3 (continued)

Elements of Comparison							
Building Type	Climatic Zone[a]	Building Image	Form	Colour	Material		Sustainability Features
		 a. *http://www.archidatum.com/articles/the-taita-sustainable-vernacular-house/* b. *http://www2.exxonmobil.com/Chad/Library/Photo_Video/web_photoalbum/english/14g.html* c. *http://www.africavernaculararchitecture.com/gallery/guinea-bissau/* d. *http://sudantribune.com/spip.php?article46395*	Circular floor plans with conical thatched roof	Earth/ red earth	Mud	Thatch roofed hut	– High thermal mass ensuring suitable indoor temperature – Overhangs of roof shade walls – Very small operable windows – Local materials
		 *https://www.citylab.com/design/2015/08/the-beauty-of-african-vernacular-architecture-captured-in-one-big-database/400778*	Rectangular with pitched roof	Earth	Bamboo	Hut system on silts	– Local materials – Huts built on silts
		 *http://www.archidatum.com/projects/ecological-house-in-gambia-virai-arquitectos*		Beige	CEB, sand, lime	Wall bearing	– Local materials – Courtyards improving microclimate, creating shades, daylighting, – Patio collects rainwater from roof and stores it

		 *http://tibarose.com/social-project.php?id=222*	Rectangular with pitched roof	Pink and blue	Wood	Wood structure	– Local materials – Pitched roof shed rain – Houses raised above the ground level
		 *http://www.malawiarchitecture.com/mwanza*	Rectangular with thatched roof	White plaster	Burnt bricks, plaster, grass	Wall bearing	– Local materials – Opening in roof's kitchen to allow smoke to escape
		 *http://www.traveladventures.org/continents/africa/mbini10.html*	Rectangular	Beige, blue	Wood	Wood structure	– Local materials
		 *https://atcnews.org/2015/02/20/creole-architecture-all-part-of-reunions-attractions/*		Vivid pink			

(continued)

Table 15.3 (continued)

Elements of Comparison							
Building Type	Climatic Zone[a]	Building Image	Form	Colour	Material		Sustainability Features
		 *https://www.britannica.com/place/Zaria-Nigeria*		Vivid yellow and orange	Adobe (clay)	Wall bearing	
		 *http://www.middleafrica.com/ blog/togo/ togo-architecture*		Earth			
	Tropical rainforest climate “Af”	 *http://www.wondermondo.com/Comoros.htm*	Rectangular	Crème, shades of coral lime	Basalt, plaster, coral lime	Wall bearing	– Maze of narrow streets – Local materials
		 *http://www.seychellesnewsagency.com/articles/1652/The+traditional+Creole+architecture++An+aspect+of+the+Seychelles+heritage+in+need+of+preservation*	Rectangular with pitched roof	Brown	Wood	Wood framing	– Local materials – Natural ventilation and daylighting with the presence of large openings, high-pitched roof and large balconies
	Tropical monsoon climate “Am”	 *http://vintagemauritius.org/vintage-houses/old-colonial-creole-style-houses-mauritius-part-1/*		Blue, orange, grey	Rocks, wood, metals	Wall bearing	– Local materials – Pitched roof to shed rain

		 a. *http://www.monitor.co.ug/Magazines/HomesandProperty/The-hut%2D%2Dthe-first-building-in-Uganda/689858-1522996-r03yg4z/index.html* b. *http://www.visualgeography.com/pictures/guinea_2_1.htm* c. *http://www.ibike.org/bikeafrica/sierra-leone/essay/09-Tiwai.htm*	Circular hut forms	Earth	Natural, mud, straw and tree residual	Hut system	– Local materials – Dwellings are cool and easy to maintain – Traditional architecture creates a cooler and more liveable environment and is economically affordable
	Humid subtropical "Cfa"	 *https://www.citylab.com/design/2015/08/the-beauty-of-african-vernaculararchitecture-captured-in-one-big-database/400778/*	Rectangular with hut roof		Wood, rock, mud plaster	Wooden structure	– Local materials

(continued)

Table 15.3 (continued)

Elements of Comparison							
Building Type	Climatic Zone[a]	Building Image	Form	Colour	Material		Sustainability Features
	Dry-summer subtropical (Mediterranean Climate) "Csa"	 *http://endangeredsites.leadr.msu.edu/ghadames/*	Compact rectangular	White	Plaster, mud	Wall bearing	– Local materials – Overhangs creating shaded pathways
		 *https://i.pinimg.com/564x/7c/6a/45/7c6a45124b127f895 4340d734d35ae65%2D%2Dunusual-houses-star-wars.jpg*		Earth	Adobe mud		– High thermal mass so the shelter is cold during arid weather
	Oceanic subtropical highland climate "Cwb"	 *http://www.travelmadagascar.org/CITIES/Antananarivo-surroundings.html*	Rectangular with pitched roof	Red earth			– Local materials – Pitched roof to shed rainfall
		 *http://www.designindaba.com/articles/point-view/ hut-decorating-competition- could-give-urban-homes-run-their-money*	Rectangular with hut roof	Earth-brown		Wall bearing and roof hut system	
	Marine west coast "Cfb"	 *https://commons.wikimedia.org/w/index.php?curid=3797723*	Round or oval		Stones, tree limbs, thatched roof	Roofing structure is made of pole	– Local materials

		 *http://naturalhomes.org/african-vernacular*	Cone (round)	Earth	Bamboo	Bamboo plaited hut	– Pointy top dome to shed heavy rainfall – Local materials
Educational	Mid-latitude steppe & desert "Bsh"	 *https://www.designcauseinc.org/single-post/2016/04/12/Blending-traditional-and-modern-to-create-successful-humanitarian-architecture*	Rectangular	Light brown	Unbaked bricks, recycled steel bars	Wall bearing, Steel roof	– Local materials – Recycled materials – Double-layer roof for natural ventilation – Low embodied energy – Daylighting
	Tropical Savanna climate "Aw"	 *https://www.archdaily.com/372709/umubano-primary-school-mass-design-group*			Wood	Wood	– Local materials – Clerestory windows allowing natural ventilation, and daylighting
		 *https://www.archdaily.com/144527/educational-building-in-mozambique-andre-fontes-sixten-rahlff*		White, beige, brown	Wood, mud, sand bags, concrete	Truss roof, concrete room	– Local materials – Roof extension provides shade and collects rain water – Trusses allow natural ventilation
	Tropical monsoon climate "Am"	 *https://www.aecom.com/projects/*		Earth	Block work, plaster, timber	Wall bearing and timber trusses roofs	– Local materials

(continued)

Table 15.3 (continued)

Elements of Comparison							
Building Type	Climatic Zone[a]	Building Image	Form	Colour	Material		Sustainability Features
Cultural	Tropical Savanna "Aw"	 *https://www.designcauseinc.org/single-post/2016/04/12/Blending-traditional-and-modern-to-create-successful-humanitarian-architecture*		Brown	CEB, baked clay	Wall bearing	– Hallway porch provides shelter from rain and sun – Blinder controls access – Sloped overhang roof – Ventilation and daylighting
	Mid-latitude steppe and desert "Bsh"	 *https://www.dezeen.com/2017/01/25/toshiko-mori-compressed-earth-bamboo-thatch-cultural-centre-senegal-africa-architecture*		Earth, white	Bamboo, CEB		– Roof overhangs provides shades and helps in collecting rainwater – Roof perforated sections allow natural ventilation, – Use of recycled materials
Hospitality	Tropical & subtropical desert "Bwh"	 *http://freevst.x10.mx/sahara/algeria/ethnography/algeria%2008.jpg*	Rectangular	Earth (red)	Red adobe	Wall bearing	– Local materials – Compact structure – Courtyard

Religious	Tropical & subtropical desert "Bwh"	 *http://looklex.com/tunisia/nefta04.htm*	Rectangular with dome	White	Plastered mud brick		– Local materials
	Mid-latitude steppe & desert "Bsh"	 *https://www.citylab.com/design/2015/08/the-beauty-of-african-vernacular-architecture-captured-in-one-big-database/400778/*	Rectangular	Earth, brown	Sun-dried mud		
	Tropical Savannah climate "Aw"	 *http://www.dailyherald.com/article/20151129/news/311299977*			Local red brick		
		 *http://www.thedigitalglobetrotter.com/country-185196-republic-of-congo-a-weddin g-in-a-great-african-family/*	Unusual shape (native + European inspirations)	Green, brown	Green tiles, copper		
		 *https://www.wmf.org/project/larabanga-mosque*	Towers and buttresses	White	Timber, mud	Towers and buttresses	– Local materials – Horizontal timber sections, – Triangular perforations over entry portals.

[a]All climatic data is based on www.weatherbase.com (Accessed on 01.04.2018)

transportable huts, rectangular or circular huts and pitched roofs (Box 15.5). These shapes were simple to be constructed by local people. Regarding the colours, vernacular architectures were described to be using vivid and earth colours as shown in Box 15.6, and the structure elements used in African vernacular architecture are presented in Box 15.7. The environmental features used in local architecture in Africa are summarized in Box 15.8. These vernacular architecture examples simply exhibit the essence of environmentally friendly buildings.

Box 15.5: Forms

- Traditional residential dwellings are termed Agudo, Basoth, Taita, Toukouls and Tukel,
- Typical vertical architecture or single storey,
- Branch-framed, transportable huts,
- Round or oval plans with pointy domes,
- Circular floor plans with conical thatched roof,
- Branch-framed, transportable huts,
- Round or oval plans with pointy domes,
- Circular floor plans with conical thatched roof,
- Pitched roof and
- Rectangular or circular huts.

Box 15.6: Colours

- Shades of blue set against white and muted creams,
- White colour paint,
- Vivid colours (pink and blue),
- Green tiles,
- Vibrantly and vivid yellow and orange coloured symbols and
- Red earth dwellings.

Box 15.7: Structure

- Thatch roofed huts or huts on silts,
- Roof is supported by poles,
- Wooden poles, rocks fill the walls with mud plaster,
- Strong grass roof lasting 20–30 years,
- Pitch truss roof,
- Horizontal timber, pyramidal towers and buttresses and
- Load bearing walls.

Box 15.8: Environmental Features

- Clustered houses resembling a honeycomb,
- Characteristic maze of narrow streets,
- Overhangs covering alleys between houses creating shades,
- Hallway porch providing shelter from rain and sun,
- Courtyards improve microclimate and create shades and daylighting,
- Blinder controls access,
- Cool shelter during arid weather,
- Small operable windows in arid climate,
- Large openings allowing natural ventilation and daylighting,
- Clerestory windows allowing natural ventilation and daylighting,
- Roof trusses structure with double layers allows and maximizes natural ventilation in hot-humid climate,
- Roof keeps comfortable indoor microclimate,
- Roof extension provides shade from summer sun and collects rain water,
- Roof openings in kitchens to release smoke and perforated sections allow natural ventilation,
- High domes collect the hot air thus providing efficient cooling
- Use of recycled materials in construction,
- Glass-bottle wall for letting in light and keeping dust out and
- Unbaked bricks which do not need a large amount of embodied energy.

Synergies in African Vernacular Architecture

The following section highlights the synergies among different vernacular buildings which are drawn and presented in Table 15.3 based on the review examples depicted in Table 15.2. Such analysis is important in order to draw the patterns on how the building types and different climatic zones affect the architectures' characteristics, which include five main criteria as shown in detail in Table 15.3:

a. Form (Shape);
b. Colour;
c. Materials;
d. Structure and
e. Sustainability Features.

Middle East

Vernacular architecture of the Middle Eastern countries will be reviewed, assessed and presented in Table 15.4. The Middle East region is a geographical region encompassing the following 16 countries: Bahrain, Cyprus, Iran, Iraq, Palestine, Israel, Jordan, Kuwait, Lebanon, Oman, Qatar, Saudi Arabia, Syria, Turkey, United Arab Emirates and Yemen[2] as illustrated in Fig. 15.9.

[2] https://www.worldatlas.com/webimage/countrys/me.htm (Accessed on 20.02.2018).

Table 15.4 List of Vernacular architecture in the Middle East countries

Middle east country	Building image	Building type	Materials, form and elements
Bahrain	 Al Jasra House *http://culture.gov.bh/en/visitingbahrain/destinations/Name,10553,en.html%23.WtJK9S5ubIU*	Residential (currently used as a Museum)	– Use of local materials such as coral stones and palm leaf trunks[a]
Cyprus	 Vavla Rustic Retreat—Cyprus *https://forkandfoot.com/vavla-rustic-retreat-cyprus/*	Hospitality	– Eco-Awarded Traditional Guest House, – Local materials used (stone and timber), – Presence of courtyard and skylights, – Uses renewable energy (solar power), – Rubbish recycling scheme[b]
Iran	 Mehraban house in Bushehr *https://commons.wikimedia.org/wiki/File:Mehraban_house2.jpg*	Residential	– Adapt to hot and humid climate, – Prevent heat gain by façade bright colours, – Exclude direct sunlight by: covering verandas, courtyards, – Allow natural ventilation by high ceiling, wide windows and windows on both sides (Soleymanpour et al. 2015)
	 Avadis house—Rasht city *https://ajbes.eiph.co.uk/index.php/ajbes/article/download/37/160*	Residential	– Adapt to temperate humid climate, – Rain disposal by roof extending over balconies, – South orientation to receive solar heat, – Prevent moisture absorption by having ground floor's slab above the ground level, – Natural ventilation by many openings (Soleymanpour et al. 2015)

(continued)

Table 15.4 (continued)

Middle east country	Building image	Building type	Materials, form and elements
	 Hedayat House—Urmia city *https://ajbes.eiph.co.uk/index.php/ajbes/article/download/37/160*	Residential	– Adapt to cold climate, – Rain disposal by gentle sloped roofs, – Allow heat absorption by having ground floor lower than a natural ground level and thermal mass materials, – Introverted building with courtyard, – Heat gain by large southern windows, – Windows allowing cross ventilation (Soleymanpour et al. 2015)
Iraq	 Inner Court in Baghdad House *http://earsiv.cankaya.edu.tr:8080/xmlui/bitstream/handle/123456789/347/Ayyash%2C%20Ahmed%20Basil.pdf?sequence=1&isAllowed=y*	Residential	– Rooms open on central courtyard, – House surrounded by high walls to provide privacy, – Ventilation by windows and three air-scoops "Bad-Geer" facing—wind catcher, – Daylight is present, – Local materials (clay brick), – Bent entrance for privacy, as outsiders can't see behind the entrance[c]
Jordan	 Dana guest house in Jordan *http://www.khammash.com/projects/dana-guesthouse*	Hospitality	– Nine-room hotel made of local materials with high thermal mass, – Building elevation has a unique random nature with a raw finish, to be integrated with surroundings, – Building looks like an eroded cliff[d]

(continued)

Table 15.4 (continued)

Middle east country	Building image	Building type	Materials, form and elements
Kuwait	 The courtyard of Beit Khalid *http://www.traveladventures.org/continents/asia/beit-khalid01.html*	Residential	– Traditional house made of adobe and wood, – Wooden columns to support second floor and wooden beams sticking out of the wall holding the structure together, – Rooms arranged around the courtyard, – Enhanced natural ventilation[e]
Lebanon	 Ghazir traditional architecture *http://www.meda-corpus.net/arb /fitxes/F1sites/eng/lb_s07.pdf*	Residential	– Traditional rural houses made of local materials, – Architecture elements include: iwan, central lobbies and arches[f]
Oman	 Nizwa Fort and Jabrin Castle *https://www.tourismoman.com.au/news/an-architectural-tour-through-oman/*	Residential	– Famous Jabrin castle has beautiful structure, detailed lattice-work, curved arches and detailing in architecture elements such as doors, – Also, town is famous for its big height rounded watch towers[g]
Saudi Arabia	 Traditional house in West (Hijaz)—Saudi Arabia *http://www.limedesign.com.sa/assets/vernacular-architecture-in-saudi-arabia-revival-of-displaced-traditions.pdf*	Residential	– Respond to hot-humid climate, – Multi-storey buildings, – Extroverted architecture as windows are placed on external façade to allow natural ventilation, – Use of Mashrabyya (screens) for natural ventilation while providing privacy

(continued)

Table 15.4 (continued)

Middle east country	Building image	Building type	Materials, form and elements
	 Traditional house in East—Saudi Arabia *http://www.limedesign.com.sa/assets/vernacular-architecture-in-saudi-arabia-revival-of-displaced-traditions.pdf*	Residential	– Respond to hot-humid climate, – One- to three-storey houses massed together to create narrow passages, – Thick walls of high heat resistance, – Layered roofing system of wooden beams, and palm trunks covered with palm leafs, – Installation of wind catchers (Badgeers) to create natural ventilation
	 Traditional house in Central (Najd)—Saudi Arabia *http://www.limedesign.com.sa/assets/vernacular-architecture-in-saudi-arabia-revival-of-displaced-traditions.pdf*	Residential	– Respond to hot-dry climate, – Houses are arranged around courtyards that regulates microclimate, – Compact layout of houses so reducing heat gain by opaque walls, – Minimal openings on exterior walls allowing air ventilation and privacy, – Triangular shape decoration on parapet
Syria	 Syria's Beehive Houses *https://syriaphotoguide.com/beehive-houses-%D8%A7%D9%84%D9%82%D8%A8%D8%A8*	Residential	– Known as "beehive houses" – Rounded conical shape – Local materials (dirt, mud, straw and stone) are used. – The houses are connected to each other and built around a central courtyard. – This allows the houses to be cool in the summer and warm in the winter

(continued)

Table 15.4 (continued)

Middle east country	Building image	Building type	Materials, form and elements
	 House with a Riwaq, Syria (Images source: *http://www.rehabimed.net/Publicacions/Corpus/Manual%20para%20el%20mantenimiento%20y%20rehabilitacion%20de%20la%20arquitectura%20tradicional%20del%20Libano/pdf/livret/ats_eng.pdf*)	Residential	– It consists of several rooms connected to each other via a covered gallery called "Riwaq", – Local materials are used[i]
	 Rural House with Courtyard in Syria (Images source: *http://www.rehabimed.net/Publicacions/Corpus/Manual%20para%20el%20mantenimiento%20y%20rehabilitacion%20de%20la%20arquitectura%20tradicional%20del%20Libano/pdf/livret/ats_eng.pdf*)	Residential	– Local materials are used (mud structures), – Dwelling usually consists of one main unit, covered with a cupola roof. The richer the user, the more rooms present, – Overtime, the roof shape changed to become flat and covered with local materials such as: wood, tree residues or earth, – Rooms are divided: some for day time, some for men or women only and areas for domestic animals, – Dwelling has minimum openings on the external façade[i]
	 House with a Liwan, Syria (Images Source: *http://www.rehabimed.net/Publicacions/Corpus/Manual%20para%20el%20mantenimiento%20y%20rehabilitacion%20de%20la%20arquitectura%20tradicional%20del%20Libano/pdf/livret/ats_eng.pdf*)	Residential	– Consists of three aligned rooms, – The room "liwan" is opened to the exterior and is used as a distribution space, – House has multipurpose function, i.e. used for living and as a local workshop, – The room "Liwan" is used as a shelter for domestic animals, – Local materials are used[i]

(continued)

Table 15.4 (continued)

Middle east country	Building image	Building type	Materials, form and elements
	 Urban house with Courtyard (Images Source: *http://www.rehabimed.net/Publicacions/Corpus/Manual%20para%20el%20mantenimiento%20y%20rehabilitacion%20de%20la%20arquitectura%20tradicional%20del%20Libano/pdf/livret/ats_eng.pdf*)	Residential	– Minimum openings on the external façade, while large number on inner court walls, – Courtyard is planted to provide shade and has a fountain (evaporative cooling), – Increase of shades by the presence of: façade projections, projected wooden elements on façade, roof garden improving indoor microclimate, and shaded streets, – Natural ventilation by the presence of wind catcher and openings on opposite sides[i]
Turkey	 Timber structure, Turkey (Images Source: *https://archnet.org/system/publications/contents/3944/original/DPT0472.pdf?1384777691*)	Residential	– Timber-framed structures in northern Turkey regions due to the availability of wood as a local material[i]
	 Half-timbered mud house (Images Source: *https://archnet.org/system/publications/contents/3944/original/DPT0472.pdf?1384777691*)	Residential	– Mud dwellings on the hot, dry Anatolian plateau in central Turkey regions due to the availability of mud as a local material, – Walls are made of plastered mud brick and rest on foundation (Alkhalidi 2013).
	 Stone house, Turkey (Images Source: *https://archnet.org/system/publications/contents/3944/original/DPT0472.pdf?1384777691*)	Residential	– Stone and timber houses in southern Turkey mountainous regions due to their availability as local materials (Alkhalidi 2013)

(continued)

Table 15.4 (continued)

Middle east country	Building image	Building type	Materials, form and elements
United Arab Emirates	 Leewan space in house, UAE (Alkhalidi 2013)	Residential	– Time lag materials in walls and roofs, – Use of courtyard to enhance ventilation by big opening on opposite sides while maintaining privacy by wooden screen which also allows daylighting, – Wind catcher (Barageel) facing four directions catching breeze, – Also, ventilation by "Leewan" which is shaded space and completely open to the court thus lets cold air enter house (Alkhalidi 2013)
	 *Photo credit: Author*	Dubai Castle currently used as Cultural (museum)	– Use of local material and coral stones for external wall, – Exploit of thick wall with high thermal capacity to help in delaying heat gain during summer (time lag) and use woods and palm trees leaves for roof coverage. – Use of courtyard for cooling and promote breeze. – Wind catcher (Barageel) facing four directions catching breeze, – Also, ventilation by "Leewan" is promoted in the shaded arcades and completely open to the courtyard, thus lets cold air enter spaces of the house
	 *Photo credit: Author*	Residential Currently used as Cultural (museum of Sheik Saeed House)	

(continued)

Table 15.4 (continued)

Middle east country	Building image	Building type	Materials, form and elements
Yemen	 Al Hajarayn Town, Yemen *https://misfitsarchitecture.com/tag/yemeni-vernacular-architecture/*	Residential	– Earth made dwellings, – Multi-storey buildings, – Architecture rhythm is present due to similarities of colour, window shapes, alignment, positioning or size[j]
	 Ba Jammall Mosque, Yemen *https://misfitsarchitecture.com/tag/yemeni-vernacular-architecture/*	Religious (Mosque)	– Local materials are used (mud bricks), which is painted white as it is a mosque, – Presence of environmental friendly techniques such as white colour helps in reflecting sun, thick walls and small windows (Bayoumi 2018)

[a]http://culture.gov.bh/en/visitingbahrain/destinations/Name,10553,en.html%23.WtJK9S5ubIU (Accessed on 01.03.2018)
[b]https://forkandfoot.com/vavla-rustic-retreat-cyprus/ (Accessed on 01.03.2018)
[c]http://earsiv.cankaya.edu.tr:8080/xmlui/bitstream/handle/123456789/347/Ayyash%2C%20Ahmed%20Basil.pdf?sequence=1&isAllowed=y (Accessed on 01.03.2018)
[d]http://www.khammash.com/projects/dana-guesthouse (Accessed on 02.03.2018)
[e]http://www.traveladventures.org/continents/asia/beit-khalid01.html (Accessed on 01.03.2018)
[f]http://www.meda-corpus.net/arb/fitxes/F1sites/eng/lb_s07.pdf (Accessed on 01.03.2018)
[g]https://www.tourismoman.com.au/news/an-architectural-tour-through-oman/ (Accessed on 01.03.2018)
[h]http://www.rehabimed.net/Publicacions/Corpus/Manual%20para%20el%20mantenimiento%20y%20rehabilitacion%20de%20la%20arquitectura%20tradicional%20del%20Libano/pdf/livret/ats_eng.pdf (Accessed on 01.03.2018)
[i]https://archnet.org/system/publications/contents/3944/original/DPT0472.pdf?1384777691 (Accessed on 05.03.2018)
[j]https://misfitsarchitecture.com/tag/yemeni-vernacular-architecture/ (Accessed on 05.03.2018)

Examples of Vernacular Architecture in Egypt

This section demonstrates that Egypt is rich with vernacular architecture examples. It presents the following traditional architecture:

- Aswan's Traditional Architecture—Old Nubian,
- Luxor's Traditional Architecture—Old Gourna and
- Western Desert of Egypt which includes detailed description of the traditional architecture of Bahariya, Kharga and Siwa oases.

Fig. 15.9 Middle East Map. *Source*: https://www.worldatlas.com/webimage/countrys/asia/metimes.gif

Aswan's Traditional Architecture: Old Nubian, Egypt

The old Nubian architecture demonstrates knowledge practices that have succeeded to adapt to their environment and create distinct architectural features that are environmentally, socially and culturally sustainable (Bayoumi 2018). Thus, traditional Nubian architecture, both on the building scale and on the urban scale, provides a great example to learn about sustainability principles (Momtaz et al. 2012). Such vernacular architecture combined features and elements of the Nubian culture and traditions, yet respected the microclimate depending on the available local resources. The climate in the Nuba, Aswan can be described as hot-dry (arid) climate, exceeding the thermal comfort during day and night (Momtaz et al. 2012).

Traditional Urban Layout

The traditional Nubian village, shown in Fig. 15.10, is considered a unique example due to its integration with the surrounding environment creating harmony with the nature where the village was formed according to the topography in the area.

Fig. 15.10 Nubian village merges into topography and looking at the Nile, Upper Egypt. (**a**) Nubian village on Elephantine Island. https://www.egypttoursplus.com/nubian-village/. (**b**) Aerial view of Nubian village (Momtaz et al. 2012)

Another factor that determines the village setting is the Nile River, as most of the buildings and streets' orientation are related to the Nile. Also, the village is considered a compact urban tissue and the roads are not straight—to follow the natural shape of the site (topography), which help in preventing the movement of hot air and dust to the inside of the settlement and causing cooling effect. In addition, the compact layout provides shade and blocks winds during winter, whereas it promotes air velocity during the summer. Moreover, the road size is so narrow, which is an important feature, especially in hot summer days, as it allows maximum shade and shadow. Thus, people like to sit outside in front of their houses (Moustafa 2015).

Nubian customs and traditions are also an important determinant that has been reflected on every component of the society and buildings. The Nile River has a significant role in the ancient Nubian lives and has been related to many myths, and thus it has been taken into consideration in the village planning as roads are parallel to the Nile and buildings' main elevations are oriented toward the river. Moreover, Nubians are known for their generosity and well hospitality, thus they have a communal guesthouse building termed "Al Madiafa" in the centre of the village to guest visitors from other neighbours. Finally, privacy is an important feature in the Nubian community which was achieved on the urban scale by separate buildings, as the building is surrounded by four streets from all directions, thus creating more privacy (Moustafa 2015). In addition, houses are usually clustered in groups, each group called a "nog" or "naja", who shares a common ancestor which is key social attribute as each person of this group participates in any work, building construction and food and water gathering (Momtaz et al. 2012) (Box 15.9).

Box 15.9: Environmental and Social Sustainability in Nubian Village—Urban Scale

In the traditional Nubian village, environmental sustainability exists due to the harmony of the village with the nature and streets pattern merged with topography, while cultural sustainability is present due to reflecting the following items on the village urban planning: (a) value of the Nile River, (b) hospitality and generosity of Nubians, (c) privacy and (d) houses clustering in groups.

Traditional Architecture on Building Scale

The Nubian traditional houses are built with local materials, as the foundation is usually built with stone, while sun-dried mud bricks are used in walls construction. The roofing varies between flat roofs, vaults and domes. The roof shape reflects the underneath space function, for example: vaults are used to cover spaces like guest room, while domes may appear upon entrance hall or kitchen (Moustafa 2015). The vernacular vaults are made of earth bricks and mortar which makes it ecologically sustainable, carbon neutral, durable, thermally and acoustically comfortable, economically viable and modular as it is applicable to a wide variety of buildings. Also, the elevation's colours are usually white, and openings are green or blue (Momtaz et al. 2012).

The old Nubian house consists of the following architecture elements (Fig. 15.11):

- The entrance gate: located in the centre of a high mud wall that is oriented toward the Nile,
- Entrance transition area "main entrance portico",
- Guest room (*Mandara*): it is located next to the main entrance,
- Courtyard,
- Bedrooms: two bedrooms or more depending on the economic status of the resident,
- Kitchen: consisting of two rooms and has an open vent on the top for ventilation and
- Storage room (Momtaz et al. 2012).

This architecture on building (Fig. 15.12) scale exhibits environmental and social sustainability. The environmental techniques used include building houses from sun-dried mud bricks which is a suitable material in hot-arid areas as it has high thermal mass (Moustafa 2015). The wall thickness is 500 mm and thus it maintains a cold room temperature (Momtaz et al. 2012). Also, indoor ventilation in the houses

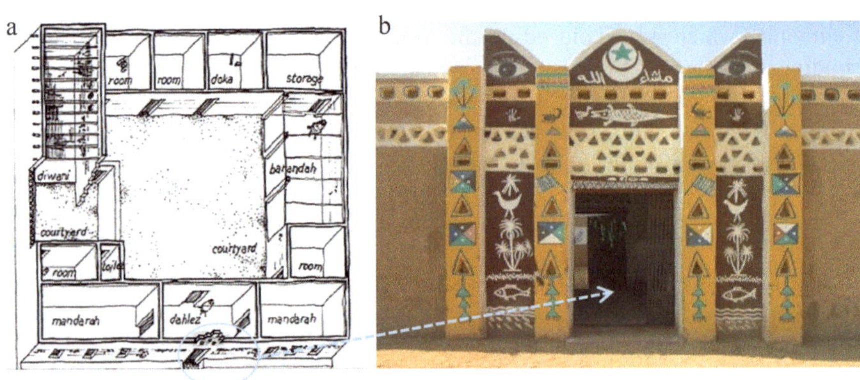

Fig. 15.11 Old Nubian building components, Upper Egypt. *Source*: (**a**) Dr. Yasser Mahgoub research "the Nubian Experience" p. 12 (Bayoumi 2018). (**b**) https://architecture.knoji.com/nubia-the-living-survivor-of-traditional-egyptian-architecture/

Fig. 15.12 Old Nubian architecture, Upper Egypt. (**a**) Vivid colours and decorations in guest house http://www.dailymail.co.uk/travel/travel_news/article-3341353/Anakato-Nubian-guest-house-oasis-colour-Egyptian-desert.html. (**b**) Minimal openings in exterior elevation. https://www.egypt-toursplus.com/nubian-village/. (**c**) Small triangular openings in a Nubian house. http://looklex.com/egypt/aswan12.htm

is achieved by the presence of the following elements: (a) interior courtyard which modifies the hot-arid climate provides daylighting, ventilation and shaded outdoor area; (b) minimal openings on exterior elevation which reduces indoor heat gain and sun glare; (c) light colour painting on external building facades to reflect sun rays and reduce heat absorbance and (d) upper openings which help in ventilation as to allow hot air to escape. In addition, the Nubian houses provide a good example of how the cultures and traditions influence it which can be referred to as social sustainability. The Nile value is reflected on the architecture as main entrances are oriented toward the Nile, and also the elevations have decorative symbols (group of triangles in refracted lines) which symbolize the Nile River (Moustafa 2015). It is clear that walls have interesting drawings symbolizing their habit and beliefs, such as palm trees, birds and crocodiles (Moustafa 2015).

Furthermore, hospitality is demonstrated by the main entrance projection and its decorations, the existence of a guest room with wide windows and decorations and by the presence of an external place (*Mastaba*) for sitting with neighbours and visitors. Also, privacy is a significant trait in Nubian houses as they have small exterior openings, presence of side entrance that is separated from guest entrance and having screen wall beside guest entrance to prevent guests from seeing the inner spaces of the house. Finally, religious beliefs influence the architectures as Quran verses or religious statements appeared on some houses' elevations (Box 15.10).

Box 15.10: Social and Environmental Sustainability in Nubian Architecture—Upper Egypt
The traditional Nubian architecture exhibits environmental sustainability which includes: using local sun-dried mud bricks and by enhancing indoor ventilation and daylighting and reducing heat gain in hot-arid climate, while cultural sustainability is present due to reflecting the following items on the houses' architecture: (a) value of the Nile River, (b) hospitality and generosity of Nubians, (c) privacy and (d) religion.

Luxor Architecture: Old Gourna

Al Gourna is the same name used for three villages that are located on Luxor's Left Bank. This section refers to the first village or "Old Gourna", which was completely designed and constructed by local residents, and thus provides a great example for vernacular architecture. The layout of the village is considered organic as the urban fabric appears to be generated naturally rather than being professionally designed (Fig. 15.13). The dwellings vary from one family to another, and the building heights range between one to two storeys. The distribution of the houses follows an irregular pattern, which is a resultant of respecting the area's mountain topography. Moreover, building decisions were influenced by social and cultural traits of the local residents. Thus, Gourniis used to examine the mountain morphology before the construction process to produce a layout that is highly adapted to their local values and traditions (Mahmoud 2016).

Fig. 15.13 Old Gourna village, Luxor, Egypt. *Source*: https://en.wikipedia.org/wiki/Kurna

A strong feature present in Old Gourna's village is that residents have psychological connectedness with their surroundings and perceive the spaces' values. The place identity is unmistakable, for example, the presence of wall openings on houses' elevations so that pigeons can fly easily inside the house. In addition, the presence of unique vernacular drawings and the use of colours, as shown in Fig. 15.14, illustrate meanings that are understood by inhabitants. Moreover, the social aspects are highly taken into consideration in the village as cultural traits affect the spaces design. The architecture is mostly introvert except of some dwellings which have another main terrace to enjoy the view. Moreover, constructing houses on levelled mountains make residents feel secure (Mohamed et al. 2015).

Also, the visual dimension is demonstrated in the village as it respected the existing topography (Fig. 15.15). Residents acquire their visual preference or satisfaction by seeing their village and having feeling that they are on top of the mountain (Mohamed et al. 2015) (Box 15.11).

Fig. 15.14 Vernacular drawings on buildings in Old Gourna and Luxor, Upper Egypt. *Source*: Mahmoud (2016)

Fig. 15.15 A religious building—The Mosque, the Old Gourna's architecture. *Photo credit*: Author

Fig. 15.16 Private and public spaces in Old Gourna, Luxor, Egypt. (**a**) Public spaces. (**b**) Private space. *Source*: Mahmoud (2016)

The lived spaces present in Old Gourna are very sensitive to their own environmental design, their visuals and aural privacy. It is important to note that the presence of vernacular architecture satisfied the residents' needs. Inhabitants enjoyed living in their open spaces whether it is inside or outside the houses as shown in Figs. 15.16. Terraces in front of the dwellings are used as sitting areas to enjoy the breathtaking view of Luxor, while the spaces that subside the houses were used for raising domestic animals (Mohamed et al. 2015).

Box 15.11: Vernacular Architecture in the Old Gourna in Luxor, Egypt
"The Old Gourna will remain a legend of vernacular architectural/urbanism and cultural heritage".

Vernacular Architecture in the Western Desert of Egypt

The Western Desert covers a large area estimated be two thirds of Egypt. It consists of five main oases: Bahariya, Farafra, Dakhla, Kharga and Siwa (Mohamed et al. 2015). This section will present selected examples of vernacular architecture in three of the oases: Bahariya, Dakhla and Siwa. The selection was based on the environmental similarities between both El-Kharga and ElDakhla regions and also between El-Farafra and El-Baharia (Walid and Omar 2014). The climate is characterized by hyper-arid climatic conditions with rare rainfall and extremely high temperature. In addition, the north western and northern winds extend from the Mediterranean over the Western Desert with decreasing speed southwards (De Filippi 2006).

The presence of vernacular architecture provides rich examples of several approaches for building earth buildings. It uses local materials as the wall materials

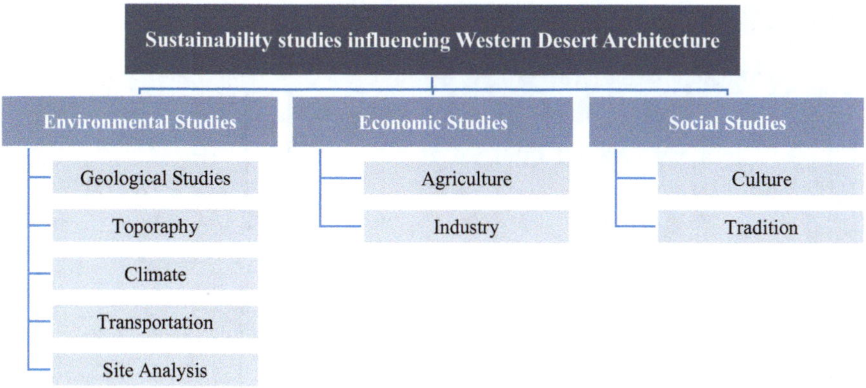

Fig. 15.17 Sustainability studies that influence the architecture in Western Desert, Egypt. *Source*: Mohamed et al. (2015)

used are mainly mud bricks with the exception of the use of salt blocks and mud (Karshif blocks) in Siwa oasis. Usually, the roof construction material is local wood either from trees or palms. In addition, reeds and palm ribs may be used as secondary construction materials.

It is important to note that using local materials has several benefits including: (a) materials can be easily recycled, either reusing old earth blocks as building material or returning it to the soil; (b) conserving energy as material has low embodied energy; (c) providing thermal insulation; (d) stabilizing indoor temperature; (e) absorbing excess humidity and (f) economically viable as material is taken from the surroundings thus eliminating the transportation or manufacturing costs (Dabaieh 2013).

The presence of arid climate plays an important role on the architectural and urban composition. In addition to the environmental sustainability, the Western Desert architecture is also influenced by economic and social sustainability (Mohamed et al. 2015), as illustrated in Fig. 15.17.

Traditional Architecture in Bahariya Oasis

The Bahariya oasis is located in the central part of the Western Desert. The presence of traditional architecture is quite similar to that present in the Nile Valley villages at the edge of the cultivation. Typically, the traditional house in Bahariya oasis (Fig. 15.18) is constructed using mud brick or stone walls, with sand stone foundation and local timber roof (Mohamed et al. 2015).

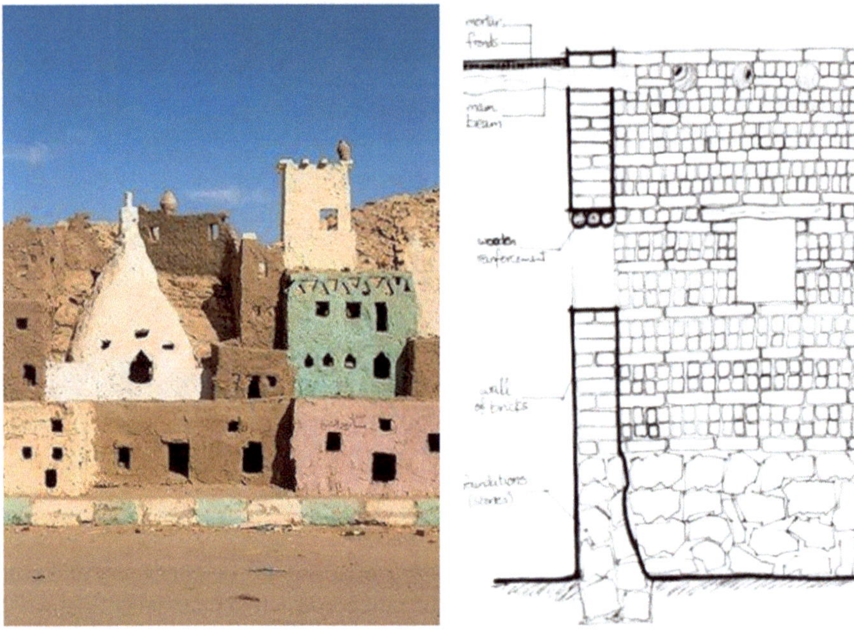

Fig. 15.18 Elevation and section in Bahariya house. *Source*: Mohamed et al. (2015)

Traditional Mud Brick Architecture in Dakhleh Oasis

Dakhleh Oasis, located in the Western Desert of Egypt, includes many ancient constructions such as the villages of Al Qasr and Balat. These demonstrate rich examples of vernacular architecture that uses ingenious methods and techniques to cope with the local harsh environment.

The layout of the region is characterized as compact mud dwellings with narrow roads separating houses. Settlements are grouped to form neighbourhoods (quarters) that relatively act as independent units (De Filippi 2006).

It is worthy to note that although the quarter is a secured closely knit group that is self-sufficient, the quarters do participate in the community and economic affairs of the city as a whole. Moreover, the traditional architecture that is present is unique and of great value as over the centuries, it was able to adapt the changes in society, defensive needs, climatic conditions and to interact with the environment (De Filippi 2006).

First, the desert defensive architecture played an important role in developing the village; this is demonstrated by the presence of the following: a compact fortress with concentric and radial connecting streets to allow internal communications within the village, walls of mud brick on all sides and closely packed buildings and connected to one another by covered narrow alleys (*zuqûq*). Streets are covered to fulfil two opposite requirements: (1) need for extra space that can be accessed from the housing dwellings and (2) pedestrian requirement to protect and shade from direct sun. In addition, the defensive wall system is achieved by the presence of the blind rear walls of the houses in the perimeter of the village. The entrance of the village was allowed by more than one gate (De Filippi 2006).

Second, the village adapted to the harsh climatic conditions by having passive techniques whether on urban or architecture scale (Fig. 15.19). The urban passive techniques include: compact layout that minimizes the exposed building facades to direct sun and narrow, often covered and shaded streets which reduce heat gain and help in ventilation (Box 15.12). While the indoor passive techniques include: minimal openings on exterior façade to reduce heat gain, indoor air ventilation by a vertical and cross stream of fresh air thanks to a system of apertures and having separate living spaces for summer or winter (day or night) to achieve human thermal comfort (De Filippi 2006).

Box 15.12: Environmental Sustainability in Western Desert Architecture, Egypt

The environmental aspects incorporated into the present local architecture included light, ventilation and thermal comfort.

This is featured by the presence of narrow shaded streets, ventilation and daylighting through courtyards, evaporative cooling from well or water feature in courtyard, small building depth (positively affecting lighting, air movement and thermal capacity), narrow frontage on shaded street, high thermal mass materials, low window-to-wall ratio and minimal openings on street while openings present on walls facing the courtyard (Ahmed 2014).

Fig. 15.19 Typological architectural formation of dwellings in Balat. *Source*: Walid and Omar (2014)

Al Qasr Housing typology

Houses are compact with minimal openings on exterior façade in almost all the Western Desert villages (Fig. 15.20). However, in the former capital of the oasis "Al Qasr", buildings are exceptionally tall. The internal configuration of the houses is designed according to the residents' needs of the members of the extended families, economic wealth and social class (De Filippi 2006).

The house investigated in Al Qasr is located in the "Mahdiy Awada" quarter, and looks onto "Al Jazarin" street, which connects the north to the south of the ancient settlement. The dwelling consists of a ground floor, two storeys and a terrace connected by a central staircase which acts as ventilation shaft as it has perforated wooden steps. The house has typical traditional housing elements, illustrated in Fig. 15.21, such as the primary entrance vestibule or passage with a right-angle turn which allows for covered area and ensures privacy of residents (De Filippi 2006).

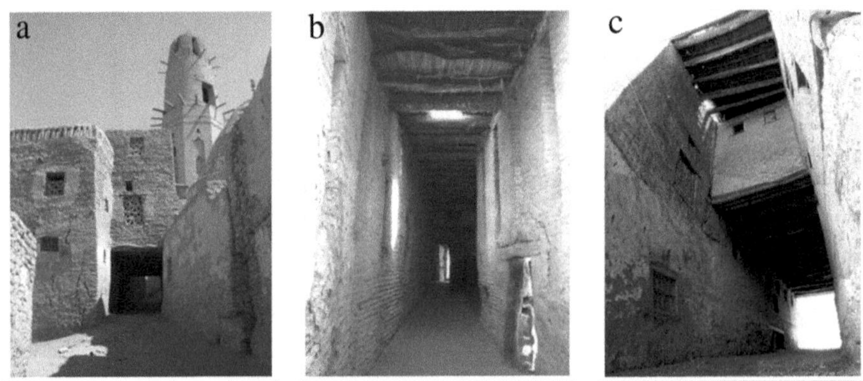

Fig. 15.20 Al Qasr architecture elements. (**a**) Al Qasr, New Valley, Egypt. (**b**) A covered street in Al Qasr. (**c**) The sabat in the old Al Qasr. *Source*: De Filippi (2006)

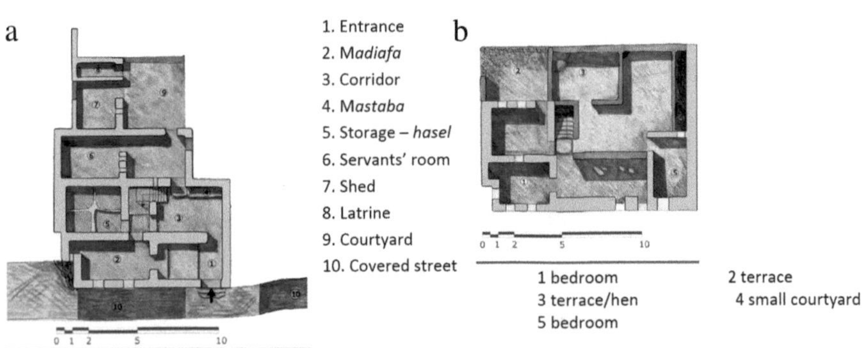

Fig. 15.21 Al Qasr house. (**a**) Ground floor plan. (**b**) Roof plan. *Source*: De Filippi (2006)

Traditional Kershef Brick Architecture in Siwa Oasis

Siwa oasis is considered to be one of the ancient oases since pharaohs' days. It is considered an agricultural oasis as it is popular for its palm and olive trees and the presence of many freshwater streams and springs. In ancient fortified dwellings, buildings normally do not exceed 5-m height, and residents used salt-mud brick "kershef" as building material (Fig. 15.22). Regarding the construction technique, salt is used which is left to dehydrate in direct sun, then fermented wet mud from the salty soil "Tlakht" is used as the filling material and left to dry for 1 or 2 weeks. Yet, the area is largely abandoned. The climate in Siwa is described as harsh conditions as it has hot-dry summer reaching 39 °C and cool winter reaching 5 °C (Ahmed 2014).

Siwa's architecture adapted to the harsh climatic conditions on both urban and architecture scale. The environmental techniques on urban scale include: compact layout shape to minimize the amount of exposed facades to direct sun before passing through the windows, having narrow alleys between houses that are usually covered and shaded to reduce heat gain and provide ventilation shaft, and the use of vegetation next to openings which cools air before passing through windows (Ahmed 2014).

While in the indoors, Siwa's architecture incorporated passive techniques such as best utilization of the available local building materials, using thermal insulation (as thickness of walls is normally 50 cm starting from the first row in the ground reaching 30-cm thick in the last row), incorporating passive cooling techniques such as wind towers and atriums and having windows in opposite directions to allow cross ventilation, and having natural sun lighting. In addition, Siwa's local residents managed to incorporate their social dimension in the local architectural practices (Ahmed 2014). Table 15.5 presents a comparison between three vernacular architecture buildings of residential/hospitality use in Siwa Oasis to highlight similarities and differences.

a b c

Fig. 15.22 Traditional architecture in Siwa (Shali). *Source*: (**a, b**) Walid and Omar (2014). (**c**) https://www.egypttoday.com/Article/9/20720/Siwa%E2%80%99s-best-kept-secret-eco-sleep-with-a-twist

Table 15.5 Comparison between vernacular architecture buildings in Siwa oasis

Criteria	Siwan house: Haj Ali	Shali Lodge and its extension Al Baben Shali	Adrere Amellal Eco-lodge
Illustration			
Use	Residential	Hospitality	Hospitality
Material	Local kershef		
Structure	– Foundation: 50-cm high concrete wall on the ground floor to protect it from ground water, – Walls are supported by palm trunks connections on roofs for straitening	– Wall bearing – Palm trees' layout was used for ceiling construction	– Wall bearing
Project description	– Healthy, low-cost, thermally comfortable and well-designed house, externally and internally	– The first kershef built lodge in Siwa, no electricity was used due to awareness, unlike other eco-lodges	– Lodge located in unique surroundings has 39 environmentally friendly rooms don't use electricity
Social sustainability	– L-shaped entrance and locating private rooms at the back of house for privacy, – Encouraging family social interaction by providing central gathering area, – Encouraging society social interaction by providing outdoor sitting area for men	– Inspired by old Siwan vocabulary as alcoves were grooved on the walls for storage or decoration, – Project gave training and helped employment of local residents as it required knowledge of local kershef building techniques	– Respecting Siwa's culture, norms and nature
Economic sustainability	– Local materials reducing costs of transportation	– Project helped in employment of over 45 Siwans	– Creating economic opportunities for local people while restoring the physical environment and marketing local products to the international market

(continued)

Table 15.5 (continued)

Criteria	Siwan house: Haj Ali	Shali Lodge and its extension Al Baben Shali	Adrere Amellal Eco-lodge
Environmental sustainability	– Use of local materials, (wood and kershef)—climatic-responsive zero carbon emission material, – External walls are painted for emitting solar radiations, – Use of built-in kershef cooker for minimizing the use of electricity	– Use of local materials – Interior walls exposed to the sun are in light colours for emitting solar radiations – Using dehydrated salt in manufacturing furniture and lighting units	– Best use of the available local building materials – Furniture is made of local materials (sand blocks, kershef and palm wood), – Skylight made of local palm wood blended with sand blocks and stone floors – Climatic-responsive materials used
	– Passive ventilation: a. Use of stair case tower as an atrium (cooling tower), b. Orienting the windows to allow cross ventilation, – Daylighting as staircase acts as lighting pipe where mirrors have been installed on the walls to reflect the sun light inside the house, – Distinct bedrooms for summer/winter; roofed rooms for sleeping in winter, while non-roofed for summer. On similar basis, another un-roofed kitchen is located on the upper floor	– Passive ventilation: a. Atrium between the hotel rooms creating passive cooling with wide opening arches are surrounding the atrium for enhancing the wind speed and creating cross ventilation 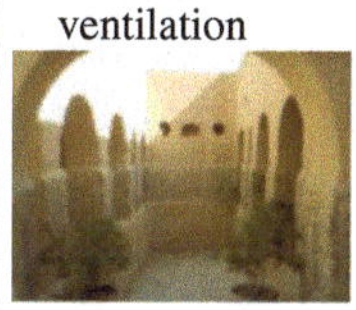 Atrium in Shali lodge b. Narrow alleys with small windows from one side and wider ones on the other side creating cross ventilation 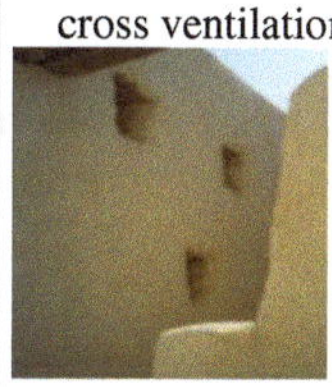 Building wall facing alley	– Passive cooling and cross ventilation were achieved for better climatic conditions – Respecting surroundings as the building merges with the mountain – Minimum openings embedded in thick walls  Small opening in exterior wall  Skylight roof details

Source: Developed by author based on Ahmed (2014)

Lessons Learned from Global, Regional and Egyptian Vernacular Architecture

The main concept that could be learnt from exploring the different vernacular buildings presented is how to blend the building with the natural surroundings using the traditional vocabularies whether on the exterior façade or through the interior vocabularies used. Moreover, vernacular architecture can be considered sustainable as it respects the economic, environmental and social factors. Finally, examining the vernacular architecture forms and solutions worldwide could help in solving problems facing present-day architecture, in particular the critical housing situation in the developing countries. Thus, lessons learned from these solutions should be implemented in the current contemporary designs.

Guidelines for Future Developments

Environmental Sustainability

- Passive design is a mean to achieve sustainability since it reduces energy consumption while achieving human comfort,
- Incorporating techniques to allow natural ventilation is recommended in hot-arid climate. Cross ventilation is required in hot-humid climate and
- The use of local materials (such as, straw mud, wood and kershef) is significant as they are considered climatic-responsive zero carbon emission materials.

Material Sustainability

- Natural resources and local materials should be used along with recycled materials to reduce the embodied energy, and hence reduce energy consumption as well as CO_2 emissions and.
- Building using local materials requires creativity in order to be able to produce structures that can withstand the environmental conditions.

Social Sustainability

- Designs should encourage society social interactions by providing liveable outdoor spaces,

- Urban planning and architecture should respect country's culture, norms and traditions to ensure that it meets the users' need, hence achieving social sustainability and
- Colours, decorative symbols and drawings can be used on buildings' elevations to symbolize meanings, habits and beliefs of a community.

Economic Sustainability

- Vernacular architecture is a mean for creating economic opportunities for local people, while restoring the physical environment and marketing local products to the international market,
- Designs should incorporate the use of local materials to reduce the costs of transportation and manufacturing costs thus become economically viable and.
- The economic activities of residents should be studied in order to provide suitable spaces that account for their needs.

Financial Sustainability

- Vernacular architecture is considered as inexpensive hence its cost is considerably low due to the use of natural resources and local materials and not using sophisticated construction methods yet built by its owners and users, thus financially viable.

Conclusions

Vernacular architecture at the global, regional and local scale was reviewed and presented. Worldwide leading historical vernacular architecture examples were reviewed with the main concern on Egypt.

One of the major points discussed is how vernacular architecture adheres to basic green principles which include: (1) energy efficiency, (2) utilization of local materials and (3) being climate responsive. Also, the chapter examined the factors influencing vernacular architecture: local climate, resources availability, construction skills, and culture and heritage through analysing regional vernacular architecture examples in Africa and the Middle East.

Local climate, resources availability and construction skills are the main factors that affect the vernacular architecture in Africa, while local climate and culture and heritage are the most dominant in the Middle East. Moreover, synergies were drawn among the presented African case studies to discuss it in terms of forms, colours, materials, structure and sustainable features present.

It is important to note that Egypt demonstrates rich examples of vernacular architecture. In order to gain clear knowledge about the vernacular architecture in Egypt, emphasis was made on the traditional architecture of Aswan (Old Nubian), Luxor (Old Gourna) and Western Desert of Egypt specifically the traditional architecture of Bahariya, Kharga and Siwa oases. In Nuba, the environmental, social and cultural factors were taken into account whether on urban layout or architecture scale. The environmental factors were included and integrating layout with surrounding topography, using local materials and enhancing indoor ventilation and daylighting, along with reducing heat gain in hot-arid climate. The social determinant is exhibited by the importance of the Nile on the village planning as it is related to many Nubian myths. Moreover, other social factors were taken such as hospitality and generosity of Nubians, privacy and religion. While the traditional Luxor architecture provides a clear example of social sustainability whether on building or urban scale as it includes: presence of unique vernacular drawings and use of colours in buildings' elevations, introvert architecture to achieve privacy and the integration of houses with existing topography which makes residents feel secure.

The traditional architecture in the Western Deserts exhibits a lot of environmental aspects as it incorporates strategies for daylighting, ventilation and achieving thermal comfort. The most distinct feature between the three presented oases: Bahariya, Kharga and Siwa is that Siwa oasis uses salt-mud brick "kershef" as their local material unlike the others which use mud. Many lessons were learned from global, regional and local vernacular architecture.

Finally, the guidelines for future developments addressing environmental sustainability, material sustainability and social sustainability as well as economic sustainability and financial sustainability are outlined.

References

Ahmed RM (2014) Lessons learnt from the vernacular architecture of bedouins in Siwa oasis, Egypt. In: Proceedings of the 31st international symposium on automation and robotics in construction and mining, ISARC 2014, pp 910–917. Retrieved from https://www.scopus.com/inward/record.uri?eid=2s2.084912569376&partnerID=40&md5=b05acc1ef2f216ddda950f3feb1b6b7d

Alkhalidi A (2013) Sustainable application of interior spaces in traditional houses of the United Arab Emirates. Procedia Soc Behav Sci 102(Ifee 2012):288–299. https://doi.org/10.1016/j.sbspro.2013.10.743

Bayoumi OAM (2018) Nubian Vernacular architecture & contemporary Aswan buildings' enhancement. Alex Eng J 57(2):875–883

Dabaieh M (2013) Earth vernacular architecture in the Western Desert of Egypt. In: Markku M (ed) VERNADOC RWW 2002

De Filippi F (2006) Traditional architecture in the Dakhleh Oasis, Egypt: space, form and building. systems. In: PLEA2006 - the 23rd conference on passive and low energy architecture, September, pp 6–8

Garcia VA (2016) Sustainable architecture in Cape Verde a new village for Chã-das-Caldeiras. Tecnico Lisboa. https://fenix.tecnico.ulisboa.pt/downloadFile/1970719973966190/extended-abstract.pdf. Accessed 14 Feb 2018

Mahmoud RA (2016) Old Gourna: the complexity of vernacular architecture/urbanism and cultural heritage. Procedia Soc Behav Sci 225:200–215. https://doi.org/10.1016/j.sbspro.2016.06.020
Mohamed N, El A, Ibrahim MA, Ayad HM, Tarek Z, Sayad E (2015) Comparative analysis of the desert and green vernacular architecture in the Oases of Egypt. Ierek Press, Alexandria, pp 1–8
Momtaz RI, Feisal Z, El A (2012) Sustainable features of vernacular architecture: a case study of climatic controls in hot arid climate. In: XVIII IAHS world congress. Istanbul, Turkey
Moustafa WFO (2015) Nubian vernacular architecture technique to enhance eco-tourism in Egypt. J Emerg Trends Econ Manag Sci 6(3):169–176
Soleymanpour R, Parsaee N, Banaei M (2015) Climate Comfort Comparison of Vernacular and Contemporary Houses of Iran. Procedia Soc Behav Sci 201:49–61
Svensson T, Ekvall E (2012) Study of a two-storey family house in the Dar es Salaam region, Tanzania. https://www.diva-portal.org/smash/get/diva2:616686/FULLTEXT01.pdf. Accessed 13 Feb 2018
Walid A, Omar F (2014) Vernacular architecture approach to achieve sustainability in informal settlements. In: World sustainable building 2014 Barcelona conference – GBCe, pp 200–207

Website Links

https://www.archdaily.com/155224/vernacular-architecture-and-the-21st-century. Accessed 28 Feb 2018
https://www.dzbreaking.com/2018/03/12/tourism-algerias-sahara-timimoun-oasis-known-red-ochre-color-buildings/. Accessed 05 Feb 2018
http://www.amusingplanet.com/2013/03/the-old-town-of-ghadames.html. Accessed 05 Feb 2018
https://decideyouradventure.com/ait-benhaddou-southern-morocco/. Accessed 05 Feb 2018
https://theculturetrip.com/africa/morocco/articles/an-architectural-tour-of-chefchaouen-moroccos-majestic-blue-city. Accessed 08 Feb 2018
http://naturalhomes.org/african-vernacular. Accessed 08 Feb 2018
http://looklex.com/tunisia/nefta05.htm. Accessed 08 Feb 2018
http://looklex.com/tunisia/nefta04.htm. Accessed 08 Feb 2018
https://www.designcauseinc.org/single-post/2016/04/12/Blending-traditional-and-modern-to-create-successful-humanitarian-architecture. Accessed 09 Feb 2018
http://www.wondermondo.com/Comoros.htm. Accessed 09 Feb 2018
http://www.visitdjibouti.dj/explorerEN. Accessed 10 Feb 2018
http://explore-eritrea.com/tour-6-discover-enchanting-culture-nature/tigrinya-tribe-family-and-hidmo-traditional-house/. Accessed 10 Feb 2018
http://www.eritrea.be/old/eritrea-barentu.htm. Accessed 12 Feb 2018
http://naturalhomes.org/african-vernacular. Accessed 12 Feb 2018
http://www.archidatum.com/articles/the-taita-sustainable-vernacular-house/. Accessed 12 Feb 2018
http://www.malawiarchitecture.com/mwanza. Accessed 12 Feb 2018
http://vintagemauritius.org/vintage-houses/old-colonial-creole-style-houses-mauritius-part-1/. Accessed 13 Feb 2018
http://www.earthauroville.com/mayotte_as_development_model_en.php. Accessed 13 Feb 2018
https://www.archdaily.com/144527/educational-building-in-mozambique-andre-fontes-sixten-rahlff. Accessed 13 Feb 2018
https://atcnews.org/2015/02/20/creole-architecture-all-part-of-reunions-attractions/. Accessed 13 Feb 2018
https://www.archdaily.com/372709/umubano-primary-school-mass-design-group. Accessed 13 Feb 2018
http://www.seychellesnewsagency.com/articles/1652/The+traditional+Creole+architecture++An+aspect+of+the+Seychelles+heritage+in+need+of+preservation. Accessed 13 Feb 2018

http://wiki.colby.edu/download/attachments/6294655/Houses+in+Somalia.pdf. Accessed 13 Feb 2018
http://sudantribune.com/spip.php?article46395. Accessed 13 Feb 2018
http://www.monitor.co.ug/Magazines/HomesandProperty/The-hut%2D%2Dthe-first-building-in-Uganda/689858-1522996-r03yg4z/index.html. Accessed 13 Feb 2018
https://www.insaka.org/pages/about. Accessed 13 Feb 2018
http://www.designindaba.com/articles/point-view/hut-decorating-competition-could-give-urban-homes-run-their-money. Accessed 13 Feb 2018
http://naturalhomes.org/african-vernacular%23southafrica. Accessed 13 Feb 2018
http://www2.exxonmobil.com/Chad/Library/Photo_Video/web_photoalbum/english/14g.html. Accessed 13 Feb 2018
http://www.endingextremepoverty.org/2010/04/an-upper-class-neighborhood-in-rural-congo.html. Accessed 13 Feb 2018
https://mobile.nation.co.ke/lifestyle/Seeking-imagination-beauty-and-identity/-/1950774/2241852/-/format/xhtml/item/1/-/u4sxfn/-/index.html. Accessed 14 Feb 2018
http://tibarose.com/social-project.php?id=222. Accessed 07 Apr 2018
http://www.everyculture.com/Bo-Co/C-te-d-Ivoire.html%23ixzz5Bvym97bo. Accessed 14 Feb 2018
http://www.archidatum.com/projects/ecological-house-in-gambia-virai-arquitectos/. Accessed 16 Feb 2018
https://www.wmf.org/project/larabanga-mosque. Accessed 16 Feb 2018
http://www.visualgeography.com/categories/guinea/houses.html. Accessed 16 Feb 2018
https://www.aecom.com/projects/liberia-schools-programme/. Accessed 16 Feb 2018
https://www.citylab.com/design/2015/08/the-beauty-of-african-vernacular-architecture-captured-in-one-big-database/400778/. Accessed 16 Feb 2018
https://archnet.org/sites/183/media_contents/10783. Accessed 16 Feb 2018
http://sainthelenaisland.info/houses.htm. Accessed 16 Feb 2018
https://www.dezeen.com/2017/01/25/toshiko-mori-compressed-earth-bamboo-thatch-cultural-centre-senegal-africa-architecture/. Accessed 16 Feb 2018
http://www.ibike.org/bikeafrica/sierra-leone/essay/09-Tiwai.htm. Accessed 16 Feb 2018
http://www.middle-africa.com/blog/togo/togo-architecture. Accessed 16 Feb 2018
https://commons.wikimedia.org/wiki/File:Rondavel_house_in_Botswana.jpg. Accessed 16 Feb 2018
https://commons.wikimedia.org/w/index.php?curid=3797723. Accessed 16 Feb 2018
http://traditionscustoms.com/people/himba-people. Accessed 16 Feb 2018
http://naturalhomes.org/african-vernacular. Accessed 16 Feb 2018
https://www.worldatlas.com/webimage/countrys/me.htm. Accessed 20 Feb 2018
http://culture.gov.bh/en/visitingbahrain/destinations/Name,10553,en.html%23.WtJK9S5ubIU. Accessed 01 Mar 2018
https://forkandfoot.com/vavla-rustic-retreat-cyprus/. Accessed 01 Mar 2018
http://earsiv.cankaya.edu.tr:8080/xmlui/bitstream/handle/123456789/347/Ayyash%2C%20Ahmed%20Basil.pdf?sequence=1&isAllowed=y. Accessed 01 Mar 2018
http://www.khammash.com/projects/dana-guesthouse. Accessed 02 Mar 2018
http://www.traveladventures.org/continents/asia/beit-khalid01.html. Accessed 01 Mar 2018
http://www.meda-corpus.net/arb/fitxes/F1sites/eng/lb_s07.pdf. Accessed 01 Mar 2018
https://www.tourismoman.com.au/news/an-architectural-tour-through-oman/. Accessed 01 Mar 2018
http://www.rehabimed.net/Publicacions/Corpus/Manual%20para%20el%20mantenimiento%20y%20rehabilitacion%20de%20la%20arquitectura%20tradicional%20del%20Libano/pdf/livret/ats_eng.pdf. Accessed 01 Mar 2018
https://archnet.org/system/publications/contents/3944/original/DPT0472.pdf?1384777691. Accessed 05 Mar 2018
https://misfitsarchitecture.com/tag/yemeni-vernacular-architecture/. Accessed 05 Mar 2018

Chapter 16
Contemporary Roof Design Concepts: Learning from Vernacular Architecture

Judit Lopez-Besora, Helena Coch, and Cristina Pardal

Vernacular Architecture, Place, and Materials

Vernacular architecture is as old as humanity. The primitive forms of habitation were designed by humans to fulfil their basic needs such as protection and security. At the beginning, caves and natural shelters were the most suitable places for this purpose, since humans did not have to carry out major modifications to make them habitable. After an extended period, humans left their shelters in caves and started to build small constructions such as huts, which gathered in permanent settlements. To reach a minimum standard of living, they used the means they had on hand, raw materials provided by the nature. These materials might be easily available and not far from the settlements, so that the construction and repair of the houses might not be complex.

The type of materials employed in primitive architectures was closely related to climate and place, either those with a mineral origin (stone, sand or clay) or organic (vegetation). The first group has a strong relation with the tectonic and geological characteristics of place. Landscapes have been configured by geological processes and modelled by atmospheric phenomena such as rain, wind and sun for millions of years. Because of these processes, the earth has created several types of mineral materials with different environmental, mechanical and durability characteristics, which have the capacity to carry out specific functions in distinct parts of the building. The second group, composed of organic materials, has a direct relation with climate. This group is made up of vegetable elements which grow in the nature depending on atmospheric parameters such as rain or sun exposure. A wide range of organic building materials such as wood, bamboo, straw, seaweed and others are present in vernacular architecture depending on the building site. The use of one or

J. Lopez-Besora (✉) · H. Coch · C. Pardal
Architecture & Energy Research Group, Department of Architectural Technology, School of Architecture, UPC, Barcelona, Spain
e-mail: judit.lopez.besora@upc.edu; https://www.aie.upc.edu

A. Sayigh (ed.), *Sustainable Vernacular Architecture*, Innovative Renewable Energy, https://doi.org/10.1007/978-3-030-06185-2_16

another depends on their local availability, directly related to local climate. People living in places with advantageous conditions for vegetation entrust the construction of their buildings to organic materials. On the other hand, architecture in desert places mostly falls on the use of mineral materials such as clay, sand, stone or other mineral origin materials. Very often, both materials are mixed in vernacular architecture (Fig. 16.1).

In vernacular architecture, the place defines the elements to protect from and the means to protect with. Throughout history, cultures have built a set of construction knowledge around their needs and means which has been transmitted over the years (Weber and Yannas 2014). This practice has shaped a recognizable vernacular architecture adapted to a specific place and its climate. Nevertheless, vernacular architecture from distant places in the world with no apparent cultural or commercial link has evolved in an analogous way, sharing the same or similar architectural elements and solutions (Fig. 16.2). Constructions in these places look alike and remind each other despite the distance. In short, the same baseline conditions in distant places has led to a similar architecture which is beyond cultural trends or styles, proving that place and materials are in the basis of vernacular architectural design.

The Ephemeral Nature of the Roof

The primary forms of habitation consisted of a single roof with the purpose of being weatherproof. In a first stage, a roof provides the most useful protection from weather, more effective than walls or other elements. Many buildings in hot climates have subtle walls or do not even have them, since the effectiveness of the construction falls on the roof (Fig. 16.3). On the contrary, there is no evidence of buildings that are only configured with walls but with no element on top.

Fig. 16.1 Organic and mineral materials in Vietnamese vernacular architecture (Coch 1998)

Fig. 16.2 Similar vernacular architecture in parts of the world with comparable climates; Chiapas and Hanoi

Fig. 16.3 Vernacular architecture defined by roofs in Hanoi

As said in the previous section, roofs have traditionally been built with local materials such as wood, straw or mud. In general terms, natural materials are highly sensitive to the passage of time and lack durability, either those with a vegetal or mineral origin. In addition, the top of the building is the most exposed part of the construction and this makes it very sensitive to climatic agents. As a consequence, the design of the roof poses a challenge in which a long-lasting solution to keep the building in good condition is needed. The design of an appropriate roof with low-durability materials requires ingenuity and regular maintenance operations. The inhabitants of vernacular architecture have been taking care of their buildings for generations using inventiveness and carrying out regular maintenance operations such as replacing single pieces or restoring claddings. Thanks to that, the life of buildings has been extended for years in optimal conditions for living.

The preservation of roofs in good condition is key for the security of the whole building and the people inside. Obviously, weather conditions endanger the durability of building materials, but also structural stability must guarantee a long-term permanence. While vertical elements are self-supporting structures, horizontal elements have complex stress requirements. The material and technical solution applied brings about stability and at the same time it determines the spatial

Fig. 16.4 Lintelled porch in Mallorca and vaulted coverage in the Arabic Baths of Granada

dimension of architecture. In traditional architecture, two types of structure have solved the coverage of buildings: lintelled and vaulted structures (Fig. 16.4). Each solution is usually linked to specific materials: wood for lintelled structures and clay or stone for the vaulted ones. The dimension of the space below depends on the solution applied.

Both the quality of materials and structural stability set the durability of roofs. However, when maintenance lacks or the building goes under state of neglect, the roof is the first element to collapse. As the roof is the most ephemeral part of the building, few remains of primitive roofs have come to our days. Despite their importance, few roofs have lasted in the original condition and most of them have disappeared. The ruins of old buildings show this fact, where the only witnesses of old constructions are some parts of the walls and fragments of pavements (Fig. 16.5). In most cases, ruins are roofless constructions.

Roof Functions

A roof is a layer between the floor and the sky that changes the characteristics of the place below. This layer can have several shapes and can be made up of varied materials, with different degree of complexity. In any case, it essentially fulfils two functions: a physical one, which is providing a shelter, and a psychological one related with the creation of a space.

Fig. 16.5 Ruins of Ostia Antica, Italy

Fig. 16.6 Covered market in Cordes-sur-Ciel, southern France, and street awnings in Seville, Spain

Psychological Functions of the Roof

Building a roof defines a space even though it has no side limits, as it happens with the shade of a tree in a field. The tree defines a space under its crown which is protected from sunshine and rainfall, although there is no physical limit around. The space created under a tree or a roof has distinctive characteristics from the immediate surrounding, both from an environmental and perceptual points of view. There are many examples of architecture and architectural elements with a roof but without side boundaries which define a space, as some examples of textile architecture (tepees, tents and awnings) or other spaces such as open markets, bus stops, covered platforms and others (Fig. 16.6).

The space defined by a roof gives rise to meanings with a psychological background (Table 16.1). One of them lies in the sense of physical protection. People cover their head when feel that there is something threatening around them, as an instinctive

Table 16.1 Psychological functions of the roof

Psychological functions	– Definition of space
	– Sense of physical protection
	– Sense of community
	– Architectural image

reflex. The most fragile and appreciated part of the body needs protection from attacks and damage, and the roof provides a psychological association with security. The human perceptual fields (visual and acoustic) leave the top of the environment out of scope, since the visual field is limited at both sides, bottom and top, as well as the acoustic field. So, a barrier covering what is out of control contributes to increase the sense of security, especially from an occasional danger coming from above. In broad terms, the roof acts as a helmet, a hat, which keeps our head out of danger. It not only provides physical security but also a psychological sense of security.

In the same way that a roof enhances our sense of individual physical protection, it also involves a type of collective protection, the sense of community. People feel secure in the company of other people, either living close to them or sharing similar experiences in the same places. The closer type of personal relations is people sharing cultural, affective or familiar bonds, which tend to gather under the same roof. In primitive nomadic cultures, camp fires defined the place around which a community or a group congregated (Banham 1984); in sedentary settlements, the roof played a similar role on gathering the community for celebrations and public engagements. Today, community buildings fulfil the same function, since being under the same roof has a meaning of belonging to a group or community, either in a familiar or a collective context (cultural, religious and commercial buildings).

The evolution from small settlements to huge communities led to towns and cities which can be considered an addition of roofs settled in the same piece of land. In vernacular communities, the architectural image of the settlement often falls on the shape of the roof, the most regular element of the building. There are many examples, such as "*trulli*" in southern Italy or "*Musgum*" mud huts in Cameroon, where the important appearance of the roof constitutes the image of the community. However, the uniformity of roof design in towns and cities can only be appreciated in aerial views. While the design of façades includes a lot of elements (doors, windows, decoration, etc.), roofs are more functional and their shape is more uniform. Especially in old towns, the design of roofs makes equal the heterogeneity of the city seen from above, defining a mosaic where the similarity of roofs is clearly perceived (Fig. 16.7).

Physical Functions of the Roof

So far, the space created under a roof has been analysed from a psychological standpoint. Roof functions have been focused on the meaning or image given and not on their specific physical function. However, a roof is fundamentally a shelter which

Fig. 16.7 Aerial views of Helsinki and Siena

Table 16.2 Physical functions of the roof

Rain	– Avoid water infiltration (imperviousness) – Expel water outside
Snow	– Avoid water infiltration (imperviousness) – Expel snow outside – Support weight
Radiation downward	– Protection from solar exposition
Radiation upward	– Control of infrared radiation to the sky

protects from the sky. The architectural envelope offers protection from the environmental and atmospheric conditions, which in the case of roofs are rain and radiation, both coming from the sky (Table 16.2). Wind is also a parameter to take in consideration when designing a roof, but to a lesser extent, since it takes different directions depending on the place.

First of all, a shelter must provide protection from rainfall. The roof expels the rain outside of the building and keeps it as dry as possible, while avoiding water infiltration. When reached, imperviousness contributes to keep a healthy environment in the interior and to extend the life of the building. A special case of precipitation that takes place in cold places is snow. It can collect a heavy load on the roof for weeks, putting at risk the structural stability of the whole building. Apart from this, when snow starts to melt, it falls in the form of water or snow to the outside.

Secondly, radiation must also be controlled by shelters. Roofs are exposed surfaces which are subject to radiation exchanges with the environment. Their horizontal (or nearly horizontal) position makes them a sensitive part of the building in terms of heat exchange. The roof is liable to receive plenty of solar radiation during daytime, which can lead to an increase of its own temperature. The absorbed radiation can be conducted to the interior of the building in the form of heat, in a different degree depending on the thickness and conductivity of roof materials, and it eventually produces an increase of air temperature inside. On the other side, the hours without solar radiation, the main radiative exchange is produced between the roof and the sky. The sky has a much lower temperature than the roof and the surfaces around, many negative Celsius degrees, so in this case the heat flux goes towards the sky. In fact, the cumulated heat in the roof during the day is radiated to the inside and the outside of the building, but the lower temperature of the sky implies that most of the flux is emitted to the outside, especially with a cloudless sky condition (Fig. 16.8).

A roof must be protected from solar exposition when values of solar radiation are excessive and could imply an increase of air temperature in the interior of the building. But at the same time, the roof can protect other parts of the building from solar radiation, functioning as an umbrella that provides with shade around the whole building. On the other side, the infrared radiation between the roof and the sky during the night must be controlled, either to keep the heat and increase the radiation towards the interior, or on the contrary, to dissipate it to the outside.

Vernacular architecture offers protection from the environment outside, creating shelters away from the weather conditions. The roof holds up the most aggressive atmospheric agents described above, while the rest of the building complements the thermal behaviour on controlling air temperature and wind. In the case of wind, roofs are an exposed part of the building where special attention must be paid to geometry and an accurate execution. Intense winds jeopardize the stability of the building and its elements, and so the use of appropriate materials and a right execution is fundamental.

Vernacular Roofs and Climate

The typical features of roofs can be explained from a climatic base. As seen in previous sections, vernacular architecture reflects the characteristics of climate, especially roof design, even though architecture comes from distant places in the world. Vernacular architecture has been giving a solution to climatic inputs for years with the limited means coming from the nature.

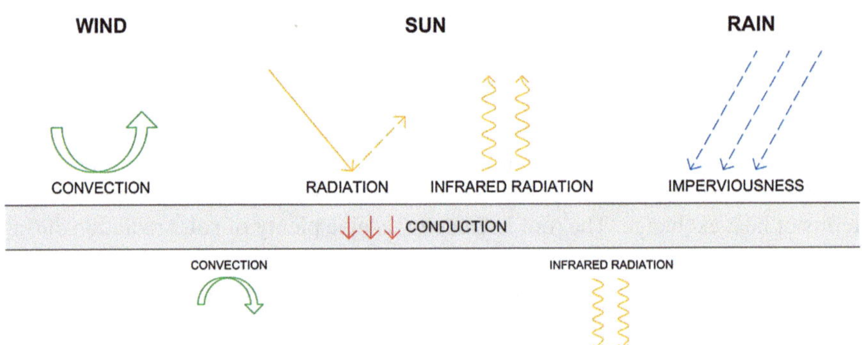

Fig. 16.8 Roof behaviour

Although the combination of climatic inputs defines the climate of a place, climates experience multiple local variations. Nevertheless, they can be divided into four main groups: cold, wet warm, dry warm and complex. The first three groups are extreme climates with a slight variation along the year. As the weather is similar throughout the year, it is easy to optimize the most suitable architectural solution. Roofs in extreme climates represent the most adapted solution considering both climate and material availability. The fourth group of climates contains some characteristics of them but with seasonal variation. Complex climates are mild, and the weather sometimes is similar to cold climates, and others to the warm ones, even though with less intensity. Vernacular architecture and roofs in complex climates must be flexible and adapted to different situations which take place simultaneously throughout the year.

Roofs in any part of the world are adapted to those particularities which can be easily defined, classified and identified by the shape and materials employed. In broad terms, the characteristics that define a roof are the pitch (from flat to steep roofs) and the degree of ventilation (ventilated or not ventilated), apart from the materials employed in their construction. A world map can then be described according to vernacular architecture roofs (Fig. 16.9).

Cold Climates

The main characteristics of cold climates are very *low temperatures*, seasonal variation with the changes of winter–summer sunshine levels, an always *pleasant solar radiation* and *aggressive winds* when they come from the direction of the

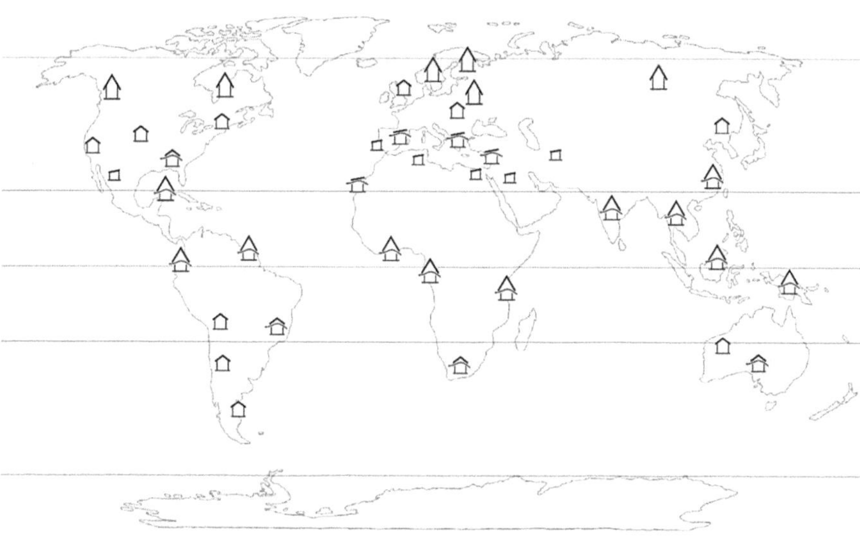

Fig. 16.9 Pitch and ventilation degree of roofs in different parts of the world

corresponding pole. These climates are typical of high latitudes or great heights in medium latitudes (Coch 1998).

The design of roofs in cold climates must deal with low temperatures as a main feature. Their function falls in materials which must guarantee high thermal insulation with the purpose of maintaining the heat inside the building. The scarce solar gaining at roof level is easily lost by transmission, radiation to the sky and convection. Aggressive winds increase convection thermal losses, while clear skies favour the radiation from the roof to the sky. Low conductivity, high thermal insulation or high thermal mass are the most suitable materials to keep the heat inside. The shape of the roof is secondary, even though roofs with a small surface exchange with the exterior perform better in cold climates. Nevertheless, the shape is mainly determined by the existence and type of rainfall. If so, pitched roofs which project the water or snow outside the walls contribute to keeping the building protected.

The expulsion of water falls mainly on the shape defined by the pitch of the roof, since the low permeability of many building materials or the need for overlapping small pieces in vernacular architecture makes materials less reliable for this function than an appropriate shape (Fig. 16.10).

Wet Warm Climates

Warm climates have significant differences depending on the degree of relative humidity. Wet warm climates have *high average temperatures* and *little day–night and seasonal variations, high humidity* and *heavy rainfall*, high and relatively *diffuse solar radiation*, and variable winds which can easily be of hurricane strength. These climates are typical of subtropical coastal regions (Coch 1998).

The main characteristic of these climates is the presence of water in the form of rainfall and high relative humidity, along with high and even mean temperatures. The shape of the roof becomes very important because it must ensure the expulsion

Fig. 16.10 Pitched roofs in Norway: Bergen and Lofoten Islands

of water outside the limits of the building while providing a high ventilation degree. With this purpose, steep pitch roofs with long eaves and a ventilated chamber or a high ventilation space below are the most common solutions. This shape usually creates intermediate spaces in the perimeter of the building, such as porches or verandas, which also help to keep solar radiation (direct and diffuse) outside the building. Intermediate spaces prevent solar gaining and the increase of air temperature in the interior caused by solar exposition.

The heat flux transmitted to the interior through the roof is low, due to the characteristics of materials and the high degree of ventilation, which helps heat dissipation by convection. In addition, the radiative exchange between the roof and the sky in climates with diffuse solar radiation is lower than in the cloudless sky climates; therefore, heat dissipation mostly depends on ventilation.

The materials employed to build roofs in wet warm climates usually have a vegetal origin, since the climatic condition is propitious for lush vegetation. These materials are not characterized by its imperviousness, so the shape and a high thickness of the roof ensures this function, to protect from both rainfall and diffuse solar radiation (Fig. 16.11).

Dry Warm Climates

Dry warm climates, though having high temperatures, perform different from the wet ones. The main characteristics are *high average temperatures* and *high temperature variations* in the daily cycle, *very low humidity* and very *directional solar radiation*, no cloud cover and practically *no rainfall*. *Dry winds* which are warm, heavy with dust and also very aggressive characterize this climate. These climates are typical of deserts close to the Equator (Coch 1998).

Dry climates are as extreme as the wet warm ones. However, though mean temperatures are high in both climates, the difference lies in a high daily variation, the strongly directional type of radiation and a much lower relative humidity.

Fig. 16.11 Ventilated roofs in vernacular architecture of Hanoi, Vietnam

The main strategy in front of these conditions consists of being as isolated as possible from the exterior, so that the roof provides some thermal insulation and high thermal mass, as well as an effective protection from strong radiation and warm winds. Shape is quite important in dry warm climates, where the surface in contact with the exterior must be as low as possible; flat roofs or domes fulfil this function (Fig. 16.12). Solar protection does not fall on ventilated roofs with long eaves, but on keeping the interior of the building as dark as possible, protected from light and warm winds. As rainfall is practically non-existent, the thermal function is the most important function of the roof. The main radiative exchanges take place at roof level, either during the day with solar gaining or at night with radiation from the roof to the sky. Variations of temperature in the day–night cycle are softened with a high thermal mass roof, using high specific heat materials and thickness.

Most materials used in dry warm climates have a mineral origin, like mud or stone, since the lack of rainfall hinders the presence of vegetation. The majority of these materials have the capacity of storing the heat during the day and releasing it during the night, when temperatures fall and the heat is needed.

Complex Climates

Complex climates display the conditions of the previous cases in their variations throughout the year, though with less intensity (Coch 1998). These climates are characteristic of intermediate latitudes and experience *seasonal variations*. The difficulty of roof design solutions in complex climates is that conditions are very changeable throughout the year.

Complex or mild climates have different features depending on their situation in the continent (distance to huge water bodies) and latitude. Secondary factors are: absolute height above sea level, topographic relief, vegetation and human action

Fig. 16.12 Flat roofs in Morocco

Fig. 16.13 Flat roofs in Lanzarote and pitched roofs with Arabic tiles in Menorca

(Coch 1998). The combination of these factors modifies the balance between rainfall regime, radiation and wind, which determine the specific characteristics of a climate to deal with.

The strategies of vernacular architecture in mild climates are based on complex solutions which employ different material and shape solutions to meet a wide variety of requirements depending on the season. The basic shapes range from flat to pitched roofs with different pitch angles (Fig. 16.13). The amount and regime of rainfall defines the best shape to evacuate water and keeping the building dry. The extension of the roof outside the walls helps to expel water outside while providing a certain degree of shadow to the perimeter walls, if necessary. However, the extension of the roof is not as large as in extreme climates or even do not exist since the characteristics of solar radiation (inclination and diffusion) and air temperature do not make it necessary.

A solution which has proved to be effective in mild climates are ventilated roofs (Dimoudi et al. 2006; Gagliano et al. 2012). Ventilation is usually produced through an opened air chamber which can have different dimensions. Sometimes, the chamber takes up an entire floor, as it happens in drying or curing rooms of rural constructions. But, it can also be thinner and span about 1 m or less. It is the case of buildings in urban environments, where the design of façades integrates small holes with the purpose of keeping the air chamber slightly ventilated (Fig. 16.14). This practice combines a shape strategy with a smart use of building materials.

Contemporary Roof Design

The evolution of architecture and construction, especially in the last century, changed the way we conceive our buildings. The emergence of new materials and new construction techniques led to a new aesthetic taste which left behind the old

Fig. 16.14 Ventilated roofs in urban environments. The small holes under the cornice are used for ventilation

Fig. 16.15 Use of concrete roofs in Joan Miró studio, Palma de Mallorca and Palazzetto dello Sport, Roma

architectural tradition. A new concept of architecture replaced the vernacular tradition and extended the limits of what was technically possible or culturally acceptable. New materials changed the shape that roofs usually had, opening a wide range of design possibilities (Fig. 16.15).

Roof Shape and Materials

Today, roof materials do not come directly from the nature. On the contrary, their production is under control in manufacturing factories, which at the same time check and prove their performance with certificates and standard tests. At the same time, contemporary roof (and building) materials are dissociated from place. Both

Fig. 16.16 Marqués de Riscal City of Wine in La Rioja, and La Cité du Vin in Bordeaux

culturally and technically, the need for proximity with the material source has become nonsense, because raw materials come from elsewhere and are processed in any part of the world. Therefore, the elements of the roof are selected depending on clients' or designers' taste and motivation, not necessarily on other criteria (Fig. 16.16).

In the last years, the environmental function of roofs has been assigned to materials, and roofs are conceived as an addition of layers, each of them fulfilling a specific function. The shape becomes then secondary, since the effectiveness of materials solves the design. This specialization of some specific layers which define a roof opened a wide range of design possibilities, today less linked to climatic conditions than before.

The combination of shape and available materials were the ingredients of vernacular architecture to create shelters for rain (snow) and radiation (downward and upward). In addition, the shape was highly determined by the characteristics of materials and their limitations. Today, the design process goes the other way around: the choice of a shape is implemented with technical solutions which involve high-performance materials. The visible layers of the roof are chosen according to aesthetic criteria, while hidden layers carry out the physical function and are expressly chosen with this purpose.

A comparison of vernacular and contemporary roofs can then be made according to the physical functions described in Section "Roof Functions", taking into account their shape and material characteristics (Table 16.3).

The protection from rainfall in vernacular architecture, either in wet warm, cold or complex climates, was carried out with steep pitch and (sometimes) long eaves. Materials were different depending on place and availability, and their performance was ensured by the creation of different layers and the overlapping of materials. On the contrary, roofs today have a waterproof layer which ensures that there is no water filtration to the inside (Brotrück 2007; Paricio 1988). The imperviousness falls today on one of the layers that compose the roof, usually hidden. Most waterproof sheets are petroleum-based products such as polypropylene or other plastics, transported in rolls which can be adapted to the shape of the roof. A right execution

Table 16.3 Comparison of shape and material characteristics of vernacular and contemporary architecture roofs

		Vernacular architecture roofs	Contemporary architecture roofs
Rain	Shape	– Steep pitch – Long eaves	
	Material	– Vegetal (wood, straw, etc.) – Mineral (slate, ceramic, etc.) – Overlapping – Superimposed layers	– Rainproof/waterproof layer – One layer
Snow	Shape	– Steep pitch	
	Material	– Vegetal (wood, straw, etc.) – Mineral (slate, ceramic, etc.) – Superimposed layers – Thickness	– Rainproof/waterproof layer – One layer
Radiation downward	Shape	– Ventilation – Long eaves	
	Material	– Vegetal (wood, straw, etc.) – Air	– Thermal insulation layer – One layer
Radiation upward	Shape	– Small surface exchange	
	Material	– Mineral (mud, stone, etc.) – High thermal mass	– Thermal insulation layer – One layer

along with an appropriate design which expels the water outside the building ensures the waterproof function.

The employed solutions to protect from radiation follow the same idea. Protection from radiation downward is especially needed in warm and complex climates. The main need in these cases is the creation of a shade. In vernacular architecture, the shade was created with a pitched roof, with different angle depending on rainfall, and usually accompanied with ventilation. The degree of ventilation was different depending on the climate and seasonal variation, sometimes with the ventilation of the space itself and other times with the ventilation of a specific space or chamber. The duplication of a roof element in mild climates proved to be successful since the function of the external roof was to create shade over the last slab, blocking the transmission of heat to the inside (Dimoudi et al. 2006; Gagliano et al. 2012).

If the radiation flux ascends upwards, the situation is slightly different. The main purpose of the roof is to preserve the heat inside and reduce the flux to the exterior. In vernacular architecture of cold, dry warm and mild climates, the choice of materials was an important part of the solution, since the performance of the roof depended on their characteristics. In cold climates, thermal insulation was reached with sealskin or other types of fur; low-conductivity materials such as wood also contributed to reduce heat loss. In dry warm climates, thick elements made up of mud, a combination of mood and straw, or even thick layers of stone served to the same purpose.

Nevertheless, the control of radiation falls today on thermal insulation layers (Brotrück 2007; Paricio 1988). In a first stage, many insulation materials were petroleum-based, though this trend has changed in the last decades. Beyond the use of polystyrene or polyurethane, other materials such as rock or mineral wool, cork, wool from sheep and others are employed today in thermal insulation. The choice of a specific material is based on their thermal transmittance which depends on thickness and thermal conductivity. Rigid and flexible materials offer a solution for any type of roof.

Learning from Vernacular Architecture

Unfortunately, the remaining roofs of vernacular architecture are not thoroughly reliable witnesses of the construction techniques of the past. These structures have suffered successive operations of maintenance in their history and have incorporated technical improvements over the years. The evolution of roof design has experimented the evolution of materials and building techniques, even though the essence of the original design has been lost in some cases. The remaining vernacular buildings along with the research in this field constitute valuable sources of information to spell out the essence of vernacular roofs, and learn some lessons for contemporary practices.

The richness of roof solutions seen in the previous sections explains the adaptability of a constructive system to climate and place. The examples show specific designs that fulfil the needs of climate with local resources. However, the evolution of building technology and the development of a global world have led to standard solutions, which are extrapolated today to any part of the world, even though with different aesthetic languages (Figs. 16.17 and 16.18).

As a result, an "international roof" design has spread around the world, encouraged by technical development and globalization. The roof solution in each case is adapted to the building and the environment by introducing small variations, such as the thickness of thermal insulation. The goodness of a shape to create a shelter is put aside by the high performance of building materials. However, the exportation of this model presents some drawbacks:

- The materials are oversized for the function they carry out.
- The features of these roofs are the same (at the most) or worse than vernacular roofs.
- In most cases, the place of manufacture is thousands of kilometres far from the site.
- The amount of embodied energy of contemporary roofs is much higher than the vernacular ones.

As the thermal or waterproof function falls in one material, and not on the combination of the whole roof design, the material usually becomes oversized. Rainproof roofs in vernacular architecture used permeable materials, air and ventilation to create an impervious but breathable solution. Roofs today cover the whole surface with a waterproof layer with sealed joints which preserve the imperviousness of the element. In glass architecture, sophisticated technical solutions cover the whole

Fig. 16.17 MuCEM and Villa Méditerranée in Marseille; Ciutat de les Arts Valencia

Fig. 16.18 Auditorium Parco della Musica in Rome

building and soften the limits between roof and façades (Fig. 16.19). The rest of materials are exempt from this function and hardly contribute to keep dry the building. Something similar happens with the control of radiation on both directions, upwards and downwards. On trusting the function on the thermal insulation element, other materials do not play any important role, and the thickness is usually oversized. In addition, building regulations in many countries require high values of thermal insulation in order to reduce the energy demand for heating. Vernacular architecture design employed the superposition of different materials along with air chambers for the same function and a similar thermal comfort performance.

Conclusion

The result is that today's roofs have reversed the balance between climatic performance and use of resources. The improvement of contemporary roofs in terms of thermal comfort and protection from rainfall is noticeable, but the resources needed

Fig. 16.19 Some examples of glass architecture where the boundaries between roof and façades are not defined; The Shard, 30 St Mary Axe and London City Hall

to fulfil these requirements have increased exponentially. Complex technical solutions offer a high performance with an additional high environmental (and economic) price. It means that high embodied energy materials are used in their construction, which have been manufactured elsewhere with controlled or uncontrolled processes. Meanwhile, common buildings are not able to implement complex solutions and their performance is affected. In these cases, low-quality contemporary materials with scarce environmental advantages and high drawbacks are used.

The lesson provided by vernacular architecture is based on the concept of effectiveness: the degree to which something is successful in producing a desired result. In the case of roof design, the success can be determined not only by the objective but also by the means employed to achieve it.

References

Banham R (1984) The architecture of the well-tempered environment. The University of Chicago Press, Chicago

Brotrück T (2007) Roof construction. Birkhäuser – Publishers for Architecture, Basel, Boston, Berlin

Coch H (1998) Chapter 4. Bioclimatism in vernacular architecture. Renew Sust Energ Rev 2:67–87

Dimoudi A, Androutsopoulos A, Lykoudis S (2006) Summer performance of a ventilated roof component. Energ Buildings 38:610–617

Gagliano A, Patania F, Nocera F, Ferlito A, Galesi A (2012) Thermal performance of ventilated roofs during summer period. Energ Buildings 49:611–618

Paricio I (1988) La construcción de la arquitectura. Institut de Tecnologia de la Construcció de Catalunya ITEC, Barcelona

Weber W, Yannas S (2014) Lessons from vernacular architecture. Routledge, London

Chapter 17
Early Design Strategies for Passive Cooling of Buildings: Lessons Learned from Italian Archetypes

Giacomo Chiesa

Introduction

The energy consumption of the building sector is responsible for almost 40% of total primary energy use (Cuce and Riffat 2016; Logue et al. 2013). Among the factors that make up this percentage, the main ones refer to space heating and cooling (Logue et al. 2013). For this reason, in order to reduce the impact that the building sector has on both energy demand and subsequent GHG emissions on the atmosphere, several strategies have been adopted and specific regulations and policies are nowadays applied in several countries. In Europe, for example, it is possible that the EPBD recast directive (2010/31/CE) towards the adoption of nearly zero energy buildings (NZEB) has been introduced. Furthermore, each country has developed its own pathway towards NZEB along with specific regulations to reduce the expected amount of consumed energy—e.g. by reducing the *U*-value of both opaque and transparent envelopes, and by increasing the efficiency of the systems (D'Agostino 2015; Ascione et al. 2016; Dall'O' et al. 2013). Nevertheless, at present, one of the rising energy consumption factors is the energy consumed for space cooling and ventilation (Isaac and van Vuuren 2009). The air-conditioner market is growing fast with a huge increase in the number of installed units, both in industrialised countries and especially in emerging ones (Chiesa and Grosso 2016). For instance, the amount of installed units in China is growing exponentially (see also Chiesa 2016) and is reached to 120 million units in warm regions alone by 2017, as was underlined by an IEA study (Koizumi 2007). Although an increase in the EER (energy efficiency ratio) of air-conditioning units is evident, especially in

With a contribution of prof. Mario Grosso, Department of Architecture and Design, Politecnico di Torino, Turin, Italy - see the Appendix: the Seven Root Sheets.

G. Chiesa (✉)
Department of Architecture and Design, Politecnico di Torino, Turin, Italy
e-mail: giacomo.chiesa@polito.it

A. Sayigh (ed.), *Sustainable Vernacular Architecture*, Innovative Renewable Energy, https://doi.org/10.1007/978-3-030-06185-2_17

industrialised contexts, constant growth levels of energy consumption in both industrialised and emerging countries are evident (GEE Directorate of Global Energy Economics 2015; Santamouris 2016).

For this reason, alternative solutions are needed to reduce energy consumption in the summer season without sacrificing good comfort quality. The study of traditional archetypes, related to the specific climate regionalism of a location, shows that valid solutions are related with the inclusion of passive cooling techniques in buildings. Nevertheless, since these systems refer to a large toolbox, including different strategies and natural heat sinks (Cook 1989; Santamouris and Asimakopolous 1996; Santamouris 2007; Kolokotroni and Heiselberg 2015; Chiesa 2013), it is important to conceive their usage from early design phases in order to maximise their potential, increase their level of integration into building shapes and choose the best solutions for each climate and location (Chiesa and Grosso 2017).

This chapter aims at translating traditional vernacular solutions for guaranteeing the indoor comfort in Italian climates into contemporary guidelines for architects.

Methodology

This study focuses on the summer season and analyses natural cooling techniques which are compatible with Italian climates. The proposed method is based on the analysis of vernacular passive cooling applications along the Peninsula and by their classification according to a recognised list of tools for passive cooling in order to organise a list of suggestions for designers to be considered from early design phases. Firstly, seven pre-fossil-era buildings are chosen from different Italian archetypes to define a representative list of traditional solutions, related to local regionalism, which were used in different Italian climatic conditions to guarantee internal comfort without consuming fossil energy. For all case studies, an analysis is developed which reports localisation, climate and functioning information (see section "Case Studies" and Appendix). The data collected in this recording activity are further analysed together with other vernacular examples in section "The Passive Cooling Toolbox for Italian Climates" to identify a list of solutions to passively cool buildings and outdoor spaces. These solutions are analysed according to their applicability in different Italian climate zones. Moreover, their relation with contemporary passive cooling strategies is defined. Finally, in section "Early Design Strategies Passively Cool Buildings in Italian Climates", this list of strategies is translated into a short guideline as to how to include passive cooling systems from early design phases, by considering, among other factors, related requirements for a performance-driven-design approach and a sample tool to study their local-climatic potential.

The proposed research, linking tradition to innovation, follows the flowchart described in Fig. 17.1.

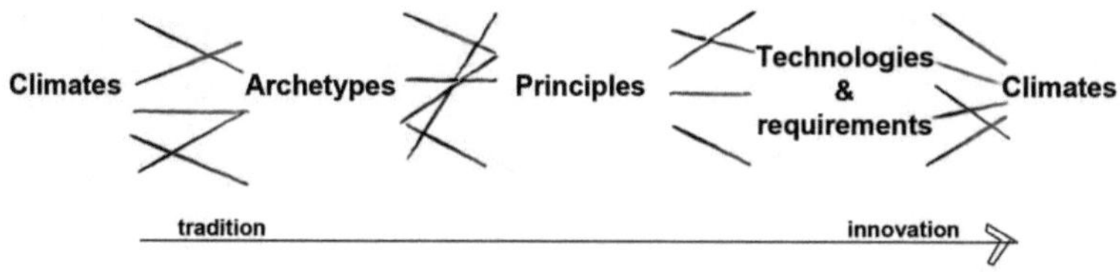

Fig. 17.1 Flowchart of the used methodology

Case Studies

The study and the renovation of Italian rural architectures is a recognised topic that involves architectural technologists, historians and design theorists. From the long list of researches into this topic, one of the first sources was the catalogue of the exhibition on Italian Rural Architecture, Triennale di Milano, edited by Pagano and Guarniero (1936). It contains a large number of photographs of traditional rural buildings all around Italy, classified according to specific architectural aspects (e.g. chimney, types of roof, shapes, etc.) and commented on according to a theoretical design approach to show the aesthetic value of rural buildings by analysing their technological, functional, cultural, climatic and geometrical aspects.

Among essential sources for studying Italian vernacular buildings are the books published in the 1920s, collections of drawings about "rustic" architecture (*architettura rusticana*) in the Italian Peninsula. Ferrari's book *L'architettura rusticana nell'arte italiana*, with 250 tables, organised according to Region contains one of the largest collections of information about traditional residential buildings in ancient times covering the Roman and Medieval ages (Ferrari 1925). Moreover, Jona's books analyse, also in this case with the help of drawings, traditional building shapes in different Italian contexts—i.e. the alpine houses of Valle d'Aosta (Jona 1920a) and the coastal houses near Amalfi (Campania) (Jona 1920b).

In this pre-fossil era, the preservation of indoor comfort conditions under outdoor stresses was principally achieved thanks to the high design quality of the envelope, while additional systems (e.g. heating devices) were limited to the essential. In contrast, today in the industrial age, the international style of buildings means that comfort issues are generally resolved by means of mechanical systems and thus allow for non-regionally specific design.

One of the first recognised studies of traditional building architectures connecting pre-fossil interactions between buildings and their environment is included in Rudofsky books (Rudofsky 1964; Rudofsky 1977) and in the analysis of roots by Olgyay (Olgyay 1963). Furthermore, Cowan's book shortly introduces, in a concise manner, structural aspects in ancient buildings (Cowan 1977).

Focusing on these approaches, seven case studies were chosen from Italian traditional architectures which represent the specific roots of local regionalism. In Fig. 17.2, the positioning of these case studies is reported, while Figs. 17.3 and 17.4 illustrate them in simple sketches. A bioclimatic approach was followed to identify the case studies, and to describe them in short sheets while analysing their

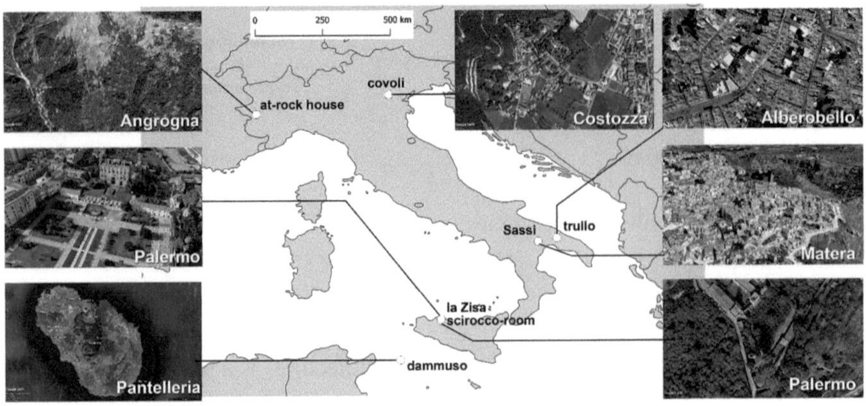

Fig. 17.2 Distribution of chosen roots throughout Italy and context views (images' source: Google)

technological and environmental functioning. Finally, Table 17.1 reports the coordinates and the climate classification of each case.

Furthermore, a specific description for each sheet was kindly provided in Appendix by Mario Grosso. The proposed sheet organisation is based on the description of three main aspects of the proposed root: its general description, the bioclimatic and regional technological principles used in the archetype and potential contemporary applications of the solution.

The Passive Cooling Toolbox for Italian Climates

In section "The Passive Cooling Toolbox for Italian Climates", main bioclimatic information concerning passive cooling solutions are extracted from the presented sheets in order to populate a toolbox of possible strategies for the natural cooling of buildings in Italian climate conditions. The passive cooling approach includes three main families of actions: heat gain prevention, heat gain mitigation and heat gain dissipation (Cook 1989; Santamouris and Asimakopolous 1996; Chiesa 2013). The first two refer to thermal control strategies, while the last includes natural cooling techniques (Grosso 2017).

Heat gain prevention is principally based on the use of shading elements to protect the environment from solar gain. Since shading systems are well-known essential techniques to prevent cooling needs in a built-up environment and reduce energy consumption in summer, they are not directly described in this chapter. Furthermore, several shading solutions are available on the market. Nevertheless, for a detailed description of shading design procedure it is possible to refer to Santamouris and Asimakopolous (1996) and Raimondo (2017). Finally, vernacular solar control solutions are present in all Italian climates, from the under-rock dwellings in the North which are exposed to the winter sun but shaded during the hottest hours in summer

Fig. 17.3 Sketches of the aspect of each considered root—Covoli (Vicenza); at-rock house (Angrogna valley); Sassi (Matera)—public domain image

Fig. 17.4 Sketches of the aspect of each considered root—Trulli (Apulia regions, e.g. Alberobello); Camera dello Scirocco (Palermo); la Zisa (Palermo); Dammuso (Pantelleria)

by the rocks to the use of wooden shading systems similar to *clastrum* and *mashrabiya* like the ones in the *Ziza* (Palermo, Sicily). Furthermore, vernacular shading architectural solutions are also represented by the use of porches and pergolas in the entire Peninsula, from ancient Roman buildings such as the *domus*, with spaces such as the *peristylium*, and porches (e.g. four-sided portico) throughout history (e.g. medieval cloisters). Porches are part of traditional building organisation since medieval times and are present in urban palaces, such as the porches in Bologna and Emilian-style

Table 17.1 Geographic coordinates of the chosen roots and their climate classification

Root	Location	Province	Lat.	Long.	Ita. climate zone	HDD_{20}	CDD_{22}
Covoli	Longare (Costoza)	Vicenza	45.48°	11.62°	E[a]	2316[a]	313[b]
At-rock house	Angrogna	Torino	44.85°	7.22°	F[a]	3527[a]	84[b]
Sassi	Matera	Matera	40.66°	16.61°	D[a]	1776[a]	295[b]
Trullo	Alberobello	Bari	40.78°	17.24°	D[a]	1644[a]	280[b]
Scirocco room	Palermo	Palermo	38.07°	13.33°	B[a]	751[a]	570[b]
La Zisa	Palermo	Palermo	38.12°	13.34°	B[a]	751[a]	570[b]
Dammuso	Pantelleria	Trapani	36.79°	11.99°	B[a]	717[a]	319[b]

[a]From DPR 412-1993 all. A—Municipality data
[b]Elaborated from interpolated TMY by Meteonorm 7.1

houses of the fourteenth to fifteenth centuries (Ferrari 1925). Even in northern areas, such as the *Pianura Padana*, the use of pergolas and porches is widespread with a prevalent effect of shading in summer, and protection in winter.

Furthermore, loggias are also widespread throughout Italy, as in the Castle of Montechiarugolo—second half of the fifteenth century, near Parma, or in the town hall of Castellarquato (Piacenza) from the same period (Ferrari 1925). Loggias are also common in southern regions, as in Amalfi's traditional houses, together with terraces covered in pergolas (Jona 1920b).

On the other hand, heat gain mitigation is generally brought about by the use of thermal masses, while heat gain dissipation refers to the usage of natural thermal sinks.

According to the chosen roots, the toolbox presented here principally focuses on dissipation techniques and on the use of thermal masses to mitigate cooling demand. Here below these techniques, adapted to the Italian context, are briefly explained.

Mitigation Techniques

In this section, some uses of thermal masses for heat gain modulation (time lag and attenuation due to thermal inertia) and the effect of buried and semi-buried buildings are explained. These solutions for passive cooling are described by defining a series of vernacular examples including some of the ones described above (section "Case Studies") as well as some new examples throughout the Italian Peninsula.

The use of massive walls and coverings, both in stone and brick, is widespread. On the one hand, this practice derives from military usage, as in Sardinian Nuragic villages (Rudofsky 1977), while on the other hand, they are an evolution of the traditional hut (Pagano and Guarniero 1936), with or without central pillars, as in Etruscan tombs (Rudofsky 1977). Other examples of massive architectures, such as the *dammusi* (Pantelleria—see the root sheet #7, Appendix) and *trulli* (root sheet #4, Appendix), are treated in sections "Natural Cooling Techniques: Thermal Sinks" and "Radiative Cooling".

Furthermore, the construction of traditional towns is in some cases linked with specific relations to rocks, as was underlined by Edward Lear when writing about small Calabrian settlements in 1852: "the houses being built on, under, and among, separate masses of rocks" (Rudofsky 1977; Berenson 1948). The use of under-rock construction, together with semi-buried buildings, is documented all along the Italian Peninsula, and allows dwellers to take advantage of the high thermal inertia of rocks and the earth. In northern regions, under-rock buildings and settlements are used to protect houses form atmospheric events, by exposing massive rock walls to the springy/autumnal and winter sun while protecting them from potential summer overheating (see the root sheet #2, Appendix). Specific examples are represented by the *barma* buildings (Piedmont, and Waldensian valleys)—where *barma* is a toponymal for under-rock shelter that in some places, as in the one here described, became real buildings. The *Barma Monastic* or *Mounastira*, at 1200 m, localised in *Valle Angrogna*, which is a small settlement of several units disposed along and below a big rock (see also Trossarelli 1998), and other small settlements in the same valley such as the *Barma d'l'Ours*. Furthermore, in Piedmont, other similar settlements are also known, such as the *Balma Boves* (see Fig. 17.5), which is made up of a small series of buildings built under a big rock in *Valle Po* (Sanfront) and which is similar to the well-known Mesa Verde cliff dwellings.

Furthermore, other "rock" buildings are described in Central Italy, such as the houses in the rocks at *Sasso* near Bologna (Ferrari 1925). Other examples are reported also in southern locations such as the houses built against rocks near Amalfi (Jona 1920b), and the already mentioned Calabrian buildings (Rudofsky 1977). Furthermore, buried or semi-buried buildings, such as the *sassi di Matera* (see root sheet #3, Appendix, and recent studies on their thermal behaviour (Cardinale et al. 2013; Cardinale et al. 2015)), the *grotto*, like the one in the *Latomia del Paradiso* in Sicily (Rudofsky 1977), and the necropolis of *Pantalica* in Sicily (Siracusa), were transformed into residential spaces during the Middle Ages (Rudofsky 1977).

The inclusion of massive thermal elements, especially in inner layers, can constitute a good solution to mitigate the effect of heat gain and can be further connected with passive dissipative techniques, such as structural ventilation (see the later section). Nevertheless, the correct positioning and dimensioning of these layers have to

Fig. 17.5 View of the Balma Boves settlement (pictures of the author)

be conceived in an integrated environmental design strategy in order to localise the masses near the designed natural airflows for direct exposure (if further dissipation is expected) and to have a defined control of heat modulations (attenuation and time lag of the thermal release). However, massive external layers may be used to increase the effect of radiative cooling dissipation.

Natural Cooling Techniques: Thermal Sinks

Natural cooling techniques focus on the use of passive thermal sinks to dissipate heat gain. Four thermal sinks are considered: air, ground, water and nocturnal sky (even if this expression in misleading since heat dissipation towards the sky arrives also during the day (Cook 1989)). Each sink defines a specific approach to the passive cooling of buildings. These approaches are, respectively: ventilation for cooling (comfort ventilation, environmental ventilation and structural ventilation), ground cooling, evaporative cooling and radiative cooling. With reference to the described vernacular architectures, devoted technologies referring to these approaches are further introduced.

Ventilation for Cooling

Cooling a space by using controlled natural airflows refers to three main strategies (see Fig. 17.6a):

- The direct physiological benefit perceived by users, which is a function of the air velocity (see Fig. 17.6b), named comfort ventilation;
- The dissipation of heat gains by substituting internal air with external low-temperature air, named environmental ventilation;

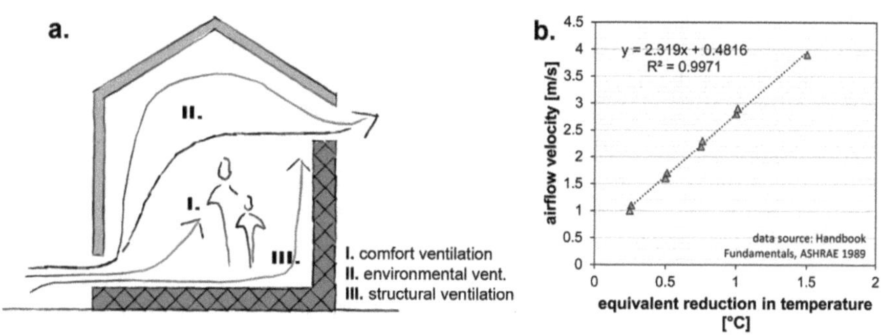

Fig. 17.6 (**a**) Sketch of the three main ventilation for cooling strategies; (**b**) relation between the expected perceived reduction in temperature due to comfort ventilation and the velocity of the airflow (re-elaboration from ASHRAE Handbook of Fundamentals 1989, p. 46—see also Chiesa and Grosso 2015a)

- And, the indirect effect due to nocturnal dissipation of stored heat by building masses due to their direct exposure to airflows, named structural ventilation.

Natural airflows are activated thanks to differences in pressure between two openings (the upper and lower part of a single-side opening system) that may be inducted by wind-driven differences (e.g. cross ventilation) or buoyancy-driven forces (e.g. stack effect—differences in height).

A possible solution to increase the airflow in stack-driven ventilation is to use solar radiation to raise the temperature of building components near an outlet opening, as, for example, is done in solar chimneys. A root for this solution can be found in some complex *Trulli* aggregations (see e.g. Ferrari 1925; Rudofsky 1977) where a few chimneys were added at the bottom of the cupola. The increase in temperature of the stones has a direct effect on the airflow, even if it is mediated by the heat capacity of the material used, that activates and guarantees a nocturnal ventilation of the mass using the chimney as extractor and the draught of the door as air inlet (Accorsi and Malaguti 2017). From a climate point of view, this solution may be applied in hot conditions or in buildings which are characterised by high heat gain (e.g. offices) when nocturnal air temperatures fall below the comfort threshold to guarantee the effectiveness of all ventilation for cooling strategies. Nevertheless, in order to be used, internal heat capacity has to be sufficiently high to make the system efficient. Furthermore, the effect of ventilation for cooling in hot climates may be increased by the joint application of other dissipative techniques, such as direct evaporative cooling or ground cooling which are able to reduce the temperature of an airflow when environmental temperatures are uncomfortable.

Considering the archetypes illustrated in section "Case Studies", the use of structural ventilation strategies is principally focused on the southern and central Italian provinces, where external temperatures may be very high during the day, and the intensity of the solar radiation is highly relevant. Nevertheless, the application of this technique, as also underlined by the majority of archetypes, means that the nocturnal environmental air temperature should fall below the comfort threshold, a condition that is not often guaranteed. Hence, in these areas, as has been already defined in the "mitigation techniques" section, traditional buildings were characterised by great inertia with depth massive opaque envelopes which were able to store solar heat during the day—solar gains were predominant over internal gains—and release it in an attenuated manner during night time when external temperatures are lower thus activating a mitigation effect. In some cases, the same effect may be used to activate structural ventilative heat dissipation. Although, in these locations, the comfort ventilative approach in outdoor or protected spaces is also used (e.g. *Camera dello Scirocco*), where the effect of ventilation is often increased by using evaporative techniques (presence of water or even exposure to marine breezes).

Furthermore, a similar effect is obtained in traditional colonnades, which are often covered in arbours or wickers, with one side positioned against buildings, where ventilation is coupled with shading effects (heat gain prevention). A further root for this technique is represented by traditional pergolas (vegetal or wicker covered) of the traditional residences in the Eolian islands (Lo Cascio 2005; Linea Blu 2018)—see Fig. 17.7—where the buildings are light coloured with flat roofs (for

Fig. 17.7 Sketch of the traditional pergolas in Eolian residential buildings interested by the double effect of shading and comfort ventilation

water collection). However, environmental ventilation solutions are mainly effective when external temperatures are below the comfort threshold even during daytime. Especially in northern regions, natural air exchanges may constitute a good solution to guarantee indoor comfort conditions and dissipate heat gain. A sample root is constituted by natural ventilation flows in several Palladian villas—see for example the representation reported in Green (1979).

The adoption of controlled natural ventilation techniques includes the careful dimensioning of openings and the definition of control systems (sensors and actuators to operate the devoted opening areas and modulate the effect) (Holzer and Psomas 2017; Allard 1998). Furthermore, the use of fans and ceiling fans to increase air movement can also constitute a good way to increase the effect and the number of hours in which such a technique may assure comfort conditions.

Ground Cooling

The ground attenuates, thanks to its great inertia, external air temperature variations, showing, at a given depth (around 10 m), a temperature behaviour which fluctuates around the average yearly external one. This effect, which follows a trend similar to the one described for the thermal mass in section "Mitigation Techniques", is increased by the depth—in a domain principally ranging from 1 to 15 m for horizontal systems (if no endogenous forces are present)—and constitutes the potential of the ground to be a heat sink. Its temperature is, in fact, higher than the external temperature in winter, and lower in summer. The effect of natural ground cooling in dissipating heat gain (for heat gain modulation in buried or semi-buried buildings see the previous section) is well known and used in several contemporary systems that may be classified according to a vertical or horizontal distribution in the soil and according to the exchanging fluid used. For a description of low-energy water-based techniques, such as borehole heat exchangers (BHE), it is possible to refer to the literature (e.g. Pfafferott et al. 2007), while a review on ground pump systems was presented in Lucia et al. (2017). Nevertheless, even if contemporary ground-source technologies are principally water-based, air-based ground cooling technologies, which are directly related with the described Italian archetypes, are analysed here. In particular, a contemporary technology that functionally refers to *Covoli* is the earth-to-air-heat exchanger (EAHX). Such a technique is composed of horizontal buried pipes, whose depth is generally less than 5 m—in which an external airflow is treated—pre-heated in winter and pre-cooled in summer—before being supplied to internal spaces or directed to an HVAC system for post-treatments. Several modern applications of EAHXs were studied in Italian locations such as a residential building near Turin (Grosso and Raimondo 2008) and a large system installed at the "Media Orsini" school in Imola (Chiesa 2013; Chiesa et al. 2014; Grosso and Chiesa 2015). The applicability of EAHX in Italian locations was analysed in the literature (Ascione et al. 2011) showing high potentialities. Furthermore, a climate-related tool to estimate EAHX potential was reported in Chiesa (2017) by calculating the expected soil temperature at a given depth, which represents, as pointed out before, the heat sink temperature, during a typical year and the corresponding hourly potential in pre-treating the external air in winter and summer by reducing thermal discomfort.

Focusing on the presented roots, ground cooling applications to reduce the temperature of an airflow are considered both in northern and southern regions (see *Covoli* and *Torri dello Scirocco*), especially in those locations characterised by a relevant amount of cooling degree days.

EAHX may be composed of a single tube, generally not coupled with a further mechanical system and used for small buildings such as residences, or multi-tube systems in which the inlet air is distributed through several parallel pipes. In Fig. 17.8a, a simple sketch of an EAHX system is presented, while in Fig. 17.8b a graph showing the effect of EAHX by comparing inlet (external), outlet (treated) and ground temperatures in a 1-year-long monitoring campaign (Imola) (see also Chiesa et al. 2014) is presented. Considering Italian climates, the applicability of such systems may be effective both in winter and in summer, while for intermediate seasons EAHXs may be turned off, even in accordance with the vernacular examples.

Evaporative Cooling

Numerous historical roots are known in which water evaporation is used to reduce the temperature of an airflow thanks to the large amount of heat adsorbed during this phase change process. Evaporative cooling is used while considering its direct effect on the reduction of dry bulb temperature (DBT), and also in indirect applications to

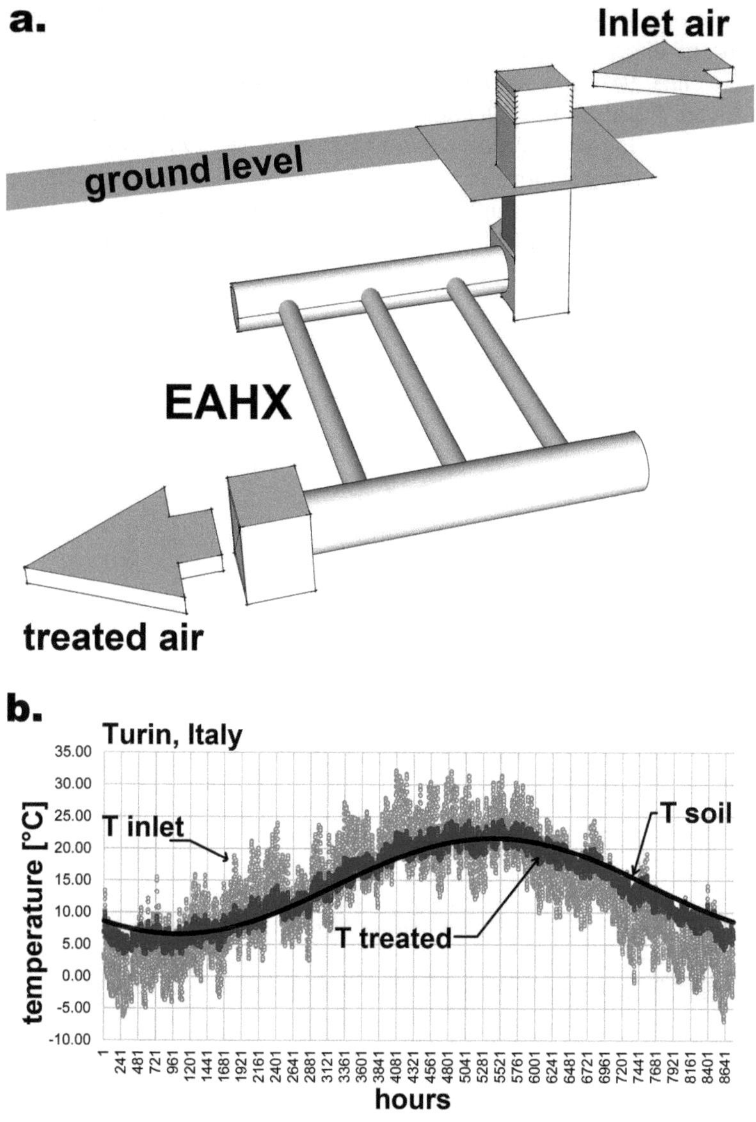

Fig. 17.8 (**a**) Sketch of a EAHX system; (**b**) external (inlet), treated (outlet) and undisturbed ground temperature profiles (3 m in depth—clay soil) for 1 year in Turin, Italy (calculated on the TMY from Meteonorm 7.11—see Chiesa 2017)

reduce the temperature of a massive element which is able to absorb heat gain. Furthermore, indirect evaporative systems may also be used to reduce the temperature of a secondary airflow coupled with a direct system in a sensible heat exchanger. The local potential of direct evaporative processes is a function of wet bulb depression, which is calculated by subtracting the wet bulb temperature (WBT) from the

DBT—see Fig. 17.9a. Several simplified expressions to calculate the temperature reduction of direct evaporative systems were presented in the literature (Givoni 1994; Givoni 1997) and compared in Chiesa and Grosso (2015b). Several of these formulas are based on the calculation of the effectiveness (ε) of the evaporative process in covering wet bulb depression (see also Cook 1989; Watt 2013). Direct evaporative cooling solutions in vernacular architectures refer to several applications, such as the effect of fountains and artificial/natural water basins, small canals and rainfall. Several historical villas, starting from the *Zisa*, collect small water

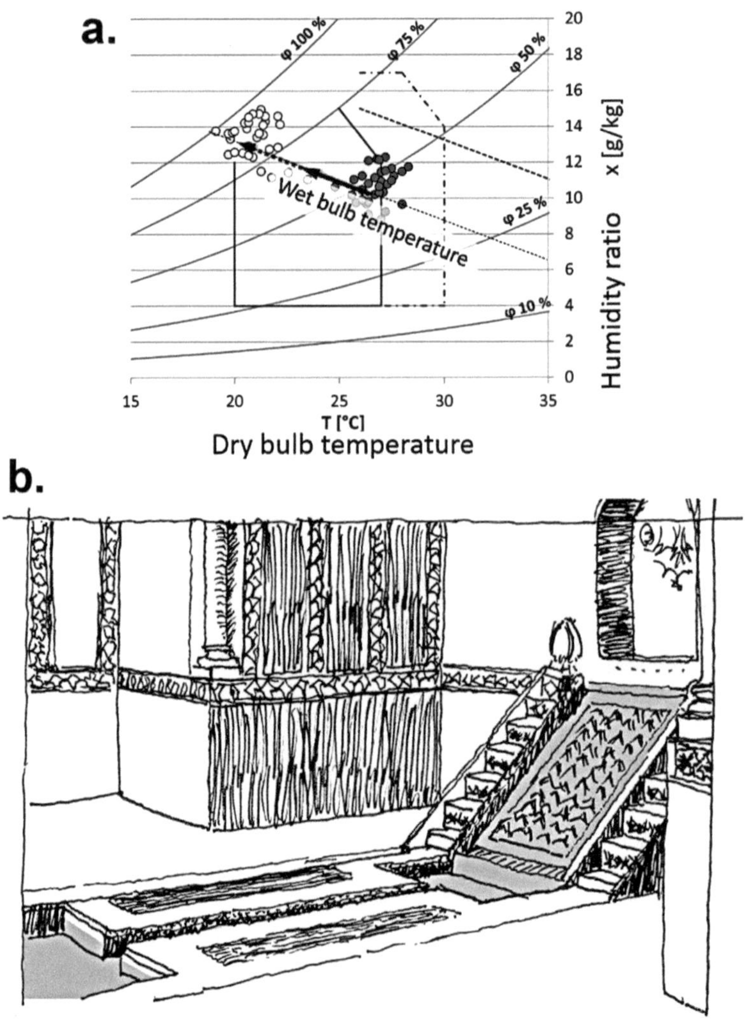

Fig. 17.9 (**a**) Sample representation on a carrier psychrometric chart of monitored data from a DEC tower (Turin, Italy), and (**b**) sketch of the internal evaporative system in the Zisa, Palermo

rainfalls into a marble or rock inclined support linked to a small canal in indoor spaces to naturally refrigerate the spaces—see for example Fig. 17.9b.

Furthermore, the use of water basins, small canals and water fountains is a traditional part of ancient Roman villas both in the *domus* of rich people as well as in larger structures like *Villa Adriano* at Tivoli, with special regard to outdoor environments. The “canòpo” of this latter archetype, with its water basin and water games (fountains), was also used for banquets, suggesting the good comfort quality reached near this space. The use of fountains is also widespread, linked with aesthetic aspects, to guarantee outdoor comfort in Renaissance and Baroque Villas such as *Villa d’Este* (Tivoli). The use of porous surfaces, humidified by water drops or flows, is also common in several ancient central and southern buildings and spaces such as Grottos where water evaporation is often linked with the effect of semi-buried spaces (see section “Mitigation Techniques”). Grottos were often used in renaissance palaces (Hwang et al. 2006). In addition, in the previously mentioned small alpine settlement of *Balma Boves*, a rainfall is localised near the “under-rock” series of buildings, probably with the intent of reducing the temperature of prevalent summer wind-driven airflows thanks to direct evaporation before reaching the settlement. Furthermore, in the example of *Torre dello Scirocco* at *Villa Naselli*, Sicily, the buried chamber and the tunnel have a water drain, also named *quanat*, connected to a water source (wellhead) to increase the cooling effect of ground and ventilative solutions. Evaporative cooling systems can be used to cool outdoor and indoor air including simple applications with passive downdraught evaporative towers or conventional systems such as desert cooler (direct evaporative cooling) or evaporative chiller (indirect), which indirectly treats a fluid used for cooling spaces (mechanical systems).

Radiative Cooling

The effect of heat dissipation from exposed surfaces (radiator) towards the sky, whose average temperature is expected to be below the surface one, is well known and used in traditional building to dissipate during the night, when the effect of sun radiation is absent, the heat stored in thermal masses, even if thermal exchange due to heat transfer between surfaces and the visible sky arrives both during day and night time (Cook 1989). Nevertheless, it is when the air temperature is low and solar radiation absent, that the radiative effect will maximise its effect in dissipating the stored heat gains. A correct design of the thermal inertia of exposed elements is essential to maximise the effect of these passive cooling solutions coupling the heat mitigation effect of the masses (time lag and attenuation factor) with the heat gain dissipation through the visible portion of the sky. The heat dissipation through exposed surfaces will reduce their temperature and consequently indirectly allow a decrease in the internal temperature thus increasing comfort sensations. Several applications were studied in hot climates, from Persian archetypes (Bahadori 1978), including roof ponds connecting the use of water evaporation on the outermost layer (indirect radiative cooling) with diurnal shading, and night radiative cooling (e.g. Pearlmutter and Berliner 2017). Nevertheless, the use of large massive constructions with exposed surfaces to the sky is also very common in Italian locations, as demonstrated by the root sheet #4 (*trulli*) and #7 (*dammuso*), see Appendix. These two examples show compact single-space buildings or aggregation of multiple units characterised by a large exposed massive roof surface to guarantee an increased radiative cooling effect especially during the night when the absorbed accumulated heat is released by the building masses. In the *trullo* shape, which is similar to the ancestral hut, there is a difference between a shaded and a direct exposed portion of the dome to direct solar radiation which creates potential natural airflow movements. Even if *trulli* are principally to be found in the central and southern Apulia region (such as the Alberobello site, part of the UNESCO world heritage list, or those in Martina Franca and Locorotondo (Ferrari 1925; Castellano 1964)), similar vernacular building typologies are also reported in the same region (e.g. in locations such as Barletta and Trani) with the name of "*casella*", where the shape is both similar to a haystack or made of small terraces a bit like a pyramid, and in some Calabrian locations where this peculiar dome shape is used in both dwellings (as in Nuova Siri, Castelluccio and S. Fili di Cosenza (Pagano and Guarniero 1936)) and small shepherd houses/shelters (as in the rural areas near Cosenza and Catanzaro (Pagano and Guarniero 1936)). Independent traditional building shapes were reported in several sites in hot regions along the Mediterranean basin, as in Aleppo or in Cilicia, or even in more distant regions such as some ancient Persian sites (e.g. Koum) or in high Turkestan villages (Ferrari 1925). The *trullo* shape is both originated by a round base and by rectangular base, especially in more complex aggregations. An evolution of this shape is the use of pitched roofs in several vernacular manor farms in the same regions (Pagano and Guarniero 1936).

Together with *trulli*, a very widespread vernacular radiative building typology is represented by the use of domed roofs, pavilion-domed roofs as well as cross-vaulted and barrel-vaulted houses. The use of domed roofs is reported throughout the Peninsula, from Liguria (e.g. Varigotti (Pagano and Guarniero 1936)) to Apulia (Ferrari 1925), as well as in Naples and Amalfi in Campania (Pagano and Guarniero 1936; Jona 1920b), and Sicily with the *dammuso* example in Pantelleria. These examples are especially related to aggregated units with limited exposed vertical surfaces to reduce solar gains, even if this means a lack of privacy, but with massive roofs, that in some cases are in the form of flat terraces, sometimes covered or adjacent to pergolas as in some Amalfi dwellings (see Jona 1920b and Fig. 17.10) or typical Eolian houses (Brancati et al. 1951), where the terrace adjacent to the house, called *bagghiu*, is generally covered by a pergola supported by cylindrical columns in the traditional white/light colouration. The materials used are, in fact, characterised by a low adsorption index, to reduce heat gain adsorption during daytime, as in modern cool roof materials.

Fig. 17.10 A typical dwelling with radiative roof in the Amalfi's coast (freely inspired from Jona 1920b)

Early Design Strategies Passively Cool Buildings in Italian Climates

In this section, the previously described technologies are analysed following a performance-driven approach to define a list of requirements for their application in building design from early design phases. The usage of a performance-drive approach to building design is well recognised in the Italian context since the 1970s (Cavaglià et al. 1975) and based on a series of UNI standards, from the UNI 7867-4:1979, now substituted by the UNI 103838:1999 (thesaurus), to the UNI 8290 part 1:1981 (technological system definition), part 2: 1983 (list of principal requirements) and part 3:1987 (requirement motivating agents), including the standard UNI 8289:1981 with the classification of the needs of final users and the UNI 11277:2008, which introduces different classes of needs including building lifespan without canceling the old ones.

Even if a series of existing requirements—where a requirement is the technical transposition of needs connected with the building process (UNI 8290-2:1983)—can be related to the specific usage of passive cooling solutions as is reported in Table 17.2, an extended list of new requirements is needed, especially for the usage of heat sinks, as partially suggested at the end of the same table.

Table 17.2 List of existing requirements from UNI 8290-2:1983 that can be used for passive cooling solutions (heat gain prevention, mitigation and dissipation)

Requirement (UNI 8290–2:1983)	Passive cooling solutions
– Solar factor control – Integration (aptitude to functional and dimensional connection)	Heat gain prevention
– Thermal inertia control	Heat gain mitigation
– Ventilation (aptitude to obtain air exchanges naturally or mechanically) – Adjustability (aptitude to be intentionally varied by an operator through technical devices) – Integration – Flow rate control – Aptitude to the integration in plants (aptitude to functionally complement building elements which are not part of a plant with plant building elements) And, if opportunely modified: – Heat dissipation control through transmission (present definition is only for winter, devoted to preventing/reducing any dissipation) – Heat dissipation control through air renewal (present definition is only for winter, devoted to preventing/reducing any dissipation)	Heat gain dissipation

Other requirements, not specifically devoted to passive cooling, even if connected to them, are not considered here—i.e. maintenance opportunity, aptitude to be cleaned, fire resistance, etc.

In this sense, it is possible to refer principally to the class of needs defined as "rational use of climatic and energy resources" and to "perceptive and sensory well-being"—see UNI 11277:2008 and Chiesa and Grosso (2017), introducing as classes of requirements "heat gain prevention", "heat gain mitigation" and "heat gain dissipation by thermal sinks". Especially for the last in the list, new potential requirements could be:

- "Exploitation of natural heat sinks according to local climate" (aptitude to maximise the local exploitation of the potential in heat dissipation through heat sinks), and in addition:
- "Control of heat gain dissipation through transmission in the summer season" (convection, conduction and especially radiation), as the aptitude to dissipate the stored heat gains by exposing external masses to the night sky. Furthermore, it can include the aptitude to activate the building thermal masses for night ventilation;
- "Control of heat gain dissipation through airflows in the summer season" by activating natural or fan-forced flows, exposing the building masses to these flows, and allowing the passage of airflows from outside to inside, through internal spaces and from inside to outside;
- "Ventilation for cooling" (aptitude to obtain naturally driven or mechanically driven airflows that are able to perform comfort ventilation, environmental ventilation if the external air has a lower temperature than the internal air and structural ventilation);

- "WBD exploitation" (aptitude to reduce the DBT of an airflow to a value which is nearer to its WBT);
- "Ground cooling exploitation" (aptitude to reduce the temperature of an airflow by using EAHX systems);
- "Mixed usage of different heat sinks" (aptitude to maximise the effect of heat sinks by mixing their usage).

Furthermore, the presented toolbox of contemporary-translated solutions for heat gain dissipation inspired by the vernacular examples is here analysed for functional, operational and climatic/site requirements to define their applicability in different Italian contexts—see Table 17.3. Nevertheless, a study to identify the main aspects to be analysed to integrate passive cooling dissipative solutions in building design is reported in a previous study (Chiesa and Grosso 2017), while a study involving the building integration of a solar-regenerated cooling system which includes heat sinks and air dehumidification is reported in Simonetti et al. (2017) and tested in an Italian context in Simonetti et al. (2016).

Furthermore, as regards climate applicability, it is possible to refer to several methods that translate the considered requirements into a performance analysis of local-climatic conditions. A simple calculation method to estimate the local potential of structural cooling was presented in the literature (Artmann et al. 2007), while a recent "ventilative cooling potential tool" was developed within IEA EBC Annex 62 (Belleri and Chiesa 2017) and is downloadable on the website of the project. Thanks to this tool, it is possible to analyse the applicability of controlled natural ventilation, evaporative cooling and fan-forced ventilation in guaranteeing indoor comfort in a space, while considering the local climate with an hourly step definition. The tool calculates the number of comfort/discomfort hours for a defined space while considering the amount of expected heat gain (internal and solar) and checking the possibility to exhaust them by using passive ventilative solutions, and can be used from early design phases. Results for six representative locations in Italy, chosen from the examples in section "Mitigation Techniques" (Longare, Angrogna, Matera, Alberobello, Palermo and Pantelleria), are reported in Fig. 17.11. A sample residential room of 16 m^2 (4 × 4 × 3 m, south-facing window of 4.2 m^2) is considered, setting the occupation time to 24 h/day. The only non-adiabatic envelope surface is facing south, fixing the *U*-values of vertical opaque surfaces and fenestrations to the limits for the 2019–2020 period defined by Italian regulation for each climate zone (Corrado 2016)—see also Appendix B, attachment 1, Chapter 4 of the GU No. 162 15/07/2015, suppl. Ord. No. 39—DM 26/06/2015. A shading control system is considered (activation for more than 40 W/m^2) together with minimum required ventilation rates in accordance with EN 15251:2007. Local Typical Meteorological Years with hourly definition were generated using Meteonorm 7.1, while the WBT is calculated according to the Stull's formula (Stull 2011). The limited amount of hours for evaporative cooling comfort, especially in locations such as Palermo, is due to the fact that this number refers only to those hours in which evaporation only may guarantee comfort conditions, but not ventilation, and does not include those in which both evaporative cooling and controlled natural ventilation may assure

Table 17.3 Early list of potential requirements (functional, operational and climatic) for the exploitation of the considered thermal sinks

Heat gain dissipative system	Main functional requirements	Main operational requirements	Main climatic/site requirements
Controlled natural ventilation	Avoid conflict between stack and wind-driven flows; allow cross-flowing inside the building if double-side ventilation is expected; allow the exploitation of stack-driven ventilation by positioning inlet openings below the neutral plane; consider the use of devoted openings to be automatically controlled; etc.	Control of the net opening areas (manually or automatically); allow to directly exhaust the air from spaces with air pollutant sources; allow temperature control to avoid overcooling during occupation; consider the use of sound barriers or filters; consider the use of ceiling fans or exhaust fan to increase the potential; etc.	Exposition to prevalent winds to increase the wind-driven potential; exploitation of nocturnal reduction in temperature to potentially activate night cooling ventilation; avoid positioning inlet openings in wind-wake-core areas; absence of outdoor airflow barriers; avoid exposing inlet opening to outside pollutant sources, etc.
Direct evaporative cooling	Correct dimensioning of the PDEC tower system (height, diameter,…); correct design of the inlet and outlet shapes; choose the correct DEC typology (shower, misting, wetPad towers or porous media); consider the inclusion of drift eliminator to reduce the risk of legionella; possibility to choose the nozzle (level of pressure, amount of sprayed water, spray angle, size of water droplets, type of cone, number of orifices, technology used); …	Control of the amount of spayed water (e.g. separate control of on/off cycle for each nozzle); treat the used water for legionella and mould prevention; consider water recirculation; consider activating a fan support; consider maintainability and nozzle substitution; consider mixing use with untreated air for reducing RH%; …	Local presence of a relevant WBD; local presence of a relevant differences between WBT and comfort temperature; avoid inlet of air near outdoor pollutant sources; consider local water presence; …
Ground cooling (EAHX)	Consider the need for free space to bury the pipe system; correctly define the number of parallel pipes according to the air velocity and the amount of air to be treated; correctly define the pipe length to minimise the pressure drops and reach an acceptable level of exchange; avoid mutual influence among tubes; etc.	Allow the elimination of potential condensation water; allow maintainability; allow the inclusion and substitution of dust filters; include grid protection on the inlet vents to prevent objects and animal entry; allow to mix treated air with external air according to environmental conditions; allow modulation cycles; etc.	Local presence of a deep ground-water level; control of the thermal transmission of the soil and tube depth to optimise the cooling effect and prevent condensation; avoid inlet of air excessively near the ground to prevent dust; consider spraying water on the ground surface to increase the cooling effect by surface evaporation; avoid the inlet of air near outdoor pollutant sources; etc.

(continued)

Table 17.3 (continued)

Heat gain dissipative system	Main functional requirements	Main operational requirements	Main climatic/site requirements
Indirect radiative cooling	Maximise the massive area exposed to the sky; consider the use of light colours to reduce heat absorption; adjust the heat attenuation and the delay effect of massive elements; etc.	Allow the cleaning of light surfaces; etc.	Guarantee sky access to thermal masses; potential use of diurnal shading systems; consider the dissipation due to nocturnal winds and breezes; consider the potential effect of coupled indirect evaporative cooling; etc.

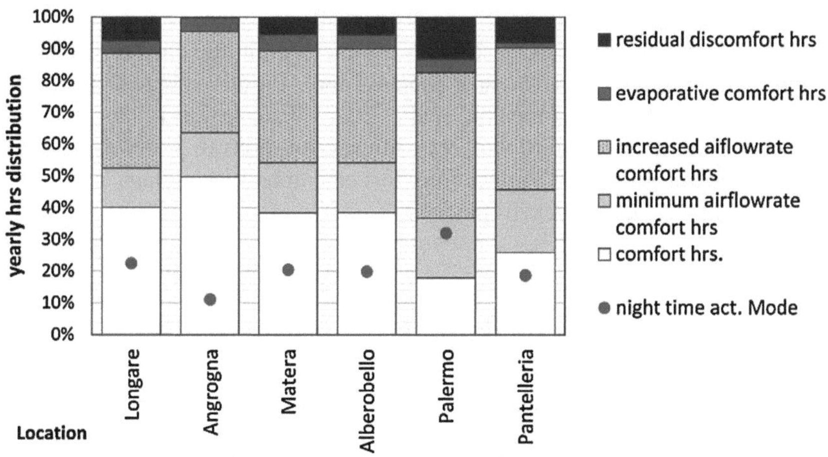

Fig. 17.11 Distribution of the amount of comfort hours, potential comfort hours with minimum airflow rate, potential comfort hours with increased airflow rate, potential comfort hours with evaporative cooling, potential comfort hours with night ventilation and residual discomfort hours

comfort. The high number of hours in which night cooling can be effective in Palermo explains the diffusion of vernacular archetypes which use this technique, combined with diurnal comfort ventilation supported by evaporative and ground cooling to cover the residual discomfort. Similarly, in Longare the use of *covoli* can cover the residual discomfort, while in the other locations radiative cooling may help. In Angrogna, the high original comfort suggests the use of ventilation for cooling and shading systems.

Finally, other climatic related methods to analyse the local cooling potential of passive systems are reported in the literature (e.g. Chiesa 2016; Santamouris and Asimakopolous 1996; Chiesa and Grosso 2015a; Chiesa 2017).

Conclusions

The chapter describes the main vernacular strategies for passive cooling of buildings throughout the Italian peninsula by presenting a list of possible solutions which could be translated into contemporary environmental and technological design choices. The toolbox of passive cooling systems is very large and ranges from heat gain prevention to modulation and dissipation. Section "Early Design Strategies Passively Cool Buildings in Italian Climates" defines a performance-driven approach to include passive cooling systems in early design phases, although further analyses are needed to develop a method for detailed design phases for building construction. As is demonstrated in this section, several strategies are adapted for the Italian climates. Nevertheless, these technologies are, in several cases, very local specific, for example, evaporative cooling, which strictly depends on local wet bulb depression and the need for to avoid an excess of relative humidity in the treated airflows. In this study, these limits are only partially analysed, even if several references to devoted international researches are included in the chapter, including a simple demonstration of a recent elaborated tool. Further researches are under development in order to further the studies into vernacular environmental design aspects, with special regard to temperate climate locations (e.g. under-rock buildings) and to relate ancient applications to contemporary buildings with special regard to climate changing environments.

Appendix: The Seven Root Sheets (by Mario Grosso)

The descriptions of the presented root sheets are principally based on Grosso (2017), while for the case studies *La Zisa* and *Camera dello Scirocco*, Ref. Tumino (2015) was also used.

Root sheet #1 | "Covoli", *Costoza (VI), Veneto Region, XVI century*
Keywords: *buried pipes; ground cooling; Palladian villas*

Description of the Archetype

The "Covoli" is the first known passive ground cooling system of Europe. Such a system is also called "Canadian well". It is composed of channels, called "covoli" (wind ducts in Venetian language), both naturally created by underground water streams and constructed by ancient romans as a transportation way from lime quarries. During the sixteenth century, in the zone where Palladian villas were built, an innovative entrepreneur thought to erect villas just at the lower end of those channels and connect them to their basements and caves through a controllable opening (door).

By doing that, it was possible to exploit the cold air present in the channels as a natural cooling sink for the villas in the summer period. In fact, the ambient temperature (more than 30 °C in the hottest month of the year) higher than the one of the air inside the covoli, which nearly approaches the annual air temperature of the location (around 12 °C in Covoli's area), causes air sliding on the tilt channel and air pressure increasing with negative altitude gradient. Hence, when the covolo/basement door is open, an outward flow to the villa's basement is generated. Once the cold air has entered and filled the basement, the pressure difference due to outdoor/indoor temperature gradient induces an airflow from the basement to the rooms at the upper floors through openings (decorative grids) in the ceiling slabs, and from the rooms to outside. A transition state in the villa brings the average indoor air temperature in a range 17 ÷ 23 °C, from the basement to the upper floor rooms, given the high thermal mass of the construction exterior and interior walls. This range is entirely within the summer comfort zone, with a lower limit even too cold. But, air temperature range can be brought at a higher level (e.g. 20 ÷ 26 °C) by alternatively opening and closing the covolo/basement door. It is not effective as a set-point temperature control of an HVAC system, but is consistent to the adaptive comfort approach typical of naturally ventilated buildings (ASHRAE 2004). The "covoli" system could work also in winter, when ambient temperature is colder than the covoli's inner air. In the periods when this condition applies, air warmer than external air could be theoretically exploited to save heating energy need, but it could not be done naturally, since air temperature from the covoli is not at comfort level. In addition, connection between covoli and villa's basements should be made at the upper end of the channels to follow the natural upward airflow inside them. This is a location set for the villas completely different from the one realised in the sixteenth century in Costoza.

Bioclimatic and Regionalist Technological Principles

The "Covoli" system does not affect architecture, since it can be applied to any architecture type and style. It is more a bioclimatic archetype of an HVAC plant rather than an architectural archetype.

However, its meaningfulness as an archetypical application of a regionalist technological principle is related to its location-bound characteristic, particularly, the availability of underground sources of air cavities or, in any case, of a terrain free of above-grade constructions, under which buried air ducts could be built on purpose.

Contemporary Potential Applications

- Earth-to air heat exchanger (EAHX) system

It is composed of horizontal buried pipes, within which air driven by a fan flows. Several EAHX applications are present in Europe, mostly in Germany for industrial and large commercial buildings.

Root sheet #2 | At-rock house, *Angrogna Valley (TO), XVIII, XIX centuries*
Keywords: *Stone building; thermal control by inertia; attachment to rocks*

Description of the Archetype

The houses under, or attached, to rocks in the Angrogna valley were built from the beginning of the eighteenth to the first half of the nineteenth century during a period of harsh persecution of the Valdensian people by the Savoia kingdom ruling on Piedmont. They were placed on almost inaccessible sites, exploiting their morph-geological characteristics whereby a plane associated to overhanging granitic rocks allowed for erecting buildings using local stones for structural walls, and attaching them to the rock. In some cases, a rock would function as the back wall, while in others a rock would function as the roof.

Bioclimatic and Regionalist Technological Principles

The bioclimatic principle of the Angrogna's "at-rock" houses is mainly related to the high thermal capacity of their stone walls as well as the almost infinite thermal mass storage capacity of the adjacent rock. This feature would allow for stabilising the indoor air temperature profile around the seasonal averages, hence reaching comfort level in summer and not so freezing conditions in winter.

From an architectural regionalism viewpoint, these "at-rock" houses represent a typical heritage of traditional mountain architecture, constructed by local material and perfectly integrated in their landscape context.

Contemporary Potential Applications

- Building construction technology, which uses high thermal capacity envelope and structure associated to controlled natural ventilation for dissipating the stored heat;
- Phase change materials (PCM) to combine heat dissipation performance and lightness of construction to facilitate the application of environmental friendly dry-assembly techniques. However, performance tests on PCM-based buildings have not yielded so far satisfactory results particularly for what stability in time of their effectiveness is concerned.

Root sheet #3 | "Sassi", *Matera*
Keywords: *hypogeum building; thermal inertia*

Description of the Archetype

The "Sassi" (literally "Stones") are a hypogeum pre-historical compound built in a calcareous natural amphitheatre in the City of Matera, which has become a UNESCO cultural heritage place (see pictures below). Abandoned for centuries, they have been inhabited since the 1950s of the twentieth century.

Sassi's builders exploited the tenderness of the calcareous rocks to create a wide variety of inner space aggregations (see schemes below). External building structures made of calcareous blocks taken from the site were erected over time.

Bioclimatic and Regionalist Technological Principles

The bioclimatic principle of the "Sassi" is based on the all-year heat exchange attenuation typical of any hypogeum structure. This is due to the quasi-infinite thermal inertia of the rock mass wherein the rooms are drawn. No openings are present, hence creating poor ventilation and daylight problems. However, these problems are currently being solved through artificial lighting systems and controlled mechanical ventilation.

Contemporary Potential Applications

- Hypogeum building development

However, a debate is still ongoing on which can be the most sustainable future urban development related to the demographic exponential increase in the planet Earth: verticality (skyscrapers increasingly higher) or underground?

Root sheet #4 | "Trullo" building, *Puglia*
Keywords: *massive building; solar control; natural ventilation, evaporative cooling*

Description of the Archetype

The "Trullo" is a traditional residential construction of the Murge plateau, located in the central part of the Region Puglia. Originally, the "Trullo" was a seasonally used building with prevailing agricultural function, made of dry-assembled calcareous blocks which could be easily dismantled in case of Government control. Currently is used as a permanent or vacation residential building, often transformed into commercial stores or accommodation resorts.

Its shape is compact and developed through the aggregation of round-plan modules, each with a cylindrical low base above which a much higher conical roof is placed, giving the "Trullo" its typical and most famous shape. The external walls, which have both envelope and bearing functions, are made of a thick layer of dry-assembled calcareous blocks. The conical roof is built-up as a pseudo-cupola and made of calcareous plates which have a decreasing thickness as they reach the keystone, surmounted by a pinnacle. The only openings are the entrance door and the outlet vent of the fireplace chimneystack, whereby cool air entering from cracks made on purpose in the bottom of the door can be stack-driven. A water basin is often placed in the ground floor as a mean to cool the air entering from the door bottom cracks.

Bioclimatic and Regionalist Technological Principles

The bioclimatic principle of the "Trullo" building is based on the combined cooling effects due to a massive structure able to store heat during the day, and heat dissipation through natural ventilation when ambient temperature drops below indoor air temperature (usually at night). When a water basin is present in the ground floor of the main room, the evaporative effect is further cooling the air. An additional cooling mechanism is given by high albedo coefficient of the white wall and roofing surfaces, which reduces the solar irradiation (short-wave length) absorbed by the envelope structure. Meanwhile, the high emittance (long-wave length) of the envelope surface allows for the daily absorbed solar heat being dissipated to the night sky by re-irradiation.

Contemporary Potential Applications

- Massive building envelope;
- Alternative system such as PCM (phase change materials).

These systems can be combined with controlled night natural ventilation, evaporative cooling and direct radiative cooling through white envelope cladding.

Root sheet #5 | "Scirocco" room, *Villagrazia (PA), XVI Century*
Keywords: *hypogeum; evaporative cooling; natural ventilation*

Description of the Archetype

The "Scirocco" room of Villa Naselli in the site of Villagrazia (PA) was built in 1552 as a component of a 50-m-long below-grade gallery connecting southward the Villa's premises to the underground river Ambleri. A stream of water is running in the middle of the gallery's floor, from the source chamber to the entrance to the gallery and, forward, to another gallery of the same length towards North, extending within the Villa's terrain. This channel had the function of irrigation as well as of providing domestic potable water. In addition, the water would provide evaporative cooling of the "Scirocco" room, created on purpose in the middle of the gallery as a resting space in hot summer days (when a south-westerly wind called "Scirocco"—from which the name of the room—blows). The gallery is 2.6 m wide in average and 2.25 high at the maximum, and has a barrel vault hollowed by seven cylinder-shaped chimneys, 5 in the 25-m-long Northward branch and 2 in the Southward part of the gallery, with respect to the "Scirocco" room. The "Scirocco" room has a 4.5-m-diameter circular plan and a reverse-cup-shaped wall envelope open on the top—with a diameter of 1.5 m—to allow for natural airflow as in the Iranian *badgïr*. The above-grade part of the "Scirocco" dome has a lace-trimming ridge and four rectangular 30 × 50 cm openings, oriented towards the main compass coordinates.

Bioclimatic and Regionalist Technological Principles

The bioclimatic principle of the "Scirocco" room and the whole gallery is based on a natural mechanism involving ventilation and evaporative cooling (Grosso 2017, Ch. IV, p. 143). A natural airflow exchange between external environment and the internal room and gallery spaces is driven by the differential pressure, which is established across the rooftop and room-dome openings due to both wind and temperature gradient. The airflow follows a path as indicated in the cross section above according to the two following types of basic schemes.

(a) When the "Scirocco" hot wind is blowing and the ambient temperature is higher than the room temperature, air can enter through the vertical openings of the room dome and exit from the gallery roof chimneys; however, the flow could reverse its direction sometimes due to the reciprocal relationships amid external and internal openings height (Grosso 2017).
(b) When wind is slowing down and the ambient temperature gets lower than the room temperature, air enters from the gallery roof chimneys and exits through the room-dome openings.

Usually, case (a) occurs during the day and case (b) at night, but the actual airflow path depends on the wind strength and the amplitude of temperature difference. In case (a), air flowing in contact to the water stream is cooled down by evaporation, hence cooling the indoor room space. In case (b), the airflow cooled down the room indoor space by environmental air change; however, a further cooling might occur by evaporation of water in contact to the airflow depending on the difference between water and air temperature.

Contemporary Potential Applications

- Controlled natural ventilation techniques associated to evaporative cooling

Root sheet #6 | "Zisa" building, *Palermo, XI Century*
Keywords: *massive building; evaporative cooling; natural ventilation*

Description of the Archetype

The "Zisa" construction was commissioned by Guglielmo I d'Altavilla (1153–1166) and erected in an extended garden called "genoard"—Arabian word for "heaven in earth". The construction was finished by Guglielmo II (1172-1189), who added an adjacent building as a "delight pavilion", called "Cuba". In this building, it is used an Islam architecture style combined to a Norman one. The "Zisa" was not substantially modified until the seventeenth century, when Giovanni de Sandoval and Platamone added a new staircase on the right wing of the building through the demolition of bearing walls and the original entrance stairs. In 1806, various consolidation and restoration works were commissioned, and in 1955 the building passed to the Italian State. In 1971, a restoration programme was commissioned and ended in 1991. Currently, the "Zisa" hosts the Islam Museum. Although most of the original features were restored, the overall naturally driven microclimate control approach is lost.

The whole palace is realised around a central double-storey high square space, which is its fulcrum (the main hall), on the side of which the residential rooms are placed in a perfect symmetry. The main façade is oriented eastward to capture sea breezes through three large arches and a window at the highest floor. The three arches let enter to a portico, whose lateral windows allow for further increasing the airflow. This portico is connected to the main hall, at the centre of which is placed a fountain. The water spilled from this fountain runs on a "salabil"—a slanted marble plate worked on relief to enliven the water stream thus increasing the exchange surface to humidify air—and afterwards to two tanks and a fishpond.

Two towers are placed, slightly protruding, in the middle of the short sides of the building. Their function is to drive natural airflow into the building by either stack effect (outlet flow) or wind catching (inlet flow). The same function had the several openings in the internal walls and the roof. Windows had no glass sheets but sashes made of perforated wood with various decorations, typical of the Arabian culture, called *mashrabyia*.

Bioclimatic and Regionalist Technological Principles

The bioclimatic principle of the "Zisa" Palace is based on the combined cooling effects due to a massive structure able to store heat during the day, and heat dissipation through natural ventilation when ambient temperature drops below indoor air temperature (usually at night). An additional cooling mechanism is given by the presence of a fountain and other water basins and streams, which generate air-cooling by evaporation.

Contemporary Potential Applications

- Massive building envelope;
- Alternative system such as PCM (phase change materials);
- Night controlled ventilation and direct evaporative systems.

Root sheet #7 | "Dammuso" building, *Pantelleria*
Keywords: *massive building; solar control; natural ventilation*

Description of the Archetype

The "Dammuso" is a traditional residential construction of the Pantelleria Island (the name meaning "Island of wind"), one of the two most Southward Sicilian, and Italian, locations; the other being Lampedusa, the famous immigrant-rescue island. Originally, the "Dammuso" was a seasonally used building with prevailing agricultural function. Currently is used as a permanent or vacation residential building, often transformed into a luxury multifunctional compound. Its shape is compact and developed through the aggregation of almost-cubic modules, each with a barrel-vaulted roof, which shows externally a typical breast-like shape made of a white lime covering (see picture below). The external walls, which have both envelope and bearing functions, are made of a double layer of dry-assembled stones with gravel embedded in the inner cavity. The barrel-vaulted roof is also made of stones and covered with a waterproof layer of lime mortar (see plan, section and elevation below).

Openings are the entrance door and small circular apertures in the middle of the walls ("occhi di pietra", i.e. "stone eyes"), allowing for night ventilation and daily feeble illuminance, while protecting from a frequent strong wind.

Bioclimatic and Regionalist Technological Principles

The bioclimatic principle of the "Dammuso" building is based on the combined cooling effects due to a massive structure able to store heat during the day, and heat dissipation through natural ventilation when ambient temperature drops below indoor air temperature (usually at night). An additional cooling mechanism is given by high albedo coefficient of the white roof covering, which reduces the solar irradiation (short-wave length) absorbed by the roof structure. Meanwhile, the high emittance (long-wave length) of the lime white roof surface allows for the daily absorbed solar heat being dissipated to the night sky by re-irradiation.

Contemporary Potential Applications

- Massive building envelope;
- Alternative system such as PCM (phase change materials);
- Controlled night natural ventilation;
- Direct radiative cooling through white roof covering.

References

Accorsi M, Malaguti A (2017) Exhibition "Felice età" organized by arch. In: Grosso M (ed) Il raffrescamento passivo degli edifici in zone a clima temperato. Maggioli, Sant'Arcangelo di Romagna, pp 144–146

Allard F (ed) (1998) Natural ventilation in buildings: a design handbook. James & James, London

Artmann N, Manz H, Heiselberg P (2007) Climatic potential for passive cooling of buildings by night-time ventilation in Europe. Appl Energy 84(2):187–201

Ascione F, Bellia L, Minichiello F (2011) Earth-to-air heat exchangers for Italian climates. Renew Energy 36:2177–2188

Ascione F, De Masi RS, de Rossi F, Ruggiero S, Vanoli GP (2016) Optimization of building envelope design for nZEBs in Mediterranean climate: performance analysis of residential case study. Appl Energy 183:938–957

Bahadori MN (1978) Passive cooling systems in Iranian architecture. Sci Am 238(2):144–154

Belleri A, Chiesa G (2017) Ventilative cooling potential tool: user guide V.1.0, Tool IEA EBC Programme. Available at: http://venticool.eu/annex-62-publications/deliverables/. Viewed 20 Sept 2017

Berenson B (1948) Aesthetics and history in the visual arts. Pantheon, New York

Brancati V, Mariani F, Simili M, Bonanno E (1951) Volto delle Eolie. Flaccovio, Palermo

Cardinale N, Rospi G, Stefanizzi P (2013) Energy and microclimatic performance of Mediterranean vernacular buildings: the Sassi district of Matera and the Trulli district of Alberobello. Build Environ 59:590–598

Cardinale T, Balestra A, Cardinale N (2015) Thermographic mapping of a complex vernacular settlement: the case study of Casalnuovo District within the Sassi of Matera (Italy). Energy Procedia 76:40–48

Castellano M (1964) La valle dei trulli. Leonardo da Vinci, Bari

Cavaglià G et al (1975) Industrializzazione per programmi. RDB, Piacenza

Chiesa G (2013) M.E.T.R.O. (Monitoring Energy and Technological Real time data for Optimization): innovative responsive conception for cityfutures. PhD Thesis XXVI cycle, Politecnico di Torino, Torino
Chiesa G (2016) Geo-climatic applicability of evaporative and ventilative cooling in China. Int J Vent 15(3–4):205–219
Chiesa G (2017) Climate-potential of earth-to-air heat exchangers. Energy Procedia 122:517–522
Chiesa G, Grosso M (2015a) Geo-climatic applicability of natural ventilative cooling in the Mediterranean area. Energ Buildings 107:376–391
Chiesa G, Grosso M (2015b) Direct evaporative cooling of buildings. A comparison amid simplified simulation models based on experimental data. Build Environ 94:263–272
Chiesa G, Grosso M (2016) Breakthrough of natural and hybrid ventilative cooling technologies: models and simulations. Int J Vent 15(3–4):183–185
Chiesa G, Grosso M (2017) Meta-design approach to environmental building programming for passive cooling of buildings. In: Sayigh A (ed) Sustainable energy for all – selected papers form the world renewable energy congress. Springer, Berlin
Chiesa G, Simonetti M, Grosso M (2014) A 3-field earth-heat-exchange system for a school building in Imola, Italy: monitoring results. Renew Energy 62:563–570
Cook J (ed) (1989) Passive cooling. MIT Press, Cambridge
Corrado V (ed) (2016) Il nuovo Decreto Ministeriale sul calcolo della prestazione energetica e sui requisiti minimi. Rockwool Italy, Milano
Cowan HL (1977) The Master Builders: a history of structural and environmental design from ancient Egypt to the nineteenth. Wiley, New York
Cuce PM, Riffat S (2016) A state of the art review of evaporative cooling systems for building applications. Renew Sust Energ Rev 54:1240–1249
D'Agostino D (2015) Assessment of the progress towards the establishment of definitions of Nearly Zero Energy Buildings (nZEBs) in European Member states. J Build Eng 1:20–32
Dall'O' G, Bruni E, Sarto L (2013) An Italian pilot project for Zero Energy Buildings: towards a quality-driven approach. Renew Energy 50:840–846
Ferrari G (1925) L'architettura rusticana nell'arte italiana. Dalla capanna alla casa medievale. Duecentocinquanta tavole. Hoepli, Milano
GEE Directorate of Global Energy Economics (2015) Southeast Asia energy outlook 2015. OECD/IEA, Paris
Givoni B (1994) Passive and low energy cooling of buildings. Van Nostrand Reinhold, New York
Givoni B (1997) Performance of the shower cooling tower in different climates. Renew Energy 10(213):173–178
Green KW (1979) Passive cooling. Res Des 2(3):4–9
Grosso M (ed) (2017) Il raffrescamento passivo degli edifici in zone a clima temperato, 4th edn. Maggioli, Sant'Arcangelo di Romagna
Grosso M, Chiesa G (2015) Horizontal earth-to-air heat exchanger in Imola, Italy. A 30-month-long monitoring campaign. Energy Procedia 78:73–78
Grosso M, Raimondo L (2008) Horizontal air-to-earth heat exchangers in Northern Italy: testing, design, and monitoring. Int J Vent 7(1):1–10
Holzer P, Psomas T (eds) (2017) Ventilative cooling sourcebook (Annex62). IEA, EBC, Department of Civil Engineering, Aalborg University, Aalborg
Hwang I, Sakamoto T, Ferré A et al (eds) (2006) Verb natures, architecture boogazine. Actar, Barcelona
Isaac M, van Vuuren DP (2009) Modeling global residential sector energy demand for heating and air conditioning in the context of climate change. Energy Policy 37:507–521
Jona C (1920a) Architettura rusticana Valle d'Aosta. Crudo, Torino
Jona C (1920b) Architettura rusticana. Costiera amalfitana. Crudo, Torino
Koizumi S (2007) Energy efficiency of air conditioners in developing countries and the role of CDM. IEA information paper, OECD/IEA, Paris Cedex 15

Kolokotroni M, Heiselberg P (2015) IEA EBC Annex 62 ventilative cooling. State-of-the-art review. Department of Civil Engineering, Aalborg University, Aalborg
Linea Blu (2018) L'architettura eoliana, video, 3′10″, Rai. http://www.rai.it/dl/RaiTV/programmi/media/ContentItem-28cd660d-f07b-45b3-820c-e3374ca21f92.html?refresh_ce. viewed Feb 2018
Lo Cascio G (2005) Dell'architettura nelle isole Eolie. Aton, Roma
Logue JM, Sherman MH, Walker IS, Singer BC (2013) Energy impacts of envelope tightening and mechanical ventilation for the U.S. residential sector. Energ Buildings 65:281–291
Lucia M, Simonetti M, Chiesa G, Grisolia G (2017) Ground-source pump system for heating and cooling: review and thermodynamic approach. Renew Sust Energ Rev 70:867–874
Olgyay V (1963) Design with climate. Princeton University Press, Princeton
Pagano G, Guarniero D (1936) Architettura rurale italiana, Quaderni della Triennale. Hoepli, Milano
Pearlmutter D, Berliner P (2017) Experiments with a 'psychrometric' roof pond system for passive cooling in hot-arid regions. Energ Buildings 144:295–302
Pfafferott J, Walker-Hertkorn S, Sanner B (2007) Ground cooling: recent progress. In: Santamouris M (ed) Advances in passive cooling. Earthscan, London, pp 190–227
Raimondo L (2017) il controllo termico della radiazione solare. In: Grosso M (ed) Il raffrescamento passivo degli edifici in zone a clima temperato. Maggioli, Sant'Arcangelo di Romagna, pp 229–274
Rudofsky B (1964) Architecture without architects: a short introduction to non-pedigreed architecture. Doubleday, New York
Rudofsky B (1977) The prodigious builders: notes toward a natural history of architecture with special regard to those species that are traditionally neglected or downright ignored. Harcourt Brace Jovanovich, New York
Santamouris M (ed) (2007) Advances in passive cooling. Earthscan, London
Santamouris M (2016) Cooling the buildings – past, present and future. Energ Buildings 128:617–638
Santamouris M, Asimakopolous D (eds) (1996) Passive cooling of buildings. James & James, London
Simonetti M, Chiesa G, Grosso M, Fracastoro GV (2016) NAC wall: an open cycle solar-DEC with naturally driven ventilation. Energ Buildings 129:357–366
Simonetti M, Gentile V, Chiesa G, Nigra M (2017) Preliminary study of the hybrid solar DEC "NAC wall" system integration in building façades in urban context. Energy Procedia 134:588–597
Stull R (2011) Wet-bulb temperature from relative humidity and air temperature. J Appl Meteorol Climatol 50(11):2267–2269
Trossarelli L (1998) Vita e morte a Barma Mounastira. Comune di Angrogna, Angrogna
Tumino G (2015). Archetipi bioclimatici di raffrescamento passive degli edifici nel territorio siciliano. Degree Thesis, BoA in Architecture, Politecnico di Torino, Academic Year 2014/15, Supervisor: Prof. Mario Grosso, Castello del Valentino, Torino, 23/07/2015
Watt JR (2013) Evaporative air conditioning, 2nd edn. Springer, Berlin

Chapter 18
Traditional Buildings Back to the Future. Adaptive Energy Efficiency in Reuse

Michele Morganti, Emanuele Habib, Edoardo Currà, and Carlo Cecere

Vernacular Buildings and Energy Retrofit

Nowadays, to place the issue of the importance of any specialist analysis of a vernacular building—even if significant for a local community or a region—it is not foreseen. It means to recognize elements, significances and values of such kind of buildings in relation to local culture, tradition and climate. Therefore, to introduce *Civita di Bagnoregio* as case study and one of the most representative buildings, *Casa Medori*, represents a value judgement in itself. It requires not only to identify in its building elements the main attributes—age, integrity and significance—and the physical characteristics of its envelope but also to recognize it as a model of a particular cultural heritage.

In fact, we argue that it is crucial at first to understand which aspects qualify it as a *house*: a complex "manufacture" dating back to the pre-industrial era and, for this reason, the result of a pre-capitalist society. A *house*, therefore, not a *building type* that requires to "definitely transform the man in a fixed scheme" as Bruno Taut claimed (Taut 1991). It is not even a construction *good,* a minor construction erected by workmen "disappeared in the ordinary daily life"—as stated by Jean-Pierre Garnier—that build useful goods for a local community, whereas a house is a sort

M. Morganti (✉)
SOS Urban Lab, Faculty of Engineering, DICEA Department, Sapienza University of Rome, Rome, Italy

Architecture and Energy, School of Architecture, Department of Architectural Technology, Universitat Politècnica de Catalunya, Barcelona, Spain
e-mail: michele.morganti@uniroma1.it

E. Habib
Faculty of Engineering, DIAEE Department, Sapienza University of Rome, Rome, Italy

E. Currà · C. Cecere
SOS Urban Lab, Faculty of Engineering, DICEA Department, Sapienza University of Rome, Rome, Italy

A. Sayigh (ed.), *Sustainable Vernacular Architecture*, Innovative Renewable Energy, https://doi.org/10.1007/978-3-030-06185-2_18

of *tool*, more expensive and complex than others, which "meets needs, desires, pleasures and faith of everyone" (Garnier 2016).

In this study, therefore, we present Casa Medori in Civita as an exemplary model of vernacular architecture. First of all, it is important to understand the meaning of the term *vernacular* as intended in Latin and quoted by Ivan Illich: "the opposite of goods", something precisely homemade: a dialectal architecture, considering its language, just of vernacular gender (Illich 1984). Among the several and possible interpretations of the vernacular architecture, some seem to be particularly risky because of the complexity of the building. The energy interpretation, for example, reveals a different profile of operational energy, typical of a pre-industrial building, compared to a modern one or to a building designed by an architect. The "adaptive" interpretation shows in the different construction technologies the capability to waste nothing and to embrace a logic based on transformation and re-use, later esteemed by contemporary architecture and also by Vittorio Gregotti.

The architecture based on the above-mentioned logic is not the same as a *resilient architecture*, nor the same as the architecture *conviviality* of alternative uses. The resilience of the built environment, i.e. the ability to recover from change, gets to modify our lifestyle in order to survive in the awareness on the limits to growth. The conviviality, proposed by Habraken or Herzberger, offers to everyone the wider opportunity to enrich the environment with the result of its own making (Habraken 2000).

In fact, today the characters of unity, complexity and vernacularity of *Casa Medori* are at risk. Its own physical integrity and its "outliving" is at risk. Further investigation on history and later transformations, urban morphology and typology of this building in relation to use are needed. Several questions on the historical evolution, on the construction and on the role of each building element are still partially open.

The present study seeks to address these questions with the aim to understanding, once and for all, conditions and situations which led to its physical elements. Indeed, any energy retrofit process undertaked on such kind of traditional building requires to face with tailored conservation principles.

To "get back to the future" *Casa Medori*, means to go beyond conventional retrofit approach for historic buildings, looking for a mature and integrated approach; to overcome energy retrofit criteria based not only on physical interventions but also including criteria aiming at identifying the most suitable non-physical ones, e.g. building use, occupants behaviour, indoor temperature set-point for heating and cooling, etc. Facing actual urban renovation processes, to operate on traditional buildings through innovative approaches based on less intrusive criteria, is crucial in order to preserve the physical integrity of cultural heritage.

The Italian Cultural Heritage

The above-mentioned arguments demonstrate, especially in the case of Italy, the complexity related to that fundamental element of the cultural heritage. In particular, the main part of the Italian building stock is composed of urban textures based on vernacular buildings, sharing regional values, features, construction materials and building elements, which distinguish the landscape. Approximately one-third of the national buildings goes back to the period before 1945 and more or less the 20% dates back maximum to 1919 (ISTAT 2017). The greater part of this historical heritage is represented by building used for residential, working and commercial purposes and not for institutional or representative purposes.

Facing with the actual challenges related to the built environment the attention of researchers, policymakers and practitioners focuses more and more on the existing built environment and their thermal performance, especially in southern Europe. In particular, the EU objectives of reducing the greenhouse gas emissions of the building stock by 95% by 2050 raise the big challenge of the energy retrofit of the existing buildings (European Climate Foundation 2010). In fact, the mean primary energy demand of buildings is around 210 kWh/m^2 (Gynther et al. 2015) and requires particular attention in relation to significance and historical narrative of the cultural heritage of our cities and villages. In the last years attitudes on this topic have significantly shifted. Carbonara compares what happened in the past in the relationship between renovation and structural consolidation with the actual approach on energy retrofit of historical buildings (Carbonara 2015). Whereas the European legislation on energy efficiency viewed the energy retrofit in the historic building as a potential threat for conservation and protection of its narrative, nowadays, it is seen as an opportunity to protect these buildings (Webb 2017). For this reason, a vivacious debate including several disciplines (restoration and renovation, building design, building physics, urban planning, structural design and consolidation) is ongoing on energy retrofit strategies and criteria for this relevant building stock.

Several research projects on energy efficiency of traditional buildings sponsored by the EU illustrate the most diffused approaches (Vieites et al. 2015; Egusquiza et al. 2016). Current research carried out on this topic assumes that energy retrofit for traditional buildings is a complex balance of multiple criteria, e.g. conservation, energy consumptions, building fabric, systems and occupant behaviour. These criteria—derived from newer building—produce strategies that have mainly concentrated on technical improvement and may cause damage to physical elements, especially to the significance of our cultural heritage. Moreover, it has been demonstrated that factors that do not directly affect the building fabric have a relevant impact on the energy performance (Baker and Steemers 2000; Ben and Steemers 2014; Geva 2015). According to existing literature, for this reason, non-physical retrofit strategies play a major role in the effectiveness and diffusion for heritage buildings.

In this framework, Italian villages and towns underwent a renovation process in which building use transformation is often driven by economic and social factors, without taking into account possible effects on energy performance. This process reduces retrofit strategies into an application of efficient technology, in order to support new uses and users comfort. This lack of assessment calls for further specific investigation on this aspect respecting the balance between conservation and energy efficiency.

The present study aims to explain to what extent building use has an impact on thermal and energy performance of traditional buildings and how their transformation could represent a consistent retrofit criterion, comparable to other physical and non-physical criteria.

Civita di Bagnoregio and Casa Medori

Like many centres of *Tuscia* region, the territory of *Civita di Bagnoregio* is rich in signs and remains of human activity, far before the Etruscan and Roman presence. The ancient village is located on the historic route between the fords of the Tiber and the lake of *Bolsena*. As two poles of a single historical settlement, *Civita and Bagnoregio* align on this route crossed with other longitudinal paths of southern Etruria. Inter alia, *Civita* "from its origins to the time of the first decisive telluric movements, which have progressively separated its geographical boundaries, has been an integral part of the city of Bagnoregio, constituting its civic centre, then also a bishopric, as well as a diocesan one" (Gargano 1988).

La Civita has settled on top of a hill between two rivers, the *Rio Chiaro* (Clean brook) and *Rio Torbido* (Murky brook). The superficial soil is pyroclastic, with a layer of ignimbrite on layers of stratified tuff. The landslide depends largely on the sedimentary substrate on which higher levels of volcanic origin are placed. It consists of marine clays of the Pleistocene and is constantly eroded by both the waterways and the rainwater. Moreover, the area is seismic and the earthquake of 1695 has contributed to its gradual decline. In addition to the destruction, it also led to the deprivation of the ancient episcopal seat, depleting it of one city sign.

After many interventions, reconstructions and obvious historicist tampering, the inhabited area still preserves the traces of the first plant (Figs. 18.1 and 18.2). It is structured on a ridge path that becomes the decuman axis of an orthogonal system, in which the characteristics of the relationship between basic construction and road, typical of the whole area, are still evident. At the centre of the fabric is the square with the ancient cathedral. In many facades, and even more in plan, the sequences of single dwellings with the narrowest street front can be found. Both porticoes and

Fig. 18.1 Civita Cadastral map, 1870

Fig. 18.2 Aerial view of Civita and Casa Medori

profferli are still recognizable in the relationship between the front and the street. Historians consider *profferlum* an element of "imported" rural building in the city. It allows external access to the dwelling by disengaging it from the shops and deposits on the ground floor. Over time, as a functional element, it takes on an architectural and representative role in the finest buildings, as is also the case with *Civita*.

The case study chosen is a residential block, or rather an aggregate, located on *Via Madonna della Maestà* next to the Cathedral. The architectural organism is significant because it presents the typical features of a building fabric that has had a centuries-old evolution. First, facing the plan reading with the well-known process method—well exemplified in Strappa et al. (2016) or in Bascia et al. (2000)—it is possible to find a regular original plant, dating back to at least the pre-modern phases of the inhabited area. Later, a review of this plant was characterized above all by a progression of the front on the road (Fig. 18.3).

A second and more decisive process is the fusion of single terraced housing cells. This phase has generated horizontal dwellings, instead of the terraced verticals, preserving the distribution of servile uses on the ground floor and houses on the upper floors. The third kind of transformation is the adaptation of a part of the island into a palace building type in the sixteenth century, through an enlargement that incorporates and hides the complex volumetric and masonry stratifications of the object, covering them on the north and west sides of a unitary garment (Fig. 18.4).

This process takes place throughout the building fabric. It generates a complex heritage, in which lifestyles have changed over the centuries and with them the original building. It is important to note that despite the duration of the process, evolution of construction techniques has not suffered the modification that we found in the form and in the type. The construction material available, the tuff, according to its properties, has determined an evolution. A slow modification of the dimensions of regular rows accompanies the most important plan and volumetric modifications. Focusing on the masonry techniques, as illustrated by *Chiovelli* (…), in the

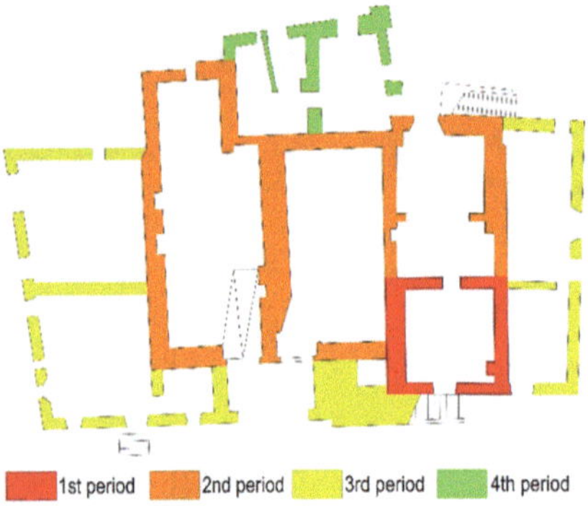

Fig. 18.3 Historical evolution of Casa Medori and external view of the main façade

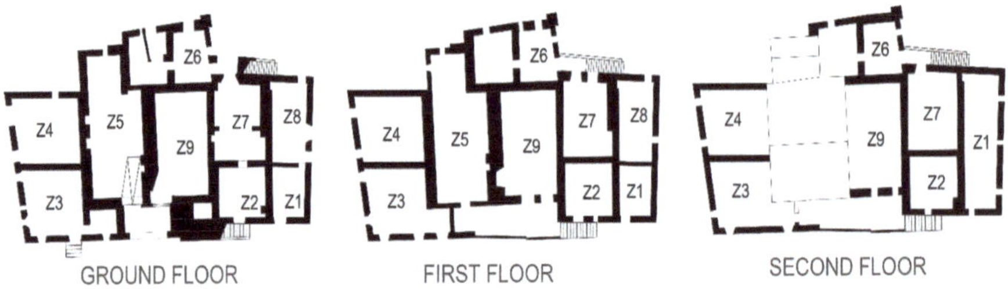

Fig. 18.4 Plans

Tuscia region and in the mediaeval *Viterbese* the technique with isometric rows prevails. He considers it an intelligent outcome of truly innovative and experimental aims at the simplification and the standardization of the great French Protogothic and Gothic shipyards, wisely mediated by local habits and materials.

Casa Medori and most of the buildings in Civita are therefore predominantly built with hewn tuff walls. They testify that "poor housing [...] from the 12th and 13th centuries were gradually replaced by masonry constructions that still partly survive throughout the regional area" (Guidoni 1980).

Thermal and Energy Performance Analysis

This section presents the research methodology applied to the case study in order to appropriately model both thermal and energy performance analysis. A comparison between different building uses with similar occupation profiles has been performed. The input parameters of building construction profiles are shown in Table 18.1.

Thermal and energy performance of Casa Medori has been evaluated through direct simulation of the whole building for different use profiles and indoor climate control. The mathematical model of the building is described by the following equations (Table 18.2).

Table 18.1 Construction profile for case study

Drawing	Construction	Description
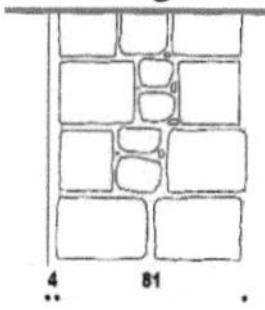	External wall	Mainly ashlar masonry of tuff blocks laid in parallel courses and plastering mortar on the internal surface
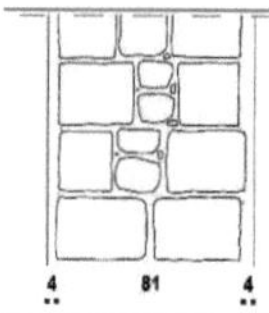	Party wall	Mainly ashlar masonry of tuff blocks laid in parallel courses and plastering mortar on the internal and external surface
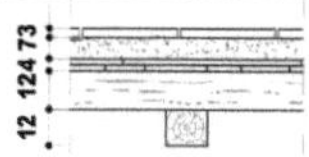	Wooden floor	Clay-tile floor, wooden floorboards with two layers and exposed wooden beams
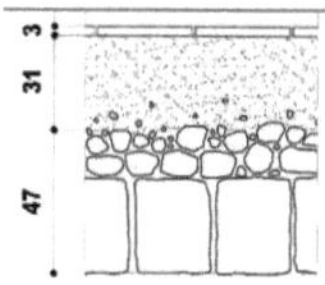	Stone vault	Clay-tile floor with hydraulic lime-based screed, vault in hewn tuff blocks
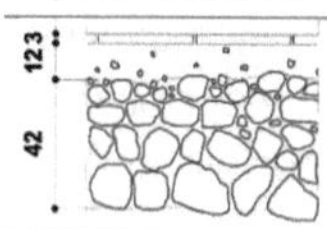	Ground floor	Clay-tile floor and stones
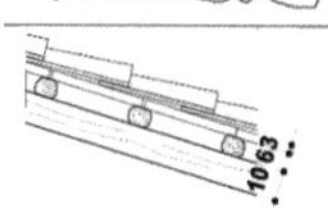	Roof	Pitched timber roofs with clay sub-roof tiles, screed and curved roof tiles
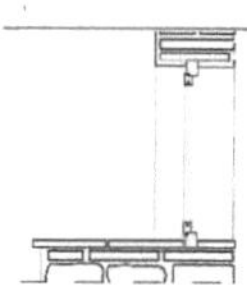	Windows	Single glazing with wooden frame
	Doors	Wood

$$c_a \rho V_i \frac{dT_{A,i}}{dt} = \sum_j h_{c,i,j} S_{i,j} \left(T_{i,j,1} - T_{A,i}\right) + G_{a,i} c_a \left(T_e - T_{A,i}\right) + Q_{C,i} \tag{18.1}$$

$$C_{i,j,k} \frac{dT_{i,j,k}}{dt} = \frac{T_{i,j,k} - T_{i,j,k-1}}{R_{i,j,k}} + \frac{T_{i,j,k+1} - T_{i,j,k}}{R_{i,j,k+1}} \tag{18.2}$$

$$T_{i,j,0} = T_{A,i} + \frac{Q_{R,i} S_{i,j}}{\left(h_{c,i,j} + h_{r,i,j}\right) \sum_j S_{i,j}} + \frac{h_{r,i,j}\left(T_{mr,j} - T_{i,j,1}\right)}{\left(h_{c,i,j} + h_{r,i,j}\right)} \tag{18.3}$$

Table 18.2 Nomenclature

c_a	The isobaric specific heat of air
ρ	The air density
V	The air volume of the thermal zone *i*
$T_{A,i}$	The room air temperature of thermal zone *i*
$h_{c,i,j}$	The convection coefficient for envelope surface *j* of thermal zone *i*
$h_{r,i,j}$	The radiant heat transfer coefficient for envelope surface *j* of thermal zone *i*
$S_{i,j}$	The area of surface *j* of thermal zone *i*
$T_{i,j,k}$	The temperature at mesh node *k* (1 is the innermost) of surface *j* of thermal zone *i*
$T_{i,j,0}$	The equivalent ambient temperature at surface *j* of thermal zone *i* that accounts for radiant heat transfer too
$G_{a,i}$	The mass flow rate of ventilation air
T_e	The outdoor air temperature
$Q_{C,i}$	The total convective load
$Q_{R,i}$	The total radiant load
$C_{i,j,k}$	The total heat capacity at mesh node *k* of surface *j* of thermal zone *i*
$R_{i,j,k}$	The total heat resistance between nodes *k* and *k-1* of surface *j* of thermal zone *i*

Heat transfer through walls is evaluated by an equivalent resistance-capacity network. Mesh nodes at walls are inside to outside, being node *1* inner surface. Equivalent sol-air temperature is used for heat transfer at the outermost node of building envelope.

Simulations were performed considering four different conditions of use:

1. No internal load
2. Residential—Internal loads: 1 W/m^2 when unoccupied (8 a.m. to 6 p.m.), 3 W/m^2, and one person per thermal zone (70 W) when occupied
3. Office—Internal loads: 20 W/m^2 and one person (70 W) per each 10 m^2 of the thermal zone surface when occupied (9 a.m. to 7 p.m.), 1 W/m^2 when unoccupied
4. Commercial (Shop)—Internal loads: 30 W/m^2 and one person (70 W) per each 8 m^2 of the thermal zone surface when occupied (9 a.m. to 7 p.m.), 1 W/m^2 when unoccupied

Loads were assumed to be half convective and half radiative. These load profiles are much simplified, but they are useful for a comparison of performance avoiding the disturbance of all other factors.

Simulations were performed in two conditions: with and without climate control. When climate control was considered, its maximum power was set to 50 W/m^3 in heating and 30 W/m^3 in cooling. Heating was considered from 15th October to 15th April, as by the law, and cooling from 1st May to 30th September. In residential use, heating is assumed to be running in the morning from 5 a.m. to 10 a.m. and in the evening from 3 p.m. to midnight. In office or shop use, heating is assumed to be running from 7 a.m. to 9 p.m. and cooling is assumed to be running from 2 p.m. to 10 p.m. in residential use, and from 10 a.m. to 6 p.m. in the other uses. Perfect control is assumed that means the cooling and heating are at the exact power needed to keep room air at set-point temperature (20 °C in heating, 26 °C in cooling), up to the available power limits stated before.

It is reasonable to expect that, in residential use, both heating and cooling will be running fewer hours, but it was set so in order to make comparison straight, avoiding that the choice on how long climate control systems are running will have any relevance. On the other hand, it is reasonable to expect that in office and commercial use heating would be running fewer hours and cooling more hours, too, in relation to actual activity period. An air change rate of 0.5 volumes per hour by outdoor air is set. In residential use, when the building is occupied higher ventilation rates may be expected due to window operation. In the other uses, lower ventilation rates are often observed as ventilation is limited to leakage of closed windows, except for spring and autumn if there is not too much noise outdoor. Yet any more detailed consideration of occupant behaviour or of the influence of wind on ventilation would make harder to compare results and would require more information on the outdoor environment.

Governing equations, along with boundary and initial conditions, are solved with the first-order backward scheme. A time step of 60 s was used. The code has been tested with validation data from IEA (Strachan et al. 2016) showing good performance, just like best performing commercial codes. Whole year simulations were done for each use and climate control strategy with normal year data generated from SODA tool (Soda pro 2018). Ambient free-floating air temperature at each hour is shown in Fig. 18.5 for rooms Z1 that are located at the north-west end on the building, for rooms Z4 that are located at the south-east end on the building, and for room Z9 that is the central part of the first floor.

The diagram plots clearly show that inner loads lead to an increase of the average temperature in each day and of the indoor daily temperature range. Even the low heat gains in residential use lead to an increase of room air temperature between 2.0 °C and 2.4 °C for the thermal zones of the ground floor, and between 2.3 °C and 3.5 °C for those of the first floor. Yet, in residential use, in summer, room air temperature is below 32 °C, reaching this value only on few days on the first floor. This means that summer air conditioning is not strictly required, as thermal conditions are within the comfort limits for non-conditioned rooms (Ole Fanger and Toftum 2002), except for few hours in some days. Moreover, a proactive operation of the windows and of their blinds may result in a reduction of peak temperature. In winter, room air temperature keeps almost above 10 °C (the minimum of outdoor temperature is −3.9 °C), but this is not enough for present habits; thus heating is required.

Office and commercial (shop) use of the building lead to a higher increase in room air temperature that may reach values as high as 38 °C in office and 43 °C in commercial use. This means that air conditioning is mandatory if the use of the building brings significant inner loads. In winter, the rise in temperature is not enough to make possible avoiding heating systems. It is thus significant to compare room air temperature for these thermal zones with heating and with air conditioning, as shown in Figs. 18.6 and 18.7.

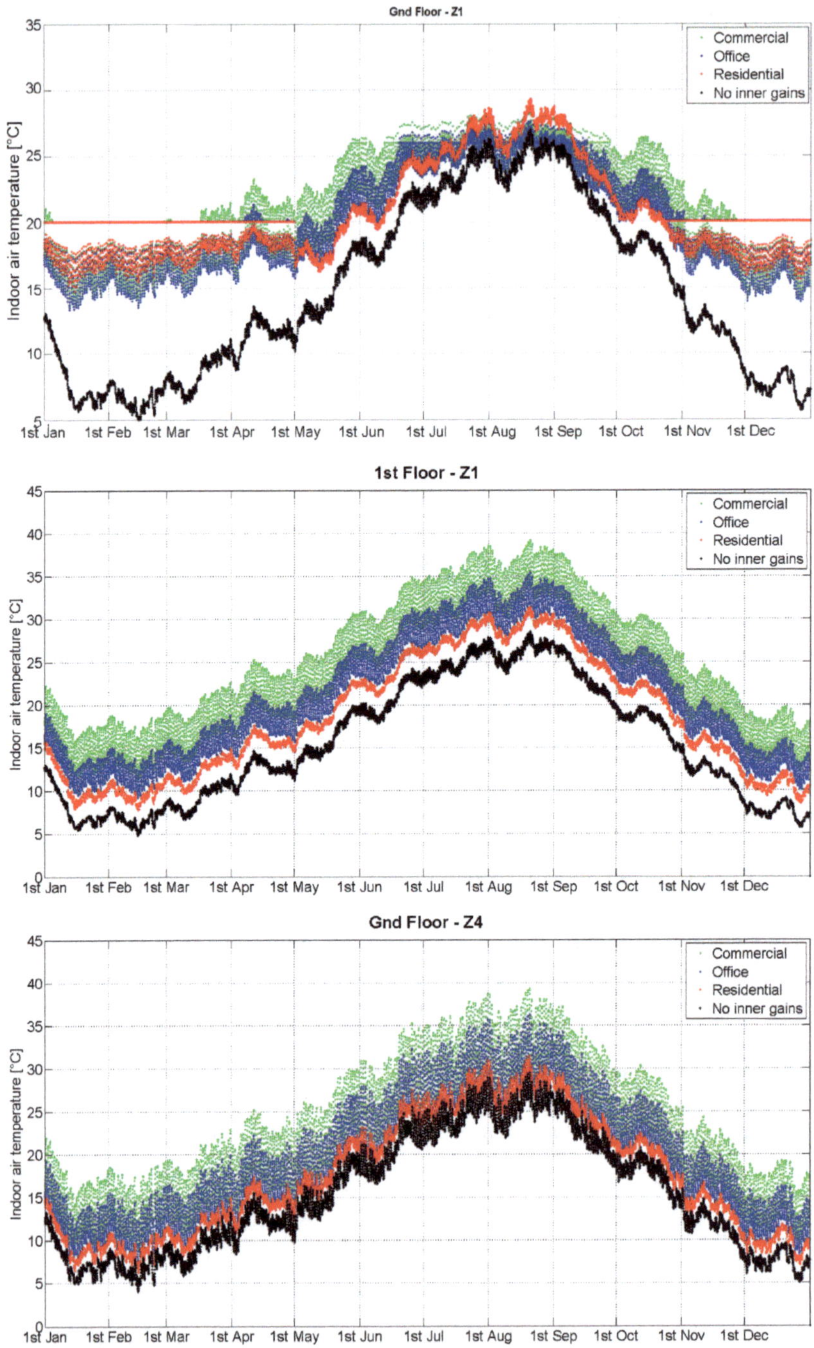

Fig. 18.5 Ambient free-floating air temperature at each hour for most representative internal zones

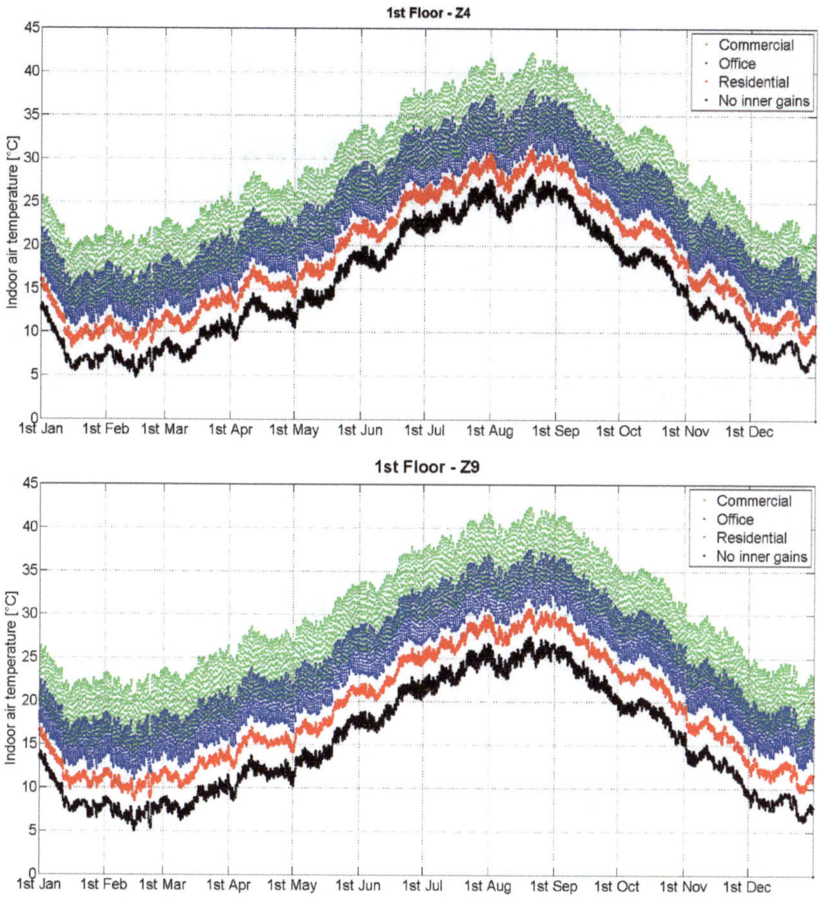

Fig. 18.5 (Continued)

The diagram plots show that even with heating and air conditioning, room air temperature has a wide daily range. This is due to the inner loads and to the heat transfer between walls and ambient. It may be better explained with a close up in Fig. 18.7.

From the diagram it is clear that even with air conditioning there are two peaks of temperatures: the first before cooling activation, while there are already the inner gains, the second just after switching off the air conditioning, the walls reject the thermal energy gained from radiant loads to air, together with the ongoing inner loads. This means that the actual operation of air conditioning is relevant for the overall performance.

As far as energy performance is considered, the demand of the whole building is shown in Fig. 18.8.

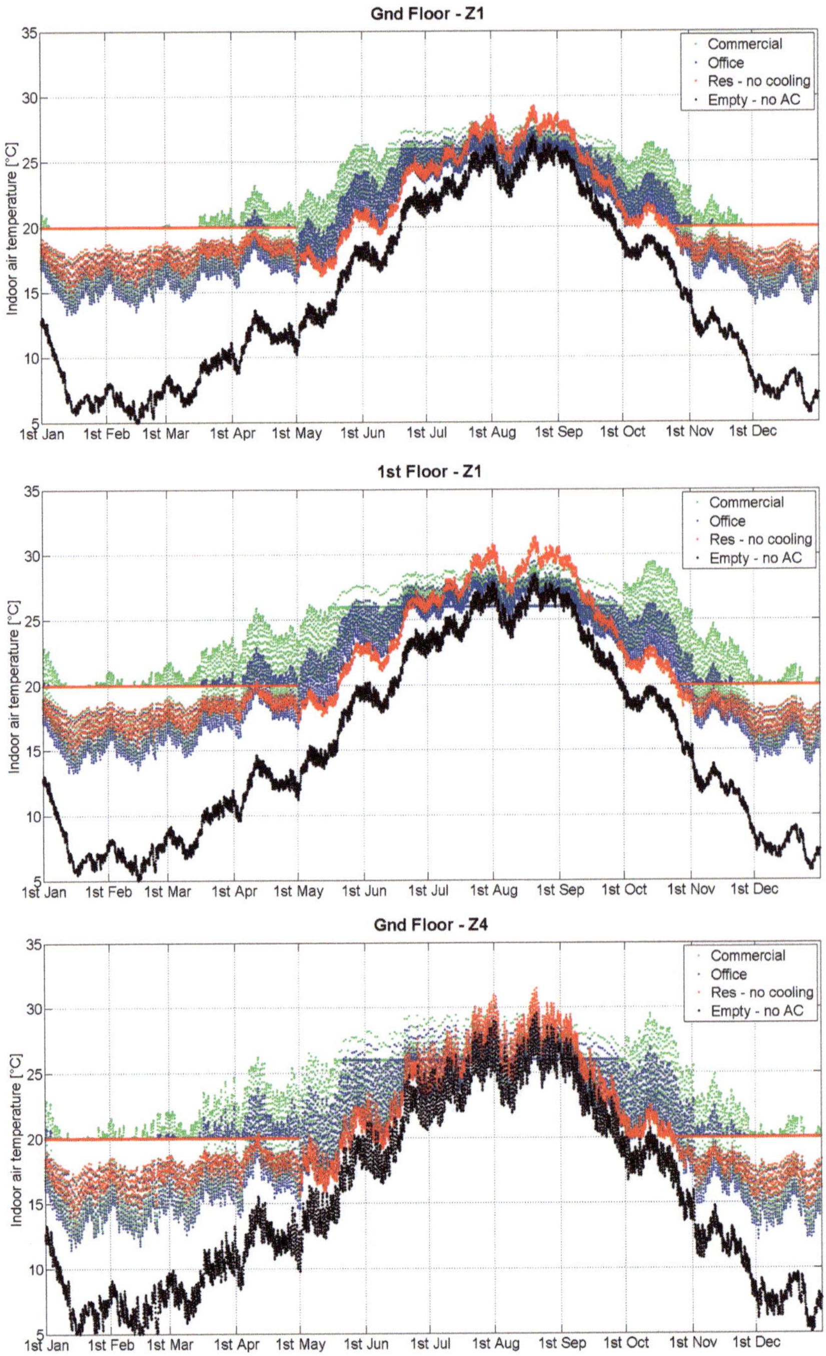

Fig. 18.6 Air temperature for most representative internal zones with heating and cooling

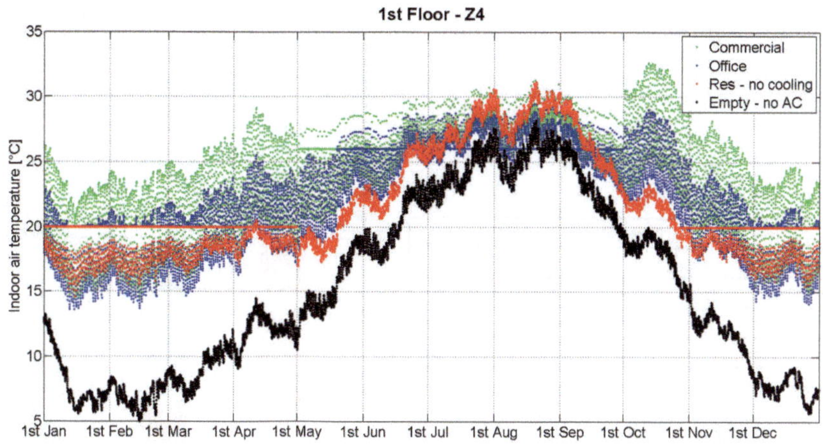

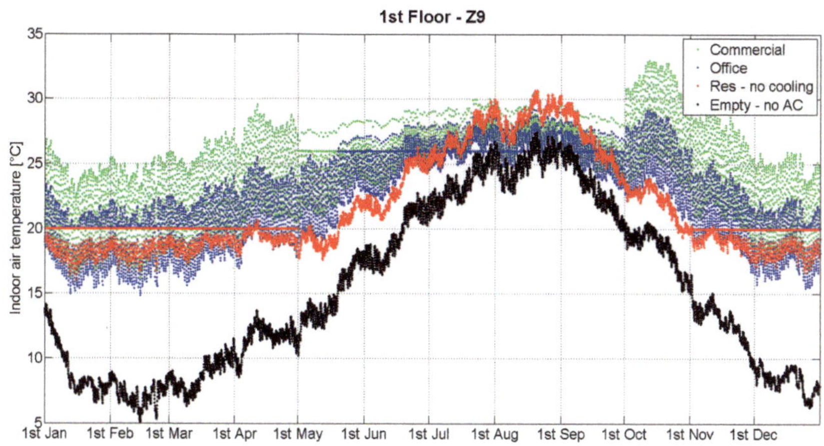

Fig. 18.6 (Continued)

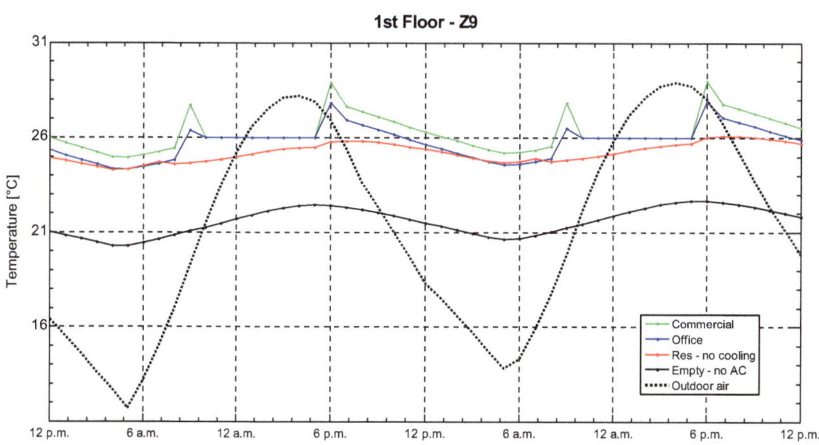

Fig. 18.7 Temperature dynamic in the first 2 days of July for Z9—first floor

Use	Energy demand [kWh/y]		
	Heating	Cooling	Total
Residential	30,290.00	2,833.00	33,123.00
Office	15,895.00	11,164.00	27,059.00
Commercial	9,903.00	18,884.00	28,787.00

Use	Energy demand [kWh/y m^{-2}]		
	Heating demand	Cooling demand	Total
Residential	54.30	5.10	59.30
Office	28.50	20.00	48.50
Commercial	17.70	33.80	51.60

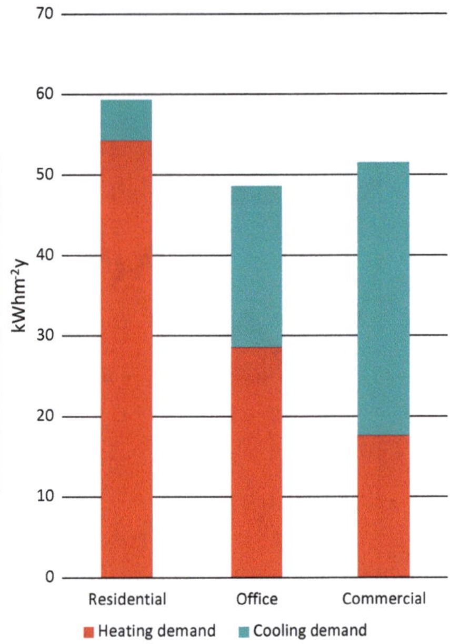

Fig. 18.8 Heating and cooling energy demand depending on building use for Casa Medori

A significant variation of energy demand exists between the three use profiles in traditional buildings. Simulations reveal a threefold variation for heating and almost sevenfold variation for cooling, while global energy performance is around 1,2 variation factor. This is due to the inner loads effect on internal air temperature in the case of office and commercial use.

The analysis of performance through whole-year simulations of the building for different uses confirms that this kind of vernacular buildings provides almost comfortable conditions in summer, due to its high thermal inertia. Yet, this is true only for residential use, as an increase in inner loads due to other activities leads to a steep increase in cooling demand.

Conclusions

The main purpose of this study is to promote renovation process aware of the energy effects of building use modifications, in order to overcome conventional approaches on the retrofit of traditional buildings and to include more suitable criteria for less intrusive interventions. Very often the energy performance improvement of heritage buildings is in contrast with conservation needs. For this reason, non-physical factors, like building use and occupant behaviour, are crucial to lead the ongoing process of renovation of historical towns. This approach allows us to better achieve energy efficiency in traditional buildings and also to develop more robust and integrated retrofit strategies.

A typical building of Central Italy—intended as a prototype for its traditional construction technique and local materials—has been analysed as case study: Casa Medori in Civita di Bagnoregio (Latium, Italy). Thermal and energy performance simulations have been performed in order to evaluate residential, office and commercial uses and considering comparable use profiles and occupations in four different use conditions.

The findings show the following:

- As expected, cooling energy demand for residential use is negligible. In office use, heating demand is higher than cooling energy demand, while in commercial use it is the reverse. The highest overall energy demand is that residential use, while the least is the office use. Since heating may be done with more efficient technologies than cooling, overall primary energy may be expected to be higher for commercial use than for the other two uses.
- To maintain the original residential use (or similar like B&B and medium level hotels) has some advantages: it is compatible with modern comfort requirements; it wants almost no cooling and a correct practice of natural ventilation could compensate some discomfort during the hot season; moreover, heating systems are much less cumbersome than cooling systems and can be introduced preserving the building; lastly the predictable seasonality of this kind of uses makes easier the compliance with the conservation issue.
- To reduce global energy demand, behavioural changes in terms of heating and cooling temperature set-point could be easily achieved in the case of residential use compared to other cases, where indoor temperature in summer reaches about 40 °C.
- Using the building as an office or a shop is efficient, as the reduction in heating demand compensates for the arising cooling demand, due to internal gains. Simulations show that inner heat gains promote a temperature increase that requires air conditioning to keep the inside usable. Yet air conditioning systems are bulky as they require heat rejection units.
- In order to limit the above-mentioned temperature increase, in case of office and commercial uses, it is crucial to accurately locate systems that produce internal loads using a design layout that favours the heat transfer with the exterior.
- At present, considering a residential use, the energy demand from simulations is 54 kWh/m^2 y. Energy efficiency may be improved combining non-physical renovation strategies with non-invasive ones, aiming at preserving the architectural value of the building. In fact, especially for residential use, these strategies are more robust and less intrusive, such as the introduction of thermal insulation on the extrados of the attic floor and mechanical ventilation with heat recovery in the attic.
- To improve the indoor thermal comfort and achieve the planned level of efficiency, users training actions need to be implemented in order to limit incorrect energy behaviours. In fact, user behaviour is an important factor for fostering less energy demand and less intrusive intervention in traditional buildings, too often neglected.

Moreover, based on the findings, it is possible to point out some consideration on the most effective actions to put into practice in order to achieve, with the ongoing urban renovation process, an effective energy retrofit based on conservation principles. Improved energy-based decision-making processes are needed: urban regulation is essential in order to support and to evaluate the renovation of historical town characterized by traditional buildings.

- Building use shall be determined taking into account as integrated values the energy performance, the construction systems and the significance of this cultural heritage. The design process has to guarantee fixed level of energy efficiency without compromising the conservation of the building. Moreover, to control the energy effect of renovation process and to revitalize communities in historical town, possible uses and related quantity shall be determined introducing specific urban regulation.
- Indoor temperature set-point both in winter (maximum level) and summer (minimum level) should be fixed by a specific regulation developed for traditional buildings. In Italy, such kind of regulatory limits has been determined for housing in the heating season, without any difference depending on the characteristics of the building stock—cultural value, period of construction, insulated/non-insulated envelope, building systems, etc.—and for these reasons, without considering the different effect on specific energy performance. A slight modification of the above-mentioned limits could help to reduce the energy demand of traditional building without compromising the indoor thermal comfort. For example, in summer a 1 °C shift could be compensated with the introduction of cross ventilation.
- As the buildings were not originally intended for heating (except for stoves and fireplaces) nor for air conditioning, the re-use of such urban textures consisting of traditional buildings may be best achieved through district heating and/or cooling systems. This implies planning of which activities might be allowed in the district and which shouldn't as well as some planning of the development.

This study contributes to heritage conservation and energy efficiency, compared with conventional approaches that mainly consider physical improvement. We envisage that emphasizing appropriate change of use in buildings through energy-based retrofit criteria would allow us to limit conventional retrofit strategies that affect building elements and appearance and better achieve energy efficiency in traditional architectures and in historical towns in order to develop more robust retrofit strategies and to inform the ongoing renovation process.

Authorship

MM conceived and designed the study; EH and MM conceived the methodology, EH performed the thermal analysis; EH and MM analysed the data; CC and MM wrote the section Vernacular Buildings and Energy Retrofit; MM wrote the section The Italian Cultural Heritage; EC wrote the section Civita di Bagnoregio and Casa Medori; EH and MM wrote the section Thermal and Energy Performance Analysis; MM and EH wrote the section Conclusions.

Acknowledgements This work was developed under Sapienza University grant AR21715C82 226BCC. The authors would like to thank Giovanni Attili for sharing documents and drawings of *Civita di Bagnoregio* and *Casa Medori*.

References

Baker N, Steemers K (2000) Energy and environment in architecture: a technical design guide. E & FN Spon, London, New York

Bascia L, Carlotti P, Maffei GL (2000) La casa romana nella storia della citta dalle origini all'Ottocento. Alinea, Firenze

Ben H, Steemers K (2014) Energy retrofit and occupant behaviour in protected housing: a case study of the Brunswick Centre in London. Energy Build 80:120–130. https://doi.org/10.1016/j.enbuild.2014.05.019

Carbonara G (2015) Energy efficiency as a protection tool. Energy Build 95:9–12

Egusquiza A, Brostrom T, Pagliula S, Hermann C, Rodwell D (2016) Strategic assessment of historic cities for energy efficiency retrofits to enable their long-term use and conservation development of the EFFESUS methodology and software tool. In: Bogusław S (curatore) Herit. Futur. Herit. Transform. Cult. Herit. Prot. XXI Century – Probl. Challenges, Predict. ICOMOS, Lublin, pp 60–81

European Climate Foundation (2010) Roadmap 2050: a practical guide to a prosperous, low carbon Europe

Gargano M (1988) Civita di Bagnoregio. In: Lattanzi F, Polci S (curatori) L'ambiente, la memoria, Progett. Testimonianze su Civita di Bagnoregio. Sugarco Edizioni, Milano, pp 29–50

Garnier J-P (2016) Architettura e Anarchia, un binomio impossibile. Nautilus autoproduzioni, Torino

Geva A (2015) Energy Simulation of Historic Buildings: St. Louis Catholic Church, Castroville, Texas. APT Bull 29:36–41

Guidoni E (1980) L'architettura popolare italiana. Laterza, Bari

Gynther L, Lappillone B, Pollier K (2015) Energy efficiency trends and policies in the household and tertiary sectors. An analysis based on the ODYSSEE and MURE databases. p 97

Habraken NJ (2000) The structure of the ordinary: form and control in the built environment. MIT Press, Boston

Illich I (1984) Il genere e il sesso. Mondadori, Milano

ISTAT (2017) Database ISTAT 2017. http://dati.istat.it/. consultato 24 agosto 2017

Ole Fanger P, Toftum J (2002) Extension of the PMV model to non-air-conditioned buildings in warm climates. Energy Build 34:533–536. https://doi.org/10.1016/S0378-7788(02)00003-8

Soda pro s.d. http://www.soda-pro.com/web-services/typical-years/normal-year-global-radiation-temperature. consultato 10 marzo 2018

Strachan P, Svehla K, Kersken M, Heusler I (2016) Reliable building energy performance characterisation based on full scale dynamic measurements. In: IEA (curatore) Rep. Subtask 4a Empir. Valid. common Build. energy Simul. Model. based situ Dyn. data. KU, Leuven

Strappa G, Carlotti P, Camiz A (2016) Morfologia urbana e tessuti storici: il progetto contemporaneo dei centri minori del Lazio. Gangemi Editore, Roma

Taut B (1991) Una casa d'abitazione. Franco Angeli, Milano

Vieites E, Vassileva I, Arias JE (2015) European initiatives towards improving the energy efficiency in existing and historic buildings. Energy Proc 75:1679–1685. https://doi.org/10.1016/j.egypro.2015.07.418

Webb AL (2017) Energy retrofits in historic and traditional buildings: a review of problems and methods. Renew Sustain Energy Rev 77:748–759. https://doi.org/10.1016/j.rser.2017.01.145

Chapter 19
Conclusions

Ali Sayigh

Vernacular architecture (VA) is associated with traditional cultures and exploits indigenous materials, creating comfort through natural concepts such as shading, night radiation and ventilation, thick walls, the use of courtyards and microclimates created by use of water and greenery.

Vernacular buildings do not normally require many mechanical devices to regulate heat and cold in summer and winter. Such architecture is not susceptible to replication because it is site dependent and varies from one climate zone to another.

The authors have demonstrated the strengths of VA in all corners of the planet. In every climate zone and all forms of economy, VA can be seen to meet the crucial demand for sustainable housing and public buildings. It is imperative that as the planet approaches ever closer to climate change point of no return, architects, economists, and politicians act now to stave off this catastrophe. Any reduction of greenhouse gas emissions, however small, is better than no reduction at all.

Architects must utilize all methods open to them via environmentally friendly building materials, natural methods for obtaining sustaining comfort temperature levels and ventilation, utilizing some of the positive elements of the VA without neglecting the need to design buildings which address both community and individual needs for adequate living space and privacy. Let us hope that architects address the challenge which face us all, and especially the next generation.

A. Sayigh (✉)
World Renewable Energy Congress, Brighton, UK
e-mail: asayigh@wrenuk.co.uk

A. Sayigh (ed.), *Sustainable Vernacular Architecture*, Innovative Renewable Energy, https://doi.org/10.1007/978-3-030-06185-2_19

Index

A. Sayigh (ed.), *Sustainable Vernacular Architecture*, Innovative Renewable Energy, https://doi.org/10.1007/978-3-030-06185-2

MIX
Papier aus verantwortungsvollen Quellen
Paper from responsible sources
FSC® C105338

If you have any concerns about our products,
you can contact us on
ProductSafety@springernature.com

In case Publisher is established outside the EU,
the EU authorized representative is:
Springer Nature Customer Service Center GmbH
Europaplatz 3, 69115 Heidelberg, Germany

Printed by Libri Plureos GmbH
in Hamburg, Germany